MW01156159

DATE DUE		
MAY 2 6 2009		
MAY 2 7 2009		
JUN 2 5 2009		
JUL 3 0 2009 BROWSED		
BROWSED		
DEC 0 9 2009		
BROWSED		
GAYLORD		PRINTED IN U.S.A.

Sink or Float?

Thought Problems in Math and Physics

© 2008 by
The Mathematical Association of America (Incorporated)
Library of Congress Catalog Card Number 2008929711
ISBN 978-0-88385-339-9
Printed in the United States of America
Current Printing (last digit):
10 9 8 7 6 5 4 3 2 1

The Dolciani Mathematical Expositions

NUMBER THIRTY-THREE

Sink or Float?

Thought Problems in Math and Physics

Keith Kendig
Cleveland State University

*Published and Distributed by
The Mathematical Association of America*

DOLCIANI MATHEMATICAL EXPOSITIONS

Committee on Publications
James Daniel, *Chair*

Dolciani Mathematical Expositions Editorial Board
Underwood Dudley, *Editor*
Jeremy S. Case
Tevian Dray
Robert L. Devaney
Jerrold W. Grossman
Virginia E. Knight
Mark A. Peterson
Jonathan Rogness
Joe Alyn Stickles
James S. Tanton

The DOLCIANI MATHEMATICAL EXPOSITIONS series of the Mathematical Association of America was established through a generous gift to the Association from Mary P. Dolciani, Professor of Mathematics at Hunter College of the City University of New York. In making the gift, Professor Dolciani, herself an exceptionally talented and successful expositor of mathematics, had the purpose of furthering the ideal of excellence in mathematical exposition.

The Association, for its part, was delighted to accept the gracious gesture initiating the revolving fund for this series from one who has served the Association with distinction, both as a member of the Committee on Publications and as a member of the Board of Governors. It was with genuine pleasure that the Board chose to name the series in her honor.

The books in the series are selected for their lucid expository style and stimulating mathematical content. Typically, they contain an ample supply of exercises, many with accompanying solutions. They are intended to be sufficiently elementary for the undergraduate and even the mathematically inclined high-school student to understand and enjoy, but also to be interesting and sometimes challenging to the more advanced mathematician.

1. *Mathematical Gems,* Ross Honsberger
2. *Mathematical Gems II,* Ross Honsberger
3. *Mathematical Morsels,* Ross Honsberger
4. *Mathematical Plums,* Ross Honsberger (ed.)
5. *Great Moments in Mathematics (Before 1650),* Howard Eves
6. *Maxima and Minima without Calculus,* Ivan Niven
7. *Great Moments in Mathematics (After 1650),* Howard Eves
8. *Map Coloring, Polyhedra, and the Four-Color Problem,* David Barnette
9. *Mathematical Gems III,* Ross Honsberger
10. *More Mathematical Morsels,* Ross Honsberger
11. *Old and New Unsolved Problems in Plane Geometry and Number Theory,* Victor Klee and Stan Wagon
12. *Problems for Mathematicians, Young and Old,* Paul R. Halmos
13. *Excursions in Calculus: An Interplay of the Continuous and the Discrete,* Robert M. Young
14. *The Wohascum County Problem Book,* George T. Gilbert, Mark Krusemeyer, and Loren C. Larson
15. *Lion Hunting and Other Mathematical Pursuits: A Collection of Mathematics, Verse, and Stories by Ralph P. Boas, Jr.,* edited by Gerald L. Alexanderson and Dale H. Mugler
16. *Linear Algebra Problem Book,* Paul R. Halmos
17. *From Erdős to Kiev: Problems of Olympiad Caliber,* Ross Honsberger
18. *Which Way Did the Bicycle Go? ... and Other Intriguing Mathematical Mysteries,* Joseph D. E. Konhauser, Dan Velleman, and Stan Wagon
19. *In Pólya's Footsteps: Miscellaneous Problems and Essays,* Ross Honsberger
20. *Diophantus and Diophantine Equations,* I. G. Bashmakova (Updated by Joseph Silverman and translated by Abe Shenitzer)
21. *Logic as Algebra,* Paul Halmos and Steven Givant
22. *Euler: The Master of Us All,* William Dunham
23. *The Beginnings and Evolution of Algebra,* I. G. Bashmakova and G. S. Smirnova (Translated by Abe Shenitzer)
24. *Mathematical Chestnuts from Around the World,* Ross Honsberger

25. *Counting on Frameworks: Mathematics to Aid the Design of Rigid Structures,* Jack E. Graver
26. *Mathematical Diamonds,* Ross Honsberger
27. *Proofs that Really Count: The Art of Combinatorial Proof,* Arthur T. Benjamin and Jennifer J. Quinn
28. *Mathematical Delights,* Ross Honsberger
29. *Conics,* Keith Kendig
30. *Hesiod's Anvil: falling and spinning through heaven and earth,* Andrew J. Simoson
31. *A Garden of Integrals,* Frank E. Burk
32. *A Guide to Complex Variables* (MAA Guides #1), Steven G. Krantz
33. *Sink or Float? Thought Problems in Math and Physics*, Keith Kendig

MAA Service Center
P.O. Box 91112
Washington, DC 20090-1112
1-800-331-1MAA   FAX: 1-301-206-9789

Preface

In his book *I Want to be a Mathematician*, Paul Halmos reveals a secret:

> ... it's examples, examples, examples that, for me, all mathematics is based on, and I always look for them. I look for them first when I begin to study, I keep looking for them, and I cherish them all.

Sink or Float? is actually a collection of examples in the guise of problems. The problems are multiple-choice, challenging you to think through each one. You don't need fancy terminology or familiarity with a mathematical specialty to understand them. They are concrete, and most are understandable to anyone.

The willingness and confidence to apply common sense to the real world can be very rewarding. It's one of the aims of this book to show that it works, that thinking can enlighten us and bring perspective to the world we live in. The range of problems to which common sense applies is almost limitless.

The example is one of the most powerful tools for discovering new things and for suggesting how to solve problems. Examples come in different degrees of concreteness and often, the more concrete, the more informative. The following tale provides an example of "concreter can be better," and is the story of what started this book.

It begins in the dentist's chair. After putting generous amounts of cotton in my mouth, my dentist commences his customary free-wheeling question and answer session. One time, things go like this:

Dentist. I've been wondering for some time, what was it that convinced you to become a mathematician?

Me. Waa-gaa ook ha!

Dentist. I see. I was sort of OK with math up through trigonometry, but when it came to calculus, the teacher went fast, I got lost and never caught up. From then on, I can't say I've liked math.

Me. Aagh hinc wah ad.

Dentist. Yes. Is there anything that you could say to give me some sense of the attraction?

Me. Eh he eee . . .

At last the cotton came out of my mouth, and so did this:

Me. You asked for some sense of the attraction in math. Here's a concrete example: 10×10 is 100, right? Now, add 1 and subtract 1 to change 10×10 into 11×9. That's 99, 1 less that 100.

Dentist (already sounding bored). Sure is.

Me. OK, now try that on 8×8, which is 64. Adding 1 and subtracting 1 from 8 changes it to 9×7. That's 63, again one less than 64.

Dentist. Funny coincidence, huh?

Me. Well, try this on 5×5, which is 25. Adding and subtracting 1 gives 6×4 which is 24, one less—again!

Dentist (whose expression has changed slightly). You're not telling me this *always* works?

Me. Try one yourself!

Dentist (possibly wanting to burst this little bubble). How about 11×11?

Me. OK, do the math . . .

Dentist. That would be 12×10. That's 120. Geez . . . 11×11 is 121, isn't it?

Me. This always *does* work!

Dentist. Well . . . not for really *small* numbers, I'll bet . . .

Me. What do you mean?

Dentist. I was thinking it wouldn't work for 1×1 . . .

Me. But it does! 1×1 is 1, while 2×0 is 0, *still* one less than 1!

Dentist (now impressed). My son is into math. I'm going to show him this!

Me. Great! By the way, it works for huge numbers, too. He can try numbers in the millions or billions on a calculator, and he'll see that it still works.

Dentist. I'll have to admit, this trick is impressive.

Me. Did I mention that it works for fractions and negative numbers, too?

Now here is this intelligent guy, one of the best dentists in town, who had become turned off from math and couldn't understand why anybody would take it up as a profession. Yet he responded positively to this little example. I decided to try it out on my classes. One was a "Technology in Teaching" course given to high-school teachers. Did anyone know that trick? Ever heard of it? Everybody is looking around. Not a single teacher had encountered it. Then I asked if they'd ever seen

$$(x + a)(x - a) = x^2 - a^2.$$

Preface

One of the more outspoken students blurted out, "*Duh! Everyone's* seen that!" I counter, "Well, then, I guess you've seen

$$(x + 1)(x - 1) = x^2 - 1^2$$

too!" Some look slightly uncomfortable, for I'm obviously up to something. "What's *that* when you make $x = 10$? Isn't it just

$$(10 + 1)(10 - 1) = 10^2 - 1^2 \;?$$

It says $11 \times 9 = 100 - 1$, which is 99.

They all immediately saw the pattern, and why the trick works for any number. In fact, it was also clear why adding and subtracting two made the number four less, and so on. The best student in class looked incredulous. She exclaimed "But ... but that's just so *basic*! How could we have missed it?" The teachers had gone part-way in concretizing, replacing *a* by a number. But going all the way? No one had ever replaced *both a* and *x* by numbers!

The story about the dentist was the germ of this book. That, followed by the experience with high-school teachers added to a growing conviction that there are significant gaps in understanding out there. Nor is it only in math—there are also serious issues with an everyday understanding of our physical world. As one example, a test given to entering freshmen at one Ivy League school revealed that a large proportion of them believed it is cold during winter because the earth is further away from the sun then. But when it's winter in the U.S., it's summer in Australia, a quick contradiction to that belief.

Given the close connection between mathematics and physics, I've included a generous selection of physics problems. This book is in a mathematical series, but separating math from science is often artificial and unproductive. The Russian mathematician Chebyshev (1821–1894) put it this way:

> **To isolate mathematics from the practical demands of the sciences is to invite the sterility of a cow shut away from the bulls.**

In addition to questions about numbers and geometry, you'll find a whole chapter devoted to the surprises hidden in Archimedes' Principle, as well as questions from the vast cultural rivers that have so richly informed mankind about the world we live in: astronomy, mechanics, electricity and magnetism. Each of these has brought understanding that has led to an intellectual or technological revolution that changed our world.

How to use this collection? After reading a question, close the book and think. Sometimes common sense leads to the answer. (Which way does our moon spin about its axis?) Other times, you will need to think longer. (A plumb line isn't plumb in New York City. Why not?) Sometimes you may need more in the way of mathematical or physics information and for that, this book includes three types of resources: chapter introductions, boxes scattered throughout the text, and a Glossary of terms. Other than such background material, however, you'll find little in the way of hints on how to solve any problem. That's the way nature operates—she doesn't hand out instruction books. Most of the time we humans simply stumble upon the right approach. However, through long experience mathematicians and scientists have discovered that certain general thinking techniques can be helpful and sometimes potent. Perhaps you ought to push to an extreme? exploit symmetry? find a pattern? Sometimes, when you get really, truly stuck, it's because your guess or conviction is wrong and you should be trying to prove the opposite! An open mind is basic in this business. At the end of each problem, the answer's page number appears in parentheses.

Truth is, most mathematicians think differently from the man in the street. They search for analogies or a clever counterexample, they try to remove exceptions, look for symmetry, rejoice over a far-reaching generalization. They'll often push to extremes, looking for the most, the least, the best, the worst. That is not how typical folk operate, nor is it the way most math students think. Most professional mathematicians have had to acquire some of these thinking techniques, too. To learn about nature, you need to think on her terms, but the rewards of doing this can be tremendous and can help us penetrate many of nature's mysteries.

In this book, you'll find that a problem usually falls into at least one of these categories:

- The problem gives a fresh or expanded appreciation of something familiar. (The dentist's problem is one example. Another: if you know the Maclaurin expansion of $\sin x$ and x is large, are you aware of the drama played out as the expansion converges?)

- The problem is an unusual or unexpected application of some principle. (Did you know that the various layers of the earth, from its dense metallic core to the lightest layers of the stratosphere, are mostly a consequence of Archimedes' Principle?)

- The problem is a celebration of math as an "information massager." The math itself may mostly consist of just arithmetic, but such massaging can put a fresh spin on available information. (On average, lightning strikes over each square mile about once every three years. So how often is lightning striking somewhere on earth?)

- Solving the problem instructs by pointing out a common misconception.

- The problem illustrates a novel use of some computational technique.

- The problem, though innocent-sounding, leads to a considerably larger idea or method in mathematics. *Up or Down?* (p. 7), *Method or Madness?* (p. 156), and *Adding Equal Weights* (p. 215) are examples.

- It can be solved with a surprisingly quick or novel argument, including ...

- Sudden death by counterexample.

I have chosen problems from ten general areas of mathematics and physics, arranged in roughly historical order. Although certain problems in the later chapters will be easier to

Preface

solve by those having more mathematical background, common sense always serves as a potent tool.

I hope this book will serve as a springboard—either individually or in a classroom—for mathematical thought, case-testing, discussions, arguments. One of my favorite historical snippets is when the Wright brothers, after watching a bird in flight, would start off with differing explanations of some aspect of flight. In searching for understanding and clarity, they'd argue. Each forced the other to defend statements. They became so good at this, they'd often reverse their original positions! But in the process each gained insight and deeper understanding, and their method goes to the heart of what scientific thinking really is. If just one problem in this book snares some young student into this dynamic and eventually leads to a life of scientific inquiry, I will have done my job.

It is a pleasure to express my gratitude to the many people who wittingly or unwittingly helped with this book:

It was the initial enthusiasm of Don Albers, MAA's Director of Publications, who fanned a few stray embers into a healthy fire that pretty much had me going for two years. He's been a wonderful sounding board and always a great source of support during the writing process.

When a "completed" manuscript was finally in hand, the Dolciani Series editor, Underwood Dudley, showed me how much better that manuscript might be. He read every word of it not once but twice, each time showing me by skillful example how to say it more simply and smoothly. My apprenticeship has been rewarding. Thank you!

Every other member of the Dolciani board, too, gave me feedback, some with several pages of thoughtful comments: Jeremy S. Case, Robert L. Devaney, Tevian Dray, Jerrold W. Grossman, Virginia E. Knight, Mark A. Peterson, Jonathan Rogness, Joe Alyn Stickles and James S. Tanton.

I owe a big note of appreciation to my thesis advisor, Basil Gordon. For many years the editor of the Anneli Lax New Mathematical Library, his sharp eye, wonderful sense of the written word and unfailing logic have once again helped me in writing a book. His many thoughtful suggestions substantially improved the manuscript.

Elaine Pedreira Sullivan (Associate Director for Publications) and Beverly Ruedi (Electronic Production Manager) both work quietly behind the scenes, but for many years their special talents have left a highly professional imprint on MAA books. I feel lucky indeed to have worked with them!

It is also my pleasure to mention some other very helpful people who led me to valuable insights or a new point of view, or made me aware of a surprising, bookworthy fact: mathematicians Annalisa Crannell and Allen Schwenk, physicists John Hofland and James Lock (both of whom read large portions of the manuscript) and two of my talented students, Allison Parsons and Jason Slowbe.

And now for the person who, throughout the project, selflessly gave me support in simply too many ways to count—my cherished wife, Joan. Thanks!! It's been quite a trip, hasn't it?

Contents

Preface	vii
What Do You Think? A Sampler	1
Geometry	9
Numbers	33
Astronomy	45
Archimedes' Principle	67
Probability	85
Classical Mechanics	105
Electricity and Magnetism	123
Heat and Wave Phenomena	143
The Leaking Tank	179
Linear Algebra	197
What Do You Think? Answers	217
Geometry Answers	227
Numbers Answers	245
Astronomy Answers	253
Archimedes' Principle Answers	267
Probability Answers	273
Mechanics Answers	285
Electricity Answers	295
Heat and Wave Phenomena Answers	301
The Leaking Tank Answers	317
Linear Algebra Answers	323
Glossary	339
References	367
Problem Index	369
Subject Index	373
About the Author	375

What Do You Think?
A Sampler

Cube-Shaped Fishing Weight

A metal cube is suspended by a vertex to make a fishing weight:

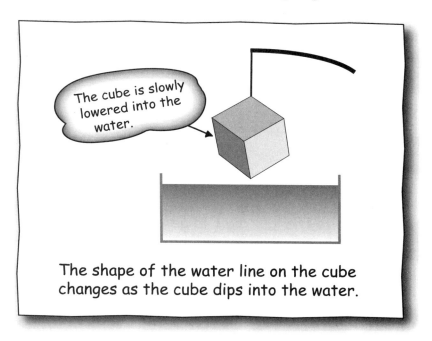

The shape of the water line on the cube changes as the cube dips into the water.

The weight is lowered into a quiet pond until it is half submerged. What is the shape of the waterline on the cube's surface?

(a) Triangle (b) Square (c) Non-square rectangle

(d) Hexagon (e) Octagon

(p. 217)

Subtracting by Adding?

This illustration shows three subtraction problems changing into addition problems. In the first step, the number being subtracted is replaced by its digitwise nines complement. For example, the nines complement of 3 is 6, because 6 completes 3 to make the sum 9. Does the method suggested by these examples generalize? Could all that time grade-school kids spend learning borrowing be eliminated?

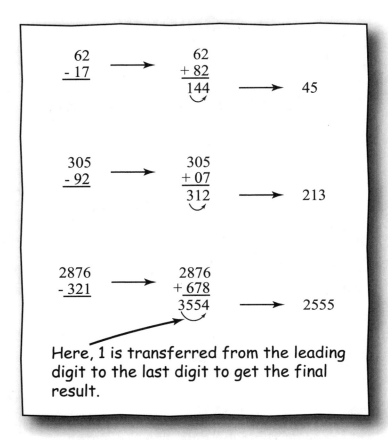

Can one always use the method suggested in the picture to subtract a smaller positive integer from a larger one?

(a) These special examples work, but there is no general free lunch here.

(b) Usually yes, but there are exceptions that could easily lead to confusion for the kids.

(c) Yes for integers, but the method doesn't work for decimals. For these, they would still need the traditional method.

(d) It always works. (p. 217)

Would This Tempt You?

You have just won $100,000 on a quiz show. You no sooner begin to recover from the shock than the emcee offers you a deal. The curtain behind you opens, revealing two baskets with 50 ping-pong balls in each. The only difference between them is that in one basket the balls are red, while in the other they are yellow. He says you can mix them up or transfer any number of them from either basket to the other. But once you have done that, you are blindfolded, the contents of each basket are stirred, and the baskets are moved around. You then randomly choose a basket and take out a ping-pong ball.

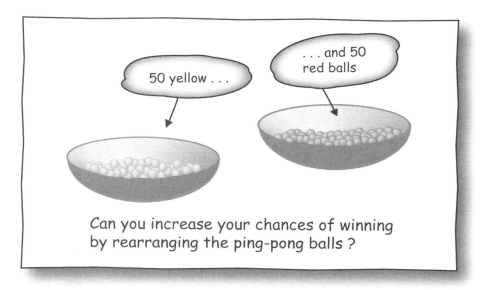

Can you increase your chances of winning by rearranging the ping-pong balls?

If you pick a yellow one, your prize is doubled. If it is red, you lose everything.

(a) All that talk about mixing up the balls and so on, is TV hype. The fact is, there are 50 balls of each color, and your chances of getting a yellow one are 50%, period.

(b) Although he stirs the contents of each basket and moves them around before you pick, you can do so much better than 50% that you are really tempted by his deal.   (p. 218)

Friday the 13th

On average—say, over a century or two—how often does a Friday the 13th occur? The answer is closest to once every

(a) 5 months   (b) 6 months   (c) 7 months   (d) 8 months

(e) Less frequently than any of these.   (p. 218)

How Turns the Earth?

From high noon on one day to high noon the next day, which of these is closest to how much the earth spins around its north-south axis?

(a) 359° (b) 360° (c) 361° (p. 218)

What Will Happen?

You are holding a chain draped over a block whose top forms two frictionless inclined planes. The block's bottom is horizontal, and the two planes are at right angles to each other:

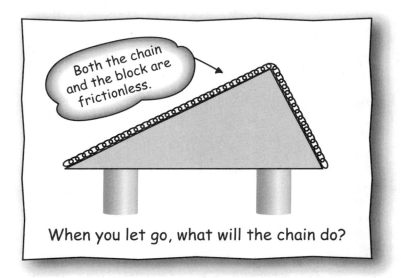

The chain, whose ends extend to the bottom of the block, is perfectly flexible and has uniform density. When you let go of the chain, it will

(a) move to the right.   (b) move to the left.   (c) just sit there.   (p. 219)

How Close?

What is the minimum distance between the graphs of $y = e^x$ and its inverse $y = \ln x$?

(a) $\frac{1}{2}$ (b) 1   (c) $\sqrt{2}$ (d) e (e) $\frac{1}{e}$ (p. 220)

Travel Time

The setup on the left is the same as the one on the right, except there are two thin horizontal wires on the left and three on the right. All five of these thin wires are identical.

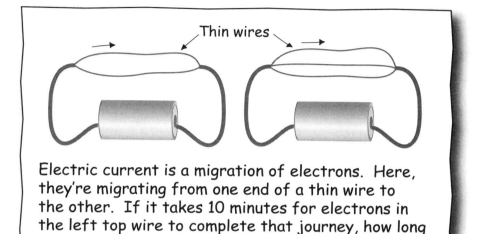

Electric current is a migration of electrons. Here, they're migrating from one end of a thin wire to the other. If it takes 10 minutes for electrons in the left top wire to complete that journey, how long does it take them to do it in the right top wire?

Electrons in an electric current move slowly. In addition to free electrons' random motion in the wire as a sort of electron gas, a battery's voltage adds a "drift speed" making them migrate from the negative to the positive terminal. If it takes electrons 10 minutes to drift through either thin wire in the left picture, then about how many minutes would it take an electron to do that in the right picture?

(a) 6.7   (b) 7.5   (c) 10   (d) 12.5   (e) 15   (p. 220)

> The electrons above may move slowly, but that's not always what they do. Within atomic nuclei, their "speed" is anything but slow, and this can turn certain atoms into tiny magnets. The enormous number of atoms, when properly lined up, can add up to a huge total magnetic effect. There are permanent neodymium magnets that can lift 1300 times their own weight, and a typical commercial-grade neodymium magnet about the size of eight stacked pennies can lift 200 lbs. Such magnets have been important in miniaturizing electronic devices: packing such strength in a small space is largely responsible for shrinking the early hard drive diameter of 16 inches. And far smaller, lighter magnets in headphones helped the Walkman initiate the personal music revolution.

Sinking 2-by-4

Archimedes' principle says that an object displacing water experiences an upward buoyant force equal to the weight of the displaced water. Suppose one end of an 8-foot long pine 2-by-4 is attached to a 50-lb anchor using nylon string, as in the picture. The stud is released into the ocean:

When the stud-plus-anchor is about halfway down its two-mile trip to the bottom of the ocean, the string breaks. What does the stud do?

(a) It goes back to the top.   (b) It stays about where it is.

(c) It sinks all the way to the bottom of the ocean.

(d) Are you kidding? It could never make it that far down without breaking the string!

(p. 220)

By the Way . . .

Is the tension in the string greater when you hold the dry assemblage before lowering it into the water, or when you hold the submerged assemblage just before releasing it on its journey? The tension is greater when it is

(a) dry.   (b) submerged.   (c) It is the same either way.   (p. 221)

Up or Down?

In this circuit, all seven resistors have the same resistance, and a battery connected to the ends makes the overall current flow from left to right:

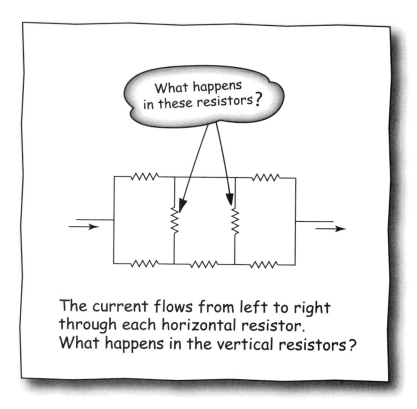

The current flows from left to right along the horizontal paths. What about the currents in the vertical resistors? If the top path had three resistors as the bottom does, then symmetry implies that no current flows through either vertical resistor. Does that missing resistor make a difference? In the vertical paths,

(a) the current goes up in both of them.

(b) the current goes down in both of them.

(c) the current goes up in the left resistor and down in the right.

(d) the current goes down in the left resistor and up in the right.

(e) No current flows in either of them. (p. 221)

Yellow Points, Blue Points

A regularly-repeating pattern of isolated points is sometimes called a *lattice*. Atoms of certain crystals, for example, define three-dimensional lattices. In the plane, all integer points (m, n) form a two-dimensional lattice, and the points can be called even or odd according to whether $m + n$ is even or odd. Painting the evens yellow and the odds blue creates a checkerboard-like pattern:

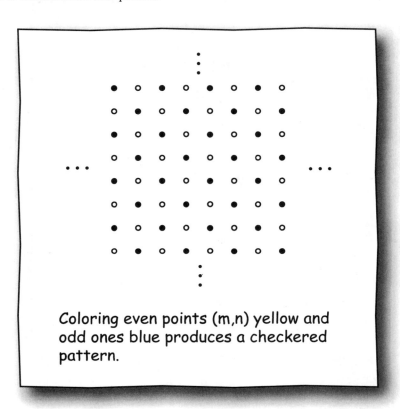

Coloring even points (m,n) yellow and odd ones blue produces a checkered pattern.

These points have the nice property that if you move one unit away from any point to another point, the new point will have the other color. Does this extend to the whole plane? Is there a way of coloring each point of $\mathbb{R}^2$ blue or yellow so that when you move one unit in any direction to a new point, that new point has the other color?

(a) Yes   (b) No   (p. 226)

Geometry

Introduction

Geometry is as old as civilization itself. Monuments from the Late Stone Age such as Stonehenge in England and the pyramids in Egypt already show considerable geometric sophistication. Just as the push toward civilization led to a need for astronomy, it did the same for geometry. High-value land next to the Nile became subdivided into parcels with property lines and taxable areas, and all sorts of tables, rules and formulas were developed to help surveyors and tax-collectors. Thus, the Nile heightened the need for geometry: the river's annual flooding kept surveyors busy re-establishing property lines.

The ancient Greeks inherited a large geometric legacy from Egypt, Babylon and Mesopotamia. By the time Euclid (ca. 325–265 BCE) came on the scene, the Greeks possessed an enormous mass of geometric information—some correct, some not, some with justifying arguments, some without. Euclid aimed to put things in order: he strove for a beautiful, aesthetic order with logical deduction as the guiding principle. Starting from a few "self-evident" axioms and postulates, Euclid developed a large, organized system of geometric facts whose truth was guaranteed through stepwise deduction. Although there were logical gaps, first filled in by David Hilbert (1862–1943) in 1899, Euclid had a keen sense of pedagogy and organization, and his *Elements* became enormously popular. Except for the Bible, the *Elements* has been the most widely disseminated book ever written.

Besides developing plane geometry for its intrinsic interest, the Greeks enriched the world at a practical level. Earlier civilizations knew how to create right angles using the (3, 4, 5)-triangle, and they also knew what is now called the Pythagorean theorem. And in the first century BCE, Heron of Alexandria—a remarkable polymath distinguished as a mathematician, scientist, engineer and inventor—proved what is now known as Heron's formula. This is a formula for the area A of any triangle in terms of its sides a, b, c, and is usually written using the semiperimeter $s = \frac{a+b+c}{2}$ as

$$A = \sqrt{s(s-a)(s-b)(s-c)}.$$

He wrote using worked-out, concrete examples, so that workers in the field could use their own numbers to find areas. His guides became the equivalent of today's best-selling how-to books.

By the time of Heron, the Greeks had a discovered a number of geometric constructions useful in surveying, construction and map-making. For finding distances and angles, they had both the Pythagorean theorem and trigonometric techniques that Hipparchus (ca. 190–120 BCE) helped develop. For areas, there were Heron's formula and $\frac{1}{2}$ Base × Height for finding triangle areas. For angles, both the $(45°, 45°, 90°)$- and $(30°, 60°, 90°)$-triangles were easy to construct, and one could also bisect any angle. There was a formula for arc length on a circle that in today's radian angle measurement is $s = r\theta$, and for the area of a sector, a formula that we write today as $A = \frac{1}{2}r^2\theta$. Both are linear in θ.

Some historians believe that ancient Egyptian geometers were known as "men of the rope" since they stretched lengths of rope to measure distances, to draw straight lines and to make circles. This would make them literally measurers of the earth, or geometers. Ropes supplied with regularly-spaced knots can be thought of as forerunners of today's tape-measures, and these knotted ropes could easily construct right angles using the $(3, 4, 5)$-triangle. As writing developed, models of these life-sized constructs were drawn on surfaces to stand for the real thing. Through time, the stake and rope morphed into compass and straight-edge. Practical Egyptian geometry grew into an aesthetic intellectual monument showcasing the power of logic and pure thought. But Euclid's first three postulates are nonetheless steeped in the rope mindset—they're about constructing a line segment, extending such a segment, and drawing a circle with given center and radius.

The ancients also developed spherical geometry (hardly a surprise when you think of how central it is to astronomy) and Hipparchus became an expert in it. Some of the main ideas of spherical geometry can be seen by seeking analogies with plane geometry. As one illustration, keep a point P on a sphere fixed, and increase the diameter of the sphere. The spherical cap around P approaches a large disk, and various spherical concepts approach corresponding planar concepts. For example, in plane geometry, the shortest path between two points is a line segment, and on a sphere, the analogue is an *arc of a great circle*, which is the shortest path between two points on the sphere. Great circles on a sphere are the largest possible circles there, and they are the intersections of the sphere with planes through the sphere's center.

Just as connecting three points in the plane by line segments creates a plane triangle, connecting three points on a sphere with arcs of great circles creates a spherical triangle. Through one point Q there pass two sides of the spherical triangle, and the spherical triangle's angle there is the angle between their tangent lines at Q. The measures of the three angles of any spherical triangle always add up to more than π, and how much more (called its *spherical excess*) is proportional to the triangle's area. If the sphere has unit radius, then the spherical excess, in radians, is the area of the triangle, a remarkable result due to Carl F. Gauss (1777–1855).

Spherical geometry brings with it a new concept not found in plane geometry: the *solid angle*. To see what it is, start with a circle on the sphere, and through each point of the circle pass a ray emanating from the sphere's center. The sphere's center is now the vertex of a cone that encloses a solid angle, a three-dimensional analogue of two rays through a point in the plane enclosing a plane angle. The circle could also be a spherical triangle or other simple curve on the sphere. The vertex of the resulting cone encloses a solid angle. Just as the radian is a natural measure of plane angle, the *steradian* is a natural measure of solid angle and as its name suggests, it is a stereo or three-dimensional analogue of radian. On a unit circle, a central angle cutting out an arc of length θ is defined to have size θ radians. Analogously, on a unit sphere, a central solid angle cutting out an area of Θ is defined to have size Θ steradians.

This leads to further analogies. For example, corresponding to $s = r\theta$ is the one-higher-dimension formula $\mathcal{A} = r^2 \Theta$ for the area on a sphere of radius r subtended by a central solid angle Θ. Corresponding to the circle's sector area formula $A = \frac{1}{2} r^2 \theta$ is the solid analogue for the volume of a sphere cut out by a solid angle Θ: $\mathcal{V} = \frac{1}{3} r^3 \Theta$. Since the area of a sphere is $4\pi r^2$, by taking $r = 1$ we see that there are 4π steradians in a full sphere. Just as the two planar formulas are linear in θ, their solid analogues are linear in Θ. Incidentally, there's yet another linear relationship on a sphere. To see it, make a fishing weight by suspending a metal ball from its north pole, and lower it to a depth D in a quiet pond. As long as some of the ball shows above the water, the area of the submerged surface varies linearly with D.

There's not just one geometry. Congruence can be made less restrictive—think, for example, of *similarity*: two objects are similar if they differ by only a uniform magnification or compression. One of the most relaxed congruence is *topological* equivalence. Two objects are topologically equivalent if there's a one-to-one mapping F of one onto the other so that both F and F^{-1} are continuous—that is, "nearby points go to nearby points" under both F and F^{-1}. Notions such as continuity, dimension and genus are studied in topology. (See **topology** in the Glossary.)

Archimedean Spirals

Suppose a point moves around a fixed point O so that its distance from O is proportional to the amount of turning. This creates a spiral, an Archimedian spiral. It was originally discovered not by Archimedes (ca. 287–ca. 212 BCE), but by his friend Conon of Samos (ca. 280–ca. 220 BCE). But Archimedes wrote an entire book on spirals in which he studied various properties of Archimedian ones, and that probably accounts for the name. The equation in polar coordinates reflects its geometric simplicity: if the center of the spiral is the origin, then the polar equation is

$$r = k\theta,$$

where $k \neq 0$ is a constant. Here's a picture of ten turns of the spiral $r = \theta$:

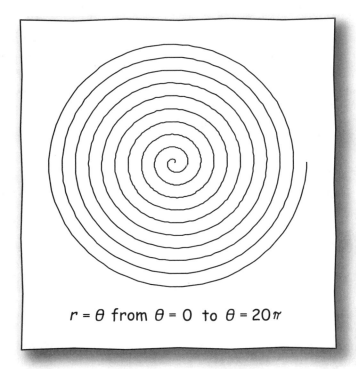

$r = \theta$ from $\theta = 0$ to $\theta = 20\pi$

The picture for $r = -\theta$ is the same, except the spiral turns clockwise instead of counterclockwise. The spiral is tighter when $|k| < 1$, and looser when $|k| > 1$.

Suppose θ increases indefinitely from 0. Then for all positive k, are the corresponding Archimedean spirals all similar to each other? (Similar means congruent up to size.)

(a) Yes (b) No (p. 227)

Spiral Length

Here's one full turn of the Archimedean spiral $r = \frac{\theta}{2\pi}$, with θ increasing from 2π to 4π:

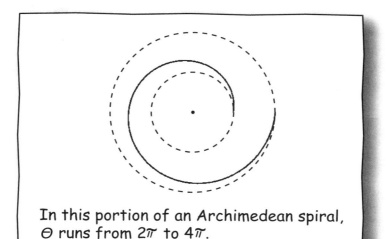

In this portion of an Archimedean spiral, Θ runs from 2π to 4π.

In this turn, r goes from 1 to 2, so the circumferences of the two dashed circles are 2π and 4π. How long is the part of the spiral between them?

(a) 3π (b) More than 3π (c) Less than 3π (p. 227)

Today, most automobile air-conditioning systems use Archimedes-spiral compressors. In the picture, one spiral orbits around the center of an identical, fixed spiral. The two spirals are always in firm contact. This divides the space between the spirals into separate crescent-shaped regions, and as the motion continues, these regions rotate and shrink in area (or expand, when run in reverse). Gas entering at the perimeter is compressed as it moves to the center. Archimedean compressors are quiet, efficient, very reliable and are increasingly replacing piston compressors in many segments of industry.

Church Window

A circular church window contains six identical smaller circular windows like this:

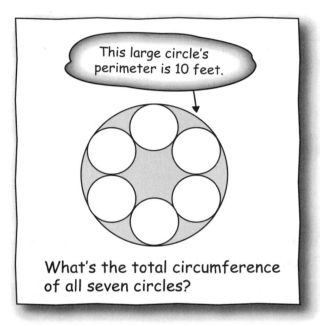

If the circumference of the large window is 10 feet, then the total perimeter (in feet) of the large circle plus the six small ones is closest to

(a) 27   (b) 28   (c) 30   (d) 31.4   (e) 34   (p. 228)

> The six small windows above form a ring of pennies, and a seventh penny fits perfectly in the center. You can extend the pattern throughout the plane, and it's not hard to show this arrangement is the closest packing of identical circles. Going up a dimension leads to a natural question, "What's the closest packing of identical spheres in 3-space?" Kepler guessed that arranging them the same way a grocer stacks oranges is densest. That is, in the closest packing of circles, make each circle the equator of a sphere and extend the circle pattern to a sphere pattern. Then stack copies of this one on top the other, letting the layers fall down as far as possible. Gauss proved that Kepler was right provided there was no denser irregular arrangement. Nearly 400 years after Kepler made his conjecture in 1611, Thomas Hales apparently completed a proof in 1998 by using a computer to analyze the thousands of possible irregular arrangements.

Geometry

Church Window Again

In the last problem, what fraction of the large circle's area is shaded?

(a) $\frac{1}{4}$ (b) $\frac{1}{3}$ (c) $\frac{\pi}{9}$ (d) $\frac{3}{7}$ (e) None of these

Which of these is closest to the percentage occupied by the central shaded area?

(a) 10% (b) 15% (c) 18% (d) 20% (e) 21%

Which has greater area, the inner shaded region or the outer shaded region surrounding the ring of six pennies?

(a) The inner (b) The outer (c) They're the same. (p. 228)

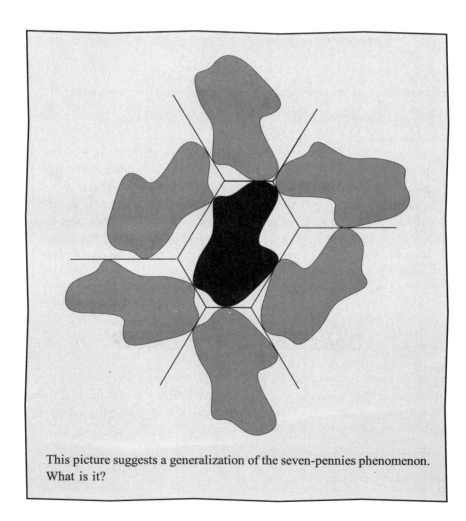

This picture suggests a generalization of the seven-pennies phenomenon. What is it?

Mastic Spreader

A tile installer used a sheet-metal hole puncher to make this tool for spreading mastic:

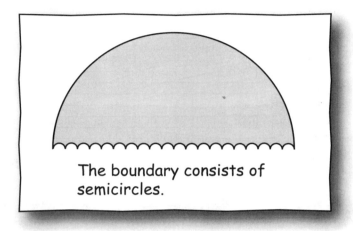

The boundary consists of semicircles.

Which perimeter is longer—the top semicircle, or the bottom serrated edge?

(a) The top   (b) The bottom   (c) The perimeters are the same.   (p. 229)

> Spreaders for tile adhesive almost always have serrated edges. This limits the amount of adhesive spread, so when you press the tile into place, excess mastic doesn't fill the channels between the tiles where grout is supposed to go. A serrated edge also creates air channels so that when the tile is put into place, air has escape paths. A bubble trapped under the tile means less adhesion area, and a large bubble can make it nearly impossible to push the tile down to make it level with the surrounding ones.

Does This Curve Exist?

Starting with a curve in the Cartesian plane, stretch it vertically so that for some $a > 0$, each of its points (x, y) moves to (x, ay). Other than horizontal lines, are there any curves for which you could accomplish this stretching without actually stretching at all, but simply by translating the curve in some direction?

(a) Yes, there are lots of curves like this.

(b) There's exactly one curve like this.

(c) No   (p. 229)

Geometry

Elliptical Driveway

In the figure below, the smaller oval is an ellipse having semi-axes 25 feet and 15 feet, and forms the inner edge of an elliptical driveway. At each point of the ellipse, a perpendicular distance of 10 feet is marked off to define the drive's outer edge.

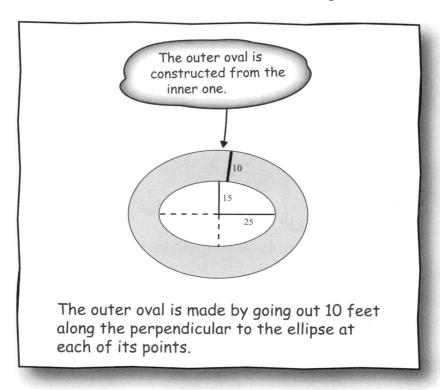

Is the outer oval an ellipse?

(a) Yes   (b) No   (p. 229)

Perpendicular at Both Ends

Suppose the edges of a driveway are two concentric ellipses, the inner having semi-axes of 25 feet and 15 feet, and the outer of 35 feet and 25 feet. A ray going out horizontally from $(\pm 25, 0)$, or vertically from $(0, \pm 15)$, meets the larger ellipse in a right angle. Can any *other* rays perpendicular to the small ellipse meet the larger one at right angles?

(a) Yes, they all do.   (b) Yes, there are some more.   (c) No   (p. 230)

Star Area

Here's a familiar image:

U.S. star in a circle

Which of these is closest to the percentage of the disk occupied by the star?

(a) 35%   (b) 40%   c) 45%   (d) 50%   (e) 55%   (p. 230)

> It's easy to think that the familiar five-pointer has always been the official star for the U.S. flag. Not so — the Continental Congress's first Flag Act states only that "… the flag of the United States (will) be made of thirteen stripes, alternate red and white; that the union be thirteen stars, white in a blue field, representing a new Constellation." Most historians now believe that the first flag was designed by Francis Hopkinson, one of the signers of the Declaration of Independence, and that it was he who chose the star shape we see today. George Washington's original pencil sketch for the U.S. flag shows six-pointed stars. Other stars were used at various times and localities: Bennington's 1777 version of the U.S. flag sported seven-pointed stars, and the Serapis 1779 and Guilford 1781 versions each used eight-pointers.
>
> Hopkinson duly billed Congress for his flag-designing efforts "a quarter cask of the public wine." Congress never paid.

Stereo Camera

In the 1950s Kodak introduced a stereo camera that took a roll of 35mm slide film:

This stereo camera manufactured in the 1950s takes one roll of ordinary 35 mm color slide film. How does it prevent double exposures without a lot of waste?

When you snap a picture, two lenses expose two standard-size rectangles on the film. The pair of shots, mounted on a cardboard frame, is then viewed through a stereo viewer.

You crank the film forward to expose fresh film, but how did Kodak avoid double exposures without wasting a lot of film? Suppose successive rectangles on the roll are numbered $1, 2, 3, \ldots$. Then if the camera exposes every other frame such as $(1,3), (2,4), (3,5), (4,6), \ldots$, most frames are double-exposed. Double exposing is avoided in a pattern like $(1,3), (4,6), (7,9), \ldots$, but this wastes a third of the roll! Which of these would you guess Kodak chose as the solution to this problem?

(a) Use some fixed advance amount to lessen waste.

(b) Appropriately vary the amount the film advances between shots.

(c) Build the lenses one frame apart, but aim them so they diverge slightly, thus achieving the necessary stereo effect.

(d) Build the lenses one frame apart, but aim them so they converge slightly, thus achieving the necessary stereo effect.

(p. 231)

Diagonals of a Parallelogram

In the parallelogram below, the side lengths are fixed. As θ varies, so do both diagonal lengths D_1 and D_2.

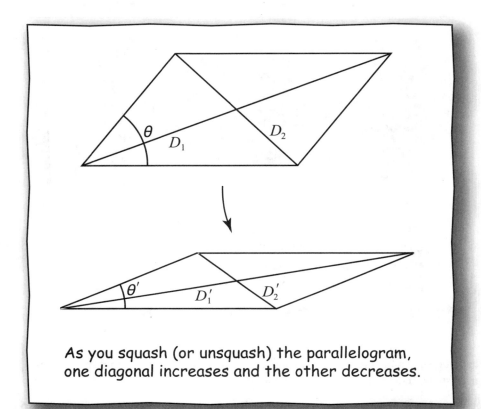

As you squash (or unsquash) the parallelogram, one diagonal increases and the other decreases.

As one diagonal gets shorter, the other gets longer. Do they offset each other in some sense?

(a) $D_1 + D_2$ is a constant. (b) $D_1 \cdot D_2$ is a constant.

(c) $D_1^2 + D_2^2$ is a constant. (d) None of the above is constant.   (p. 231)

> The standard pronounciation of a word often hides its origins. Take *parallelepiped*, for example. If the usual pronounciation were instead parallel-epi-ped, we could better see its etymology: a parallelogram on (*epi*) feet (*ped*).

Half and Half?

It's easy to show that any finite set in the line can be bisected by a point. That is, if the set consists of an even number of points then there's some point P on the line so that half the points in the set lie to the left of P and the other half lie to the right. If the set has an odd number of points, then P is in the set.

In this picture, a line similarly bisects a finite set in the plane:

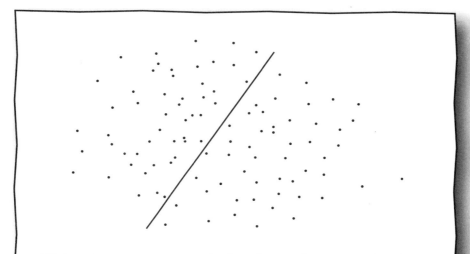

This picture was made by first drawing the line, then adding an equal number of points on each side. Can the order always be reversed?

Does this always work? That is, for any finite set in the plane, is there some line that analogously bisects it? (Assume that if the set consists of an odd number of points, then exactly one point of the set is on the line.)

(a) Yes   (b) No   (p. 232)

Half and Half Again

What about a circle analogue to the above problem? Will there always be a circle for which half the points are *in* the circle, and the rest of the points are *outside* the circle? (Assume that if the set consists of an odd number of points, then exactly one point of the set is on the circle.)

(a) Yes   (b) No   (p. 232)

A Platonic Solid

One of the five Platonic solids is a convex polyhedron with 12 identical regular pentagonal faces—the regular dodecahedron. One sometimes sees calendars made in this form. The *inscribed* sphere (the largest sphere inside this dodecahedron) touches the center of each face. Here's a picture of the reverse—the *circumscribed* sphere (the smallest sphere containing the dodecahedron):

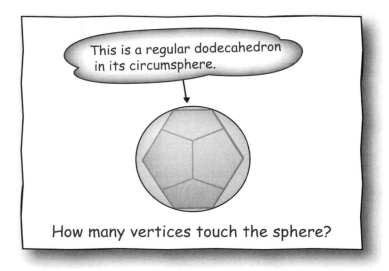

How many of the dodecahedron's vertices touch it? Choose all that apply.

(a) 4   (b) 10   (c) 12   (d) 20

(e) There's a mistake in the picture.   (p. 233)

> The ancient Greeks were familiar with the five Platonic solids, the regular polytopes of 4, 6, 8, 12 and 20 faces. But there are actually *nine* regular solids, not five. Kepler found two new stellated dodecahedra, and two centuries later Louis Poinsot discovered another regular dodecahedron (12 faces) as well as a new regular icosahedron (20 faces). It has been proved that there are no other regular solids, those for which all faces are identical regular polygons, with all corners alike. "But," you may protest, "I thought there are exactly *five* Platonic solids, not nine!" The big difference is that the Platonic solids are convex, meaning that for any two points in the set, the line joining them is entirely in the set, too. In two dimensions, a regular pentagon is convex, while the star-shaped pentagram isn't. Appropriately oriented, each of the four non-Platonic regular solids can collect rainwater. The Platonic ones cannot.

Geometry

A Chip Off the Old Block

The pyramid-shaped chip below is sawed off a cube:

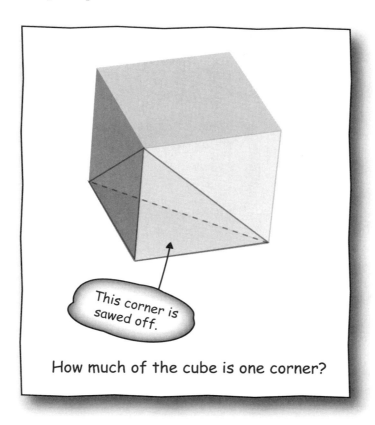

This corner is sawed off.

How much of the cube is one corner?

The volume of the chip is what fraction of the whole cube?

(a) $\frac{1}{2}$ (b) $\frac{1}{3}$ (c) $\frac{1}{4}$ (d) $\frac{1}{5}$ (e) None of these   (p. 233)

Pyramid Volumes

Compare the volume of the above *triangle*-based pyramid with that of a *square*-based pyramid whose base is one face of the cube and whose remaining vertex is the center of the cube. Which pyramid occupies the greater proportion of a cube?

(a) The chip   (b) The square-based pyramid   (c) They're the same.   (p. 233)

Ring-Shaped Fishing Weight

A metal ring is used as a fishing weight, like this:

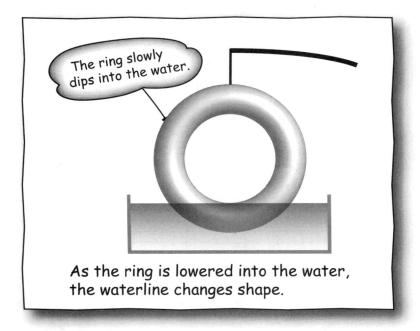

The weight is slowly lowered into a quiet pond. At first, the waterline is a little loop that grows as we continue to lower the ring. When the ring (a *torus*) is at the level shown in the figure, the waterline forms a figure 8. Here are the waterlines just before, at, and after this critical stage:

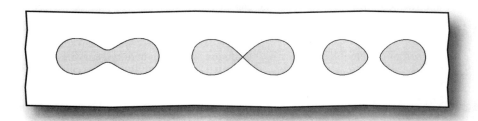

A point where a curve crosses itself is called a cross point. The figure 8 curve has exactly one.

Is it possible to hold the torus in the water so that the waterline has *two* crosspoints?

(a) Yes   (b) No   (p. 234)

Geometry

Largest Cylinder in a Sphere

Of all the rectangles inscribed in a given circle, the one with the greatest area is a square. The square in a circle can be rotated about a diameter to get a cylinder in a sphere:

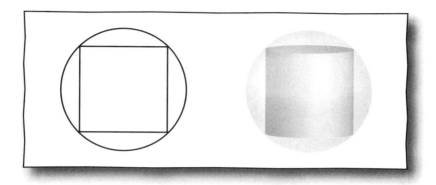

Of all cylinders inscribed in the sphere, does this one have the greatest volume?

(a) Yes   (b) No, the largest cylinder is taller.

(c) No, the largest cylinder is shorter.   (p. 235)

Largest Cone in a Sphere

Of all the triangles inscribed in a circle, the one having the greatest area is equilateral. An equilateral triangle in a circle can be rotated about a diameter to get a cone in a sphere:

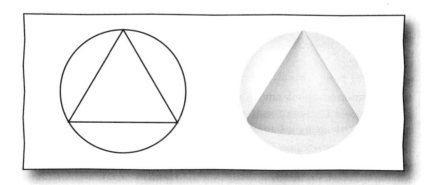

Of all cones inscribed in a sphere, does this one have the greatest volume?

(a) Yes   (b) No, the largest cone is taller.

(c) No, the largest cone is shorter.   (p. 236)

Spin It Again!

We can spin an isosceles triangle and its incircle to obtain a cone and its insphere:

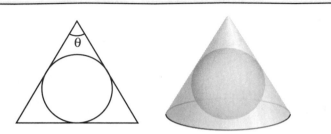

How does the circle-to-triangle ratio compare to its revolved counterpart, the sphere-to-cone ratio?

Let r denote the ratio of circle area to triangle area, and let R denote the corresponding ratio of sphere volume to cone volume. How do r and R compare?

(a) $r = R$. (b) $r < R$. (c) $r > R$.

(d) Any of (a), (b) and (c) can occur, depending on θ. (p. 237)

> One can think of a solid cone in 3-space as the union of all line segments connecting each point inside a closed plane curve with a fixed vertex outside the plane. The formula $\frac{1}{3} \times$base$\times$height gives the volume of any cone. This includes not only an ordinary right circular cone, but also any pyramid, since that's a special kind of cone. Analogously, we can generalize cylinders in 3-space by replacing a cone's vertex by a parallel copy of the base. The cylinder's volume formula can then be written as $\frac{1}{1} \times$ base $\times$ height. There's even an intermediate concept, where the (dimension 0) vertex is replaced not by a parallel copy of the base (dimension 2), but by a line segment (dimension 1) parallel to the base. To visualize an example of this shape, squash one end of a paper-towel tube from a circle down to a line segment. The volume formula for this shape is $\frac{1}{2} \times$ base $\times$ height. The denominator of the fraction is always 1 plus the difference of the two end dimensions. This works in the plane, too: the area of a parallelogram (think "cylinder") is $\frac{1}{1} \times$ base $\times$ height, and of a triangle (a cone in the plane) is $\frac{1}{2} \times$ base $\times$ height. These ideas extend to n-space.

Solid Angles of a Tetrahedron

The angles of a triangle add up to π radians, and π is half the perimeter of a unit circle. Do the solid angles of a tetrahedron add up to 2π steradians, half the surface area of a unit sphere? In at least one case, the answer is yes. In this picture, P slides down the z-axis, approaching the base of the tetrahedron:

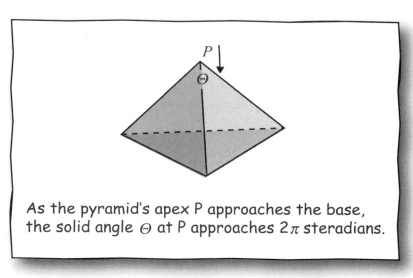

As the pyramid's apex P approaches the base, the solid angle Θ at P approaches 2π steradians.

The top angle Θ is in the limit a flat angle, and then P sees the bottom half of a sphere from its center. The flat angle Θ is therefore 2π steradians. It's easy to see that each of the other three angles of the tetrahedron approaches 0 steradians, giving a total sum of 2π.

But this case is degenerate. What about *nondegenerate* ones? Is the total sum always 2π steradians for them, too?

(a) Yes (b) No (p. 237)

Disk in the Sky

Assuming the moon is 240,000 miles from us and has a diameter of 2160 miles, what fraction of the visible hemisphere is occupied by a full moon? The answer is closest to

(a) $\frac{1}{100}$

(b) $\frac{1}{1000}$

(c) $\frac{1}{10,000}$

(d) $\frac{1}{100,000}$ (p. 238)

A Right Turn

Round duct material comes in flat rectangular sheets with interlocking grooves along the long edges. Snapping together the two long edges of a sheet to make round duct is like the reverse of cutting a paper-towel tube lengthwise and flattening it out into a long rectangle.

Now suppose you need to cut round ductwork to make a 90° turn by joining the two cut pieces to look like this:

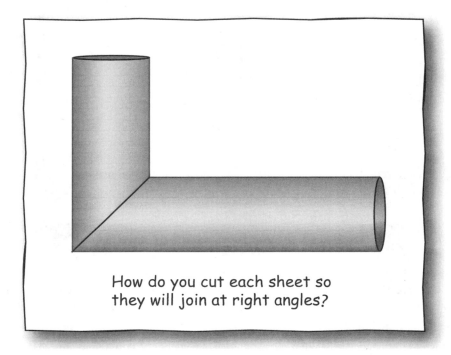

How do you cut each sheet so they will join at right angles?

You need to cut one short side of each rectangle so that after the long edges are snapped together, the two cylindrical pieces fit to make the right turn shown above. Which of the following cuts on the next page does the trick?

> When space is tight, the sharp turn can be useful. But installers usually use smoother turning elbows because they create less turbulence and are therefore quieter and more energy-efficient.

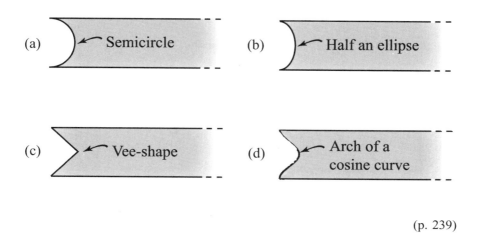

(p. 239)

A Limit — or Chaos?

Suppose the circumference 2π of a unit circle is approximated using inscribed regular n-gons. If P_n is the perimeter of an n-gon, then $P_n < 2\pi$, and we know $\lim_{n \to \infty} P_n = 2\pi$. Using circumscribed regular n-gons similarly produces perimeters Q_n, with $2\pi < Q_n$. As $n \to \infty$ the sandwich $P_n < 2\pi < Q_n$ closes down on 2π. As n gets large, where in the interval $[P_n, Q_n]$ does 2π lie? Does it approach the middle? That is, if you magnify and reposition the interval so it looks the same no matter how large n is, will 2π get more nearly centered as $n \to \infty$? Or will something else happen?

(a) 2π will get more nearly centered.

(b) 2π will approach a point to the left of center.

(c) 2π will approach a point to the right of center.

(d) 2π will jump around in a seemingly chaotic manner.   (p. 239)

What About Areas?

On a unit circle, we can play the same game as above: approximate the area π of a unit circle by the areas A_n of inscribed regular n-gons, as well as by the areas B_n of circumscribed regular n-gons. Doing this produces the sandwich $A_n < \pi < B_n$, and as $n \to \infty$, it closes down on π. As above, magnify and reposition the intervals $[A_n, B_n]$ so they look the same for any n. Will π get more nearly centered as $n \to \infty$?

(a) π will get more nearly centered.   (b) π will approach a point to the left of center.

(c) π will approach a point to the right of center.   (d) π will jump around in a seemingly chaotic manner.   (p. 241)

A Glass Skeleton

An artisan creates a cube from 12 identical glass rods:

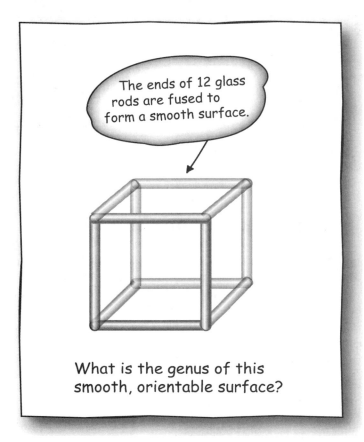

The ends of 12 glass rods are fused to form a smooth surface.

What is the genus of this smooth, orientable surface?

After the ends are fused, they're burnished so the surface is smooth everywhere. What's the genus of this surface?

(a) 3   (b) 4   (c) 5   (d) 6   (e) None of these   (p. 242)

> It was Alfred Clebsch (1833–1872) who apparently first used the term *genus* to mean the number of holes in a surface. He used it in an influential 1864 paper in which he applied abelian functions to algebraic geometry. (Uber die Anwendung der Abelschen Functionen in der Geometrie, *Journal für Mathematik*, pp. 189–243.)

Adding a Room

Suppose the artisan adds one more room to the skeleton on the previous page, using eight more rods:

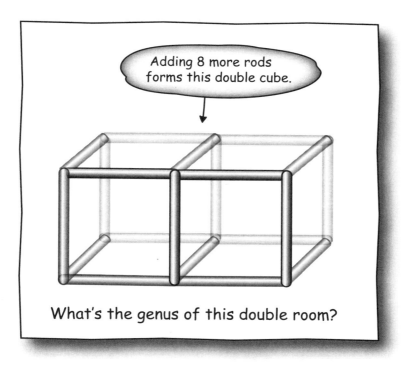

Adding 8 more rods forms this double cube.

What's the genus of this double room?

What's the genus of the double-cube?

(a) 6   (b) 7   (c) 8   (d) 9   (e) None of these   (p. 242)

> The genus makes sense not only for surfaces in 3-space, but also for ones residing in higher dimensions. For example, the set of all complex solutions to $x^n + y^n = 1$ in complex 2-space forms a surface in real 4-space. After adding a finite number of points at infinity, the surface is topologically the same as a donut with $\frac{(n-1)(n-2)}{2}$ holes. This may look innocent enough, but for large n—say $n = 1000$—the number of holes is 498,501. But plot $x^{1000} + y^{1000} = 1$ in the real plane, and all you see is a square with only the slightest rounding at the corners. The nearly half-million holes are hidden from us three-dimensioneers! Someone able to see well in four dimensions could let us know where they are. Does the surface look like a sponge? Do the holes cluster around the corners?

Genus of a 4-Cube

By induction, you can draw pictures of cubes of any dimension:

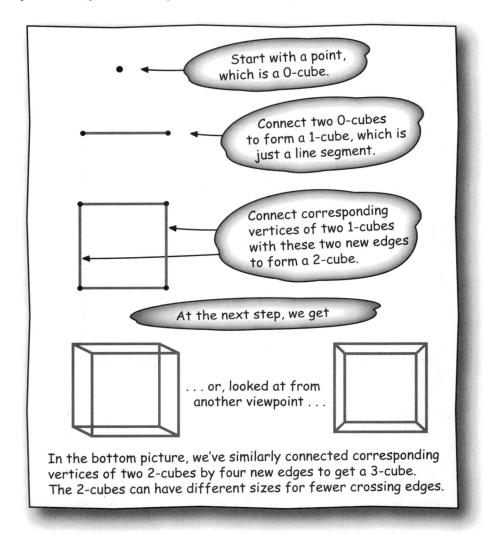

The basic idea is this: to construct the edges of an $n + 1$-cube, take two parallel n-cubes and connect corresponding vertices with straight line segments. In $\mathbb{R}^{n+1}$ itself, we can ensure that these new line segments don't cross. Making a 4-cube this way can actually be accomplished in 3-space: place a parallel, smaller model of a 3-cube inside a larger one, and connect corresponding vertices with straight edges.

What's the genus of the surface of a 4-cube constructed with glass-rod edges?

(a) 10 (b) 12 (c) 13 (d) 16 (e) 17 (p. 243)

Numbers

Introduction

People began to count long before they began to write. In the late 1930s, an archaeologist sifting dirt in what is now the Czech Republic stumbled upon the most ancient mathematical object found so far: a 30,000-year-old wolf bone with 57 notches in it. These were arranged in groups of five, much as in today's tally-method, suggesting that people may have counted on their fingers then. The relic offers no clue as to what was being counted. Was it days? food supplies? Some 25,000 years later, when trading had grown more widespread, such missing information became important, and filling it in was often done by including pictograms akin to simplified analogues of ancient cave paintings. For example, a clay token might have a stick figure animal inscribed on it to symbolize one sheep. In a sense, writing arose from business math. The original Sumerian writing system evolved from such clay tokens, and its pictograms slowly became more abstract through time. Egyptian hieroglyphs, too, are similar testimony to such evolution. Drawing upon word-sounds (spoken language developed long before writing), hieroglyphs were linked pictograms whose pronunciation could approximate a spoken word. The word itself might be abstract so that, charades-style, we could today draw a pictogram of a reed to stand for the verb "read."

Most people assume man has always counted in base ten. Not so. The earliest bases were 2, 3, 4, and especially 5, which became very widespread. Base 12 appeared in some civilizations due to its great practicality: division is a more demanding operation, but 12, with more factors, often made that easier. The base also fit in with the approximately 12 30-day months in a year. Later, the Babylonians used base 60—again, more factors meant easier division, and base 60 dovetailed with a roughly 360-day year. We still see vestiges of these other bases: one dozen is 12, there are 12 hours on a clock, 60 seconds in a minute and 60 minutes in an hour or in an angular degree—and 360 degrees in a circle. The Babylonians were greatly interested in astronomy, and base 60 had practical advantages for them.

Although by the time of the ancient Greeks base ten had become the standard, this base in no way implies a place-value system in which moving a decimal point multiplies or divides by ten. That requires introducing a zero. It's not that the ingenious Greeks failed to think of zero or "nothingness," for they in fact considered the void in some depth, saw what it seemed to imply and were terrified by it. Both the void and the infinite became regarded as ideas too dangerous to inject into mainstream thinking—either of them could insidiously eat away at the philosophical foundations of their world view. With no zero, the shortcomings of their cumbersome system survived for centuries and it wasn't until around 650 CE that the Indians at last introduced zero as a number and place-holder. But despite its superiority in computing, resistance to change was slow to die—just think of today's metric system versus the British system! Tradition can run deep: sixteenth century Italian accountants caught using the Arabic system could be jailed for it.

Any of the ancient bases of 2, 3, 4, 5, 10, 12, 60 can be modernized to a place-holding system with zero and a respective "point." Moving a number's point one place to the right or left multiplies or divides the number by the base.

How does one represent a whole number in some particular base? For those new to non-decimal bases, there is a convenient and intuitive model: money. For any base b, mint coins having values $b^0 = 1, b^1, b^2, b^3, \ldots$. It turns out that you can pay any amount, correct to the penny, provided you have $b - 1$ coins of each of the values b^n. In base 5, say, the coins would be a penny, nickel, quarter, 125-cent piece, ..., their values being successive powers of 5. With four each of these, you can exactly pay for something costing, say, 71 cents by always doling out the highest-value coin that doesn't result in overpayment. That algorithm leads to 2 quarters, 4 nickels and 1 penny. Recording this transaction from left to right gives 241, which is 71 in base 5. If you have just four coins of each denomination, then this is the *only* way to pay the bill exactly—an example of "existence and uniqueness." This monetary algorithm generalizes from base ten to any base b.

What about a monetary algorithm for converting the other way? A mechanical coin-changing machine does the trick, producing a conversion from right to left: dump 71 pennies into the machine, and let it exchange the pile for nickels. The 14 nickels slide down one chute, and down a second chute slides the remainder, the one remaining penny. That remainder of 1 is recorded in the pennies column. In the next round, the 14 nickels are dumped into the machine, and it converts 2 stacks of 5 nickels into 2 quarters, with 4 nickels sliding down the remainder chute. In the final iteration, those two quarters go into the machine, and they come out the remainder chute. The machine produces a remainder record of 241.

This machine performs long division with remainder. In the first round the dividend is 71, the divisor is 5, the quotient is 14, and the remainder is 1. In the second round, the dividend is 14, the divisor is 5, the quotient is 2 and the remainder is 4. Long division comes down to repeatedly applying the Euclidean algorithm $A = Bq + r$, with inputs A (dividend) and B (divisor) and outputs q (quotient) and r (remainder $r < B$). You also use $A = Bq + r$ every time you check a long-division problem by multiplying your answer by what you divided by, and then adding the remainder to get (you hope!) A.

One could supply the machine with a little number dial which you set to divide by any particular integer 2, 3, 4, ... After setting the dial to some base b, you pour a large number N of pennies into the machine, and it successively churns out a series of remainders—the conversion of N to base b.

Numbers

One can use the coin model to extend numbers to the right of the base point, too. In base 5, say, one could introduce coins having values $\frac{1}{5}, \frac{1}{25}, \frac{1}{125}, \ldots$. There is still the basic existence fact that for any base, any real number has a representation in that base. In terms of ever-smaller monetary units, an expansion $a_n \ldots a_0.a_{-1} \ldots a_{-k} \ldots$ converges to a real number r: given any $\epsilon > 0$, there's a j sufficiently large so that any $a_n \ldots a_0.a_{-1} \ldots a_{-k}$ with $k > j$ is within ϵ of r. That representation is unique among all representations in base b that don't terminate in an unending sequence of $(b-1)$'s.

A Curious Property

Take a whole number, then reverse its digits to get another number. Their difference is always evenly divisible by 9. For example, $83601 - 10638 = 72963$, which is 8107×9. But instead of reversing the digits, suppose you randomly scramble them. Is the difference still always evenly divisible by 9?

(a) Yes   (b) No

Suppose you take into account that a number like 35 can just as well be written as, say, 0000035. Does the reversal trick still work? What about scrambling?

(a) Yes to both

(b) Yes for reversal, no for scrambling

(c) No for both   (p. 245)

Multiplication from Another Era

Generations of school children have learned to multiply two numbers using our familiar staircase algorithm. But many more generations learned to multiply by a very different-looking method, the gelosia (pronounced jee-*low*-see-uh). It was developed in India during the middle 1000s. Leonardo Fibonacci (1170–1250) learned it as a youngster when accompanying his father on a trip there, and later introduced the method to Europe in his 1202 book *Liber Abaci*. By 1450, the method had become very popular. On the next page we see an example of how, for centuries, children multiplied numbers.

> The term *gelosia* stems from the method's resemblance to a type of iron grill put up in front of windows. These are often called "jalousa" (jealousy) gratings in various countries and became commonplace in Italy—jealous husbands installed them to prevent passersby from catching peeks of their wives.

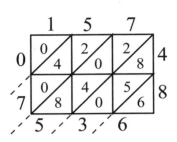

The digits of 157 × 48 are written along the top and right sides. The 2-digit products of the individual digits of 157 and 48 go in the little triangles, as shown. Add triangle values along each diagonal chute (starting with the rightmost), and carry as usual. Read the answer 07536 along the left and bottom sides.

Each square is divided into two triangles. This ingenious method works because the product of any two digits never has more than two digits. Does this method work in *all* bases? Or must we divide the square into more parts for larger bases?

(a) It works in any base.

(b) For large bases, you need to divide the squares into more parts. (p. 245)

How Often Does Lightning Strike?

On average, lightning strikes over each square mile about once every three years. (The frequency is higher over equatorial land and lower over oceans.)

Assuming the earth is a sphere of diameter 7900 miles, how often on average would you say lightning strikes somewhere on earth? The answer is closest to

(a) once each minute.   (b) 2 times each second.

(c) 20 times each second.   (d) 200 times each second.

(e) 2000 times each second.   (p. 245)

Numbers

A Never-Ending Decimal

The infinite decimal $.999\cdots$ is

(a) exactly 1.

(b) an indeterminate amount less than 1.

(c) Neither of these

(p. 246)

What is This?

If you divide 1 by 9, you get the unending sequence $.111\cdots$. Suppose you replace every other 1 by -1 to get $.abc\cdots = a \cdot 10^{-1} + b \cdot 10^{-2} + c \cdot 10^{-3} \cdots$, where $a = 1$, $b = -1$, $c = 1$, and so on. What fraction is that?

(a) $\frac{1}{9}$ (b) $\frac{1}{10}$ (c) $\frac{1}{11}$

(d) Some other fraction   (e) It's not a fraction.   (p. 246)

A Wild Sequence

Consider this sequence of numbers:

$S_o = 10\pi \approx 31.4$

$S_1 = S_0 - \frac{10^3 \pi^3}{3!} \approx -5136$

$S_2 = S_1 + \frac{10^5 \pi^5}{5!} \approx +249,880$

$S_3 = S_2 - \frac{10^7 \pi^7}{7!} \approx -5,742,765$

$S_4 = S_3 + \frac{10^9 \pi^9}{9!} \approx +76,403,121$

$\vdots$

If we continue this pattern indefinitely, the numbers on the right converge to

(a) 0.

(b) a large positive number.

(c) a large negative number.

(d) The numbers don't converge.

(p. 246)

Body-Builder

A fitness buff buys two each of 5, 10, 20, 40 and 80 pound weights for his 10-lb barbell. If each end gets the same total weight, will he be able to set up for lifting everything in 10-lb increments from 10 to 320 lbs?

(a) Yes   (b) No   (p. 247)

Decimal versus Binary

Suppose you convert digitwise from decimal to binary. For example in 57, the 5 converts to binary 101, and the 7 to 111. Concatenating these two conversions gives 101111, which is only 47 in decimal, less than 57. Or in 36, 3 converts to 11 and 6 to 110, so digitwise conversion of 36 produces 111110, and that's only 30, again less than 36. Similarly, 77 converts digitwise to 111111 which is $63 < 77$, and 777 goes to $111111111 = 255 < 777$. For positive integers, you might guess that this digitwise process always yields something less than or equal to what you started with. This conjecture is

(a) true.   (b) false.   (p. 248)

Gottfried Leibniz (1646–1716) was a man of dreams. One of his dreams was that all thought could be reduced to a series of symbols and computations—to a universal logical and mathematical language. Though he never realized this vision, it was a persistent theme running through much of his life's work. He once chanced upon a step counter that would register the number of steps as a person walked. In a blaze of inspiration he at once began to dream of a machine that could automatically perform everyday numerical calculations. When he later saw a reference to an adder built by Pascal—the "Pascaline"—he learned all he could about it, and then built a better one. Pascal's machine was only an adder; the other three basic operations of arithmetic were done by reducing them to a series of additions. Leibniz' machine could do all four operations and take square roots as well. It was in designing his brainchild that Leibniz invented the binary system and showed how it leads to simplifying calculating machines. His fellow scientists mostly ignored the new-fangled and unfamiliar binary system. Some 250 years later, however, a young electrical engineer had his own blaze of inspiration destined to continue the story...

Hexadecimal versus Binary

Suppose we write each hexadecimal digit $0, 1, 2, \ldots F$ in four-digit binary form—that is, write 0 as 0000, 1 as 0001, 2 as 0010 and so on, through F as 1111. What about digitwise conversion from hex to binary? If we try this on, say, $A9C$, we get a digitwise conversion of 1010 1001 1100; both hexadecimal $A9C$ and binary 101010011100 are 2716 in decimal. The question: Does this always work?

(a) Yes   (b) No   (p. 248)

Gelosia Again

In binary, it's not necessary to divide the gelosia-multiplication squares into triangles. (Try it!) Is base two the only base for which this is true?

(a) Yes   (b) No   (p. 249)

When nine years old, John Vincent Atanasoff (a-ta-*na*-soff) (1903–1995) developed an intense, lifelong interest in computing. His father had ordered a slide rule, and when it arrived in the mail the whole family admired but then soon forgot about it—all except for young John. It turned out his father had no real need for it, and it soon became the younster's possession and obsession. Years later, this developed into looking for ways to use the growing electronic technology to automate and speed up numerical computations. One winter day in 1937, needing a break from life's everyday problems, he sped away from Ames, Iowa in twenty-below weather. Hours later, he decided on a Bourbon before driving back home. He sipped, he mulled, and a brainchild began to grow. He aimed to make a clean break from the analog electronic computers of the day—with their inaccuracy, unpredictability and limited speed—and go digital. As part of his mold-breaking scheme, base two was to be fundamental since it dovetailed so naturally with the open/closed states of circuits that would be used. He also decided to use base-two circuitry in place of various enumeration methods found in analog computers. His new marvel would have memory, too—assemblages of capacitors whose electric charges were continually refreshed to keep any leaks from compromising data. A keyboard and screen would serve as input and output devices. Eventually he and his friend Clifford Berry transformed his dream into reality and in 1939 demonstrated to an incredulous audience the ABC—the **A**tanasoff-**B**erry **C**omputer. Today, that refrigerator-sized ABC, not the ENIAC, is legally recognized as the world's first electronic digital computer. Atanasoff is the father of the world's many millions of computers and hand-held calculators.

Do Their Behaviors Correspond?

Any real fraction always has an expansion that continues indefinitely, such as $\frac{1}{3} = .333\cdots$ or $\frac{2}{11} = .181818\cdots$. Some of these infinite expansions end in all 9s, and represent a second flavor, because they're equivalent to expansions ending in all 0s—finite expansions. For example, $\frac{1}{5} = .1999\cdots = .2$ or $\frac{3}{8} = .374999\cdots = .375$.

Is this division into two worlds absolute? Is it intrinsic to the actual fractions (ratios of one integer to another), or is it relative, depending on the base we are using? For example $\frac{1}{2}$ is $\frac{1}{10}$ in base two, and its base two expansion is 0.1—finite, just as its decimal expansion 0.5 is. And $\frac{1}{3}$ is $\frac{1}{11}$ in base two; as in base ten, its base two expansion $.010101\cdots$ goes on forever.

Do base ten and base two behaviors always correspond?

(a) Yes, any fraction having a finite expansion in either of these bases has a finite expansion in the other.

(b) A fraction having a finite expansion in base two has one in base ten, but not necessarily the other way around.

(c) A fraction having a finite expansion in base ten has one in base two, but not necessarily the other way around.

(d) Knowing only that a fraction has a finite expansion in either base two or ten tells us nothing about whether it has a finite expansion in both. (p. 249)

Base ten is an accident. For better or worse, we evolved with ten fingers, or digits, and used them to count. Base two is far more natural for electronic devices of all stripes, but it's a weak base in the sense that it takes roughly three times as many digits to represent integers, compared to base ten. Although computing devices think in base two, we humans don't. Those long strings of zeros and ones are not natural to our more conceptually-oriented brains. Base sixteen offers a convenience here: a single hexadecimal digit corresponds to four binary digits—a four-bit byte. And the double-digits, from 00 to FF, represent eight-bit bytes. Hexadecimal serves as a human-friendly link to the ultimate binary, machine base. How to write the six extra digits in base sixteen? The early 1950s saw overscores, making the 16 digits $0, 1, 2, 3, 4, 5, 6, 7, 8, 9, \overline{0}, \overline{1}, \overline{2}, \overline{3}, \overline{4}, \overline{5}$. In 1956, the Bendix G-15 computer used $0, 1, 2, 3, 4, 5, 6, 7, 8, 9, u, v, w, x, y, z$. But it was IBM who ultimately prevailed when in 1963 it introduced the current digits $0, 1, 2, 3, 4, 5, 6, 7, 8, 9, A, B, C, D, E, F$, as well as the term "hexadecimal."

Ingenious Student

A student in a physics lab is weighing an unknown mass using a pan scale looking like this:

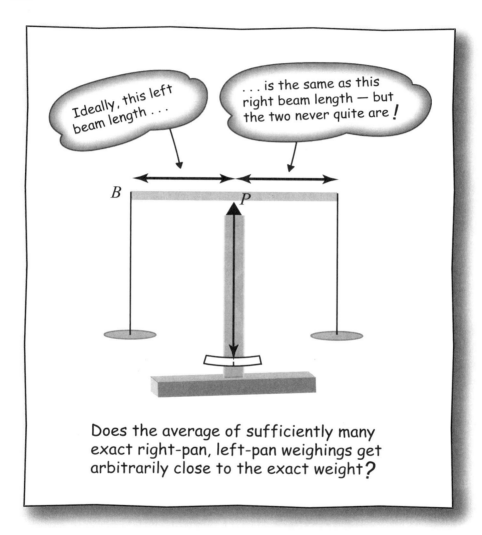

Ideally, beam B is positioned over the fulcrum P so that the left and right arms have equal length. In reality, the lengths are never exactly equal, and the student is aware of this. But he's ingenious and decides to weigh once with the unknown in the left pan, then again with the unknown in the right pan. The average, he figures, ought to be close to the actual weight, and with repeated double weighings, the errors should average out and he can get an accurate answer.

Does this work, or is there a flaw in it?

(a) By symmetry, his method works theoretically; in reality, though, he will need to take an unrealistically large number of careful weighings to get a good answer.

(b) His method works; about four or five pairs of weighings will average out most of the error.

(c) Even with a large number of weighings, his answer will be too big.

(d) Even with a large number of weighings, his answer will be too small.   (p. 249)

Gain or Lose?

A grocer using a pan scale like the above has the habit of alternating which pan gets the standard weights. Over the years, will he gain or lose? The grocer will

(a) gain.   (b) lose.   (c) break even.   (p. 250)

What Sort of Graph?

If for a sale an item is marked down a certain percent x of its usual price and, after the sale, marked up by the same percent, the item doesn't return to its original price. For example, if an article costing $100 is marked down $x = 10\%$ to $90, and then marked up $x = 10\%$ from $90, the final price is only $99. The graph of the final marked-up price versus x is an arc of a

(a) circle.   (b) parabola.   (c) hyperbola.   (d) None of these   (p. 251)

Round Trip

A motor boat makes a round trip up and down a river. Relative to the river's banks, it goes 10 miles each way. Suppose that in analogy to someone walking at a constant speed in a moving bus, the boat goes a constant 15 mph relative to the moving water. True or False: The slower the river is flowing, the better the boat's time.

(a) True. The best time would be when the river isn't flowing at all.

(b) False. For the right flow rate, the boat does better than in still water.

(c) Neither. The time lost in one direction is regained in the other.   (p. 251)

A Reality Test

The approximation $\sin\theta \approx \theta$ for small angles θ is well-known, and often can be used to approximate a nonlinear problem by a more easily-solved linear one. An example arises in the equation of a pendulum.

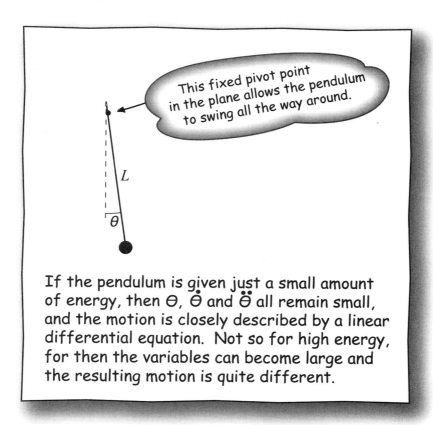

If the pendulum is given just a small amount of energy, then θ, $\dot\theta$ and $\ddot\theta$ all remain small, and the motion is closely described by a linear differential equation. Not so for high energy, for then the variables can become large and the resulting motion is quite different.

Its motion is governed by the differential equation $\ddot\theta + \frac{g}{L}\sin\theta = 0$, where the dot means differentiation with respect to time. The sine term makes the equation nonlinear, and it has no closed-form solution. Such complexity is easy to believe since pendulum behavior ranges from the familiar to-and-fro motion to swinging all the way around its pivot point when the mass is given a large initial kick. (We assume the pivot-point is like a sky hook, in that the pendulum can swing all the way around without interference from the ceiling.) The linear simplification of the equation is $\ddot\theta + \frac{g}{L}\theta = 0$, and this leads to a simple sine-and-cosine solution. It reliably predicts pendulum behavior for only small θ.

As a reality check on this approximation, how good is $\sin(.01°) \approx .01$?

(a) It's good, in the sense that the error is less than 1%.

(b) It's only moderately good—an error somewhere between 1% and 10%.

(c) It's an unsatisfactory approximation. (p. 251)

Friday the 13th Again

Can Friday the 13th occur in two successive months?

(a) Yes   (b) No   (p. 252)

Friday the 13th One More Time

Contrary to what most people think, a leap year doesn't occur every four years—it's occasionally skipped to keep the vernal equinox truly *equinox*. On a true equinox, the time from sunrise to sunset is the same as from sunset to the following sunrise. In our Gregorian calendar, a leap year is omitted in years divisible by 100 but not by 400. For example, February had only 28 days in 1700, 1800 and 1900, but 29 days in 2000. There are in this way three skips in any 400-year period, so the exact number of days in 400 years is $365.25 \cdot 400 - 3 = 146{,}097$. This happens to be evenly divisible by seven, so our calendar exactly repeats every 400 years.

With this information, try this: The probability that the 13th of any month falls on a Friday is

(a) less than $\frac{1}{7}$.

(b) exactly $\frac{1}{7}$.

(c) more than $\frac{1}{7}$. (p. 252)

What Comes Next?

What integer comes next in this infinite sequence?

$$61,\ 52,\ 63,\ 94,\ 46,\ \ldots$$

(a) 89   (b) 101   (c) 65   (d) 53   (e) None of these   (p. 252)

Astronomy

Introduction

Astronomy ranks among the oldest of sciences—hardly surprising, since astronomy played such a central role in the rise of civilization. Before the rise, we were clever nomads cooperating in small groups simply to exist, constantly searching for sustenance. Afterwards, we formed larger groups with greater organization and specialization, nomadic life being replaced by farmers planting crops and raising herds which could feed more people. Civilization gave more time for specialization and increased efficiency, and eventually culture.

The transition hinged on agriculture and that depended on good timing in planting and harvesting crops, leading to a quest for calendars. People began counting moon phases and seeing how noontime shadow lengths changed throughout the year. Such astronomical awareness and increasingly keen observation were in fact prehistoric, and by the time writing began some 5000 years ago there were horizon calendars such as the famous Stonehenge. This could predict sun positions and eclipses throughout the year. The construction material fit the time—the late Stone Age.

By the time of Aristotle (ca. 384–322 BCE) the Greeks had already reasoned that the earth was spherical, since its shadow on the moon during eclipses was circular. (This basic insight was lost by the Middle Ages.) The ancient Greeks went further: about 200 BCE, Eratosthenes estimated the earth's radius to be approximately 7000 km, respectably close to its actual radius of 6350 km. Around then, Hipparchus—Greece's most famous astronomer (ca. 190–120 BCE)—used the moon as a convenient projection screen during eclipses, and measured earth's shadow on it to be very nearly $2°$ wide. He used this to estimate the distance to the moon. He also greatly advanced spherical trigonometry which had immediate applications to astronomy, and some historians believe Hipparchus actually invented trigonometry.

In those days, people had no realistic idea of how the great celestial scene formed. Fortunately, a rough idea of our solar system's creation can be sketched using only the most basic physics, and such it allows one to explain a surprisingly wide variety of everyday facts and to answer many questions. We use just two conservation laws combined with common sense:

- **Conservation of energy.** A simple experiment, dropping a brick, shows how Nature uses this law in forming sun-planet systems and galaxies. As the brick falls and speeds up, its initial potential energy due to height ($PE = mgh$) converts to kinetic energy of motion ($KE = \frac{1}{2}mv^2$). When the brick hits the ground nearly all the kinetic energy changes to heat. Although the original potential energy has changed to other forms, the total energy stays the same. One can easily go from this terrestrial example to a celestial one by imagining a space rock falling toward a planet. Although the height is greater, it speeds up like the brick. If the rock goes into orbit around the planet, the orbit will be elliptical, the rock going faster when it gets closer to the planet, and slower when it's further away. There is a difference in kinetic energy between the slower and faster speeds, and this portion of the total energy oscillates between kinetic and potential forms. If the rock instead hits the planet, nearly all its energy converts to heat, meaning that at a molecular level, the impact has changed the rock's speed into random molecular motion.

- **Conservation of angular momentum.** The picture on the left represents some of the features from a bird's-eye view of an ice-skater pirouetting on one toe, arms outstretched. (Each m is approximately the center of mass of the arm extending along the radius.) The right picture shows that the twirling rate goes up when she pulls in her arms:

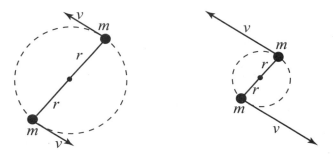

The two masses (which approximate the masses of her arms) rotate in the plane about a center, each mass having an angular momentum of magnitude equal to r times the ordinary linear momentum mv. When the skater pulls in her arms, that decreases r. The conservation of angular momentum says that each product mvr stays the same (is *conserved*). Since the masses don't change, the speed v must increase to keep rv constant, so she twirls faster. Instead of just one or two masses, we could have a celestial version with a great number of masses extending over vast distances. Over time, gravity pulls them closer together, decreasing each r and making the masses rotate faster.

Actually, angular momentum is more than just a number—it's a vector. In the picture, each mvr is the length of a vector aimed at you, perpendicular to the page. This vector is the cross product of the radial vector **r** with the ordinary momentum vector m**v**. In the celestial version, the sum of all such vectors, one for each mass, is the angular momentum of the whole system, and that vector is conserved (even if the masses happen to bump into each other). Both the length and direction of this vector remain fixed. So a disk like a solar system or a galaxy is like an enormous spinning gyroscope: it has stability and doesn't tumble around.

Astronomy

Now to the question: how did our solar system form? Gravity started working on a very large cloud of interstellar dust and gas. Over eons, the particles slowly fell toward the cloud's center of gravity, speeding up as they fell. After further eons, the falling-in process led to a large, central mass. From the conservation of energy, the average speed of its molecules increased. The motion was random, and we ended up with a very hot ball of matter—our future sun. From the conservation of angular momentum, any slight rotation of the overall cloud got magnified because over time, the average r decreased so much. Result? The sun rotates about its own north-south axis, and gravity forced the most central, fast-moving particles to crowd ever closer together into a ball, boosting temperature and internal pressure there, and finally igniting a sustained fusion reaction, one that will continue until our sun eventually runs out of nuclear fuel and burns out. Matter that was spinning around the sun fast enough or was far enough away avoided plunging into the sun. Some of it coalesced into rotating planets which, by the conservation of angular momentum, spin around the sun in the same direction that the sun rotates.

This suggests that for planets, the direction of rotation around their own axes should agree with the sun's rotation about its axis. This is generally true, but there's one big exception: Venus rotates backwards. Why should this be? Many astronomers believe that in the solar system's formative days, all planets *did* rotate in a common direction. But each planet was pummeled by a steady rain of asteroids. Though most were small, a few occasional monsters could have jolted the planet's axis, tilting it. In fact, every planet does tilt, some just slightly, others a lot. Mercury tilts by a modest 2°, the earth, by a more substantial 23.5°. Uranus' 98° tilt is so severe, it "lies on its side." The winner in this strange game is Venus, which must have been knocked hard enough—and from the right angle—to create a tilt of over 177°! At nearly 180° this makes the planet look as if it's rotating backwards with a tilt of only about 3°.

Tilts are important, for they are what give planets seasons. Here's a picture of earth orbiting the sun (the central star):

On the left, we see the earth around June 21, when the sun's rays strike the northern hemisphere most directly, and the southern hemisphere most obliquely. On the right is earth six months later, when it is summer in the southern hemisphere and winter in the northern. The topmost sphere shows earth around March 21 and the bottom one shows our position six months later. These are the vernal and autumnal equinoxes in the northern hemisphere.

It is warm during the summer and cold during the winter because of sunlight's angle of incidence with the earth. Here's an analogy: at night, a book held 20 feet from an auto's headlights can be easy to read if it is held perpendicular to the beams. Holding it parallel to the beams makes it hard to read. Easy reading corresponds to more electromagnetic radiation falling on each square inch of the page. On the earth, summer corresponds to more incident electromagnetic radiation per square mile and winter, less. Near the poles, the sun's light always hits at a more shallow angle (think of positioning the book pages so they're close to horizontal), which is the main reason it is colder in those regions.

There is another big player in the seasonal game, suggested by our illustration on the last page. In the left picture of earth, the earth's tilt means that a circle of latitude near the north pole is completely illuminated. On a more southerly latitude, the leftmost part of the circle is dark. The illuminated part is a circle minus a little arc. The arc grows as you keep going south, and therefore any northern latitude is more than half illuminated. That means northern temperate latitudes receive not only more direct illumination, but for longer periods, meaning longer days and shorter nights, and close to the north pole, the whole circle is illuminated—the sun doesn't set for about six months. In the southern hemisphere, less than half the circle is illuminated, so days are shorter and nights are longer, with a six-month night near the south pole. One more possible factor: the earth revolves around the sun not in a circle, but in an ellipse, so the distance from the sun varies during the year. How much does this contribute to seasonal temperature variation? Earth is actually some 780,000 miles *further* from the sun in June than in December! However, this doesn't amount to much because it's less than 0.84% of our approximately 93,000,000 miles from our sun.

Mars is much like us, in that it has a tilt of 24°, just a half-degree more than earth's. Jupiter's tilt of 3.2° is close to Mercury's 2°. Mercury, Jupiter and Venus are planets that have almost no tilt and therefore no seasons.

It took centuries for our everyday conception of planets spinning around the sun to take root. For a long time, peoples' views of the heavens were rife with misconceptions and falsehoods, one reason being that spaceship earth provides such a high-class ride! The earth spins around its own axis and around the sun with such silken smoothness that nobody feels its motion. No one has ever heard, felt or become seasick from it. It was only through careful observations, logical deduction and inspiration that thinkers concluded that our earth was not stationary. We were all just so close to everything that we had very limited perspective.

There is a simple but powerful symmetry principle that allows us to jump from one observation platform to another. It lets us predict what we would see on that other platform, and it serves equally well for terrestrial and heavenly motion. First, here is how it works in a line. In this picture,

P_1 and P_2 could be planets or points in a line. Let's assume they are observers looking directly at each other. They are a certain distance D apart. P_1 has a coordinate system $\{x_1\}$ with origin at P_1, and P_2 has a coordinate system $\{x_2\}$ with origin at P_2. Relative to P_1, the point P_2 has coordinate $x_1 = D$. Relative to P_2, P_1 has coordinate $x_2 = -D$.

Astronomy

The symmetry principle says that the coordinate of what one observer sees is the negative of what the other sees. More generally, P_1 might not be looking at a stationary point P_2, but one that's moving, say, $x_1 = f(t)$. The symmetry principle applies at each instant, so P_2 sees P_1 moving according to $x_2 = -f(t)$.

The symmetry principle generalizes to several dimensions. The secret is to look at things coordinate-wise. In three-space, $\{x_1, y_1, z_1\}$ are coordinates with origin at P_1, and $\{x_2, y_2, z_2\}$, a translation of them, are coordinates with origin at P_2. Suppose an observer at P_1 sees planet P_2 trace out a space curve that P_1 describes parametrically as

$$x_1 = f(t), \qquad y_1 = g(t), \qquad z_1 = h(t).$$

An observer sitting on P_2 thinks he or she is at rest, and that P_1 is moving. The observer at P_2 describes what is seen with the parametric equations

$$x_2 = -f(t), \qquad y_2 = -g(t), \qquad z_2 = -h(t).$$

The symmetry principle says that if the observers watch each other, then what one observer sees is the negative of what the other sees. If from P_1's vantage point it appears that P_2 tracing out a circular orbit around P_1, then from P_2's vantage point it appears that P_1 tracing out a circular orbit around P_2. If it's clockwise for one observer, then it is counterclockwise for the other. If one traces out a figure 8 so does the other, but in the opposite sense. If one sees the other move leftward and then suddenly make a right turn, the other sees an object moving rightward and then suddenly make a left turn. In summary:

> **Suppose two rectangular coordinate systems have their origins at points P_1 and P_2 and differ only by translation. Then in P_1's system, if P_2 has coordinates (a, b, c), then in P_2's system, P_1 has coordinates $(-a, -b, -c)$, and conversely.**

Which Way?

Suppose we look down upon the earth from above the north pole, like this:

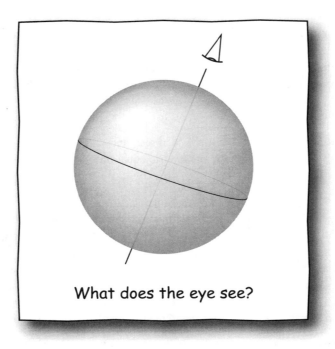

What does the eye see?

Which way will we see the earth rotating?

(a) Clockwise

(b) Counterclockwise (p. 253)

Our Spinning Moon

Our moon is gravitationally (or tidally) locked to us, meaning that the same side of the moon always faces us. (See **gravitational locking** in the Glossary for a fuller discussion.) Therefore the moon spins about its own north-south axis at the same rate it circles the earth. As in the above figure, suppose we position ourselves directly above the moon's north pole and look down. Which way will we see the moon rotating?

(a) Clockwise

(b) Counterclockwise (p. 253)

Astronomy

How High is High Noon?

In New York City, the sun is directly overhead

(a) once a day.   (b) twice a year.   (c) once a year.   (d) Never   (p. 253)

Flagpole Sundial

A vertical flagpole can be used as a rudimentary sundial. Choose the best answer: The tip's shadow moves

(a) clockwise in both Los Angeles and Sydney, Australia.

(b) clockwise in Los Angeles and counterclockwise in Sydney.

(c) counterclockwise in Los Angeles and clockwise in Sydney.

(d) counterclockwise in both cities.   (p. 254)

The Greatest Number of Stars

If comfort were no consideration, where would you be able to see the greatest number of stars during the year? Choose all that apply.

(a) At the north pole

(b) At the south pole

(c) On the equator

(d) It doesn't matter.   (p. 255)

Is This Possible?

Suppose you were standing on the moon near the center of the disk. Would it be possible to construct a sundial to record approximate time there?

(a) Yes   (b) No   (p. 256)

Besides finding the earth's radius, Eratosthenes determined the earth's axial tilt to be 23.5°. It is this tilt which is reponsible for yearly seasons. Wearing his geographer's hat, he introduced the familiar latitude and longitude system on the earth, and is today considered the father of geodesic science. In pure mathematics, he discovered the so-called Eratosthenes sieve for finding prime numbers.

Sunspots

In 1609 Europe, a toy called the telescope was all the rage. Galileo got wind of this, turned it to the heavens and in one stroke opened up a whole new world. By grinding his own lenses, he eventually reached a magnification of nearly 30 and could feast his eyes on hundreds of lunar craters, Saturn's rings, Jupiter's four large moons, ... He even demonstrated that the sun rotates by tracking sunspots. Which way did he see them go?

(a) From right to left   (b) From left to right   (p. 256)

Not a Plumb Plumb

Suppose a plumb line hangs from the top of the Empire State Building to the ground. Assume the earth is an ideal sphere and that there's no wind. Even with these assumptions, the line will not be perfectly perpendicular to the earth. The reason? Mostly because the earth spins about its axis!

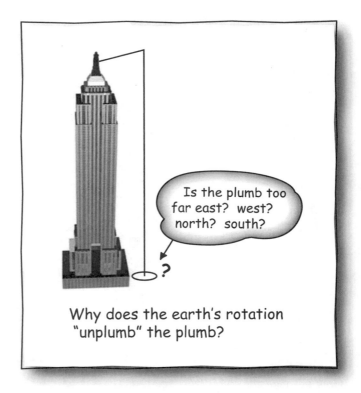

Compared to the top of the line, the bob will be displaced slightly to the

(a) east.   (b) west.   (c) north.   (d) south.   (p. 256)

Foucault's Pendulum

Leon Foucault (1819–1868) devised a pendulum that vividly demonstrates the earth's rotation. To see the gist of his brilliantly simple idea, sit on a swivel-stool and hold a plumb bob by a foot or two of string. Set this pendulum swinging directly toward and away from you. Suppose your friend then slowly swivels you and the stool 90°. Now the bob is no longer swinging toward and away from you, but instead to the right and left! The pendulum's to-and-fro direction is unaffected by the stool's rotation, and in this respect it's similar to a spinning gyroscope balanced on your finger: as the stool rotates, the direction of the spinning wheel's shaft remains unchanged.

Instead of sitting on a rotating stool, you could be sitting on a fixed chair at the north pole, holding the plumb bob by a foot or so of string. Since the earth rotates, your chair acts like the stool, slowly rotating once in 24 hours. Assuming your pendulum loses little energy as the earth rotates about its north-south axis, you'll see exactly the same phenomenon: if the pendulum is originally swinging toward and away from you, then after the earth rotates 90°, you see the pendulum swinging to your right and left. The pendulum has seemingly changed its direction. It hasn't—you have.

The pendulum is typically constructed using a wire suspended from several stories up, the other end being attached to a heavy ball near the floor. This makes it easier to maximize this phenomenon and minimize various frictional effects. Most are in museums or laboratories. In 2001, however, physicists from Sonoma University built one at the south pole in a six-story stairwell of a research station there.

Given that the pendulum never changes direction, which of these are correct? The pendulum

(a) seems to rotate faster at the north pole than in Paris.

(b) doesn't seem to rotate if you're at the equator.

(c) will seem to rotate clockwise in both the northern and southern hemispheres.

(d) will seem to rotate clockwise in the northern hemisphere and counterclockwise in the southern.

(e) will seem to rotate counterclockwise in the northern hemisphere and clockwise in the southern.

(f) will seem to rotate counterclockwise in both hemispheres.

(p. 257)

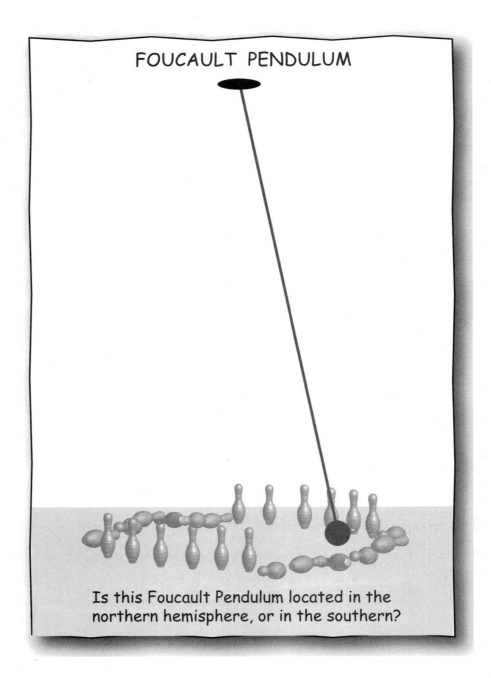

> The history of science might have been quite different if Foucault (pronounced "Foo-*coe*") had lived during the time of Aristarchus in the sixth century BCE. The Greek scientist thought that the earth makes a daily 360° rotation, and that the stars consequently only *seem* to rotate. This novel idea fell on deaf ears—it was quite apparent that *terra firma* was exactly that—a firm, solid, unmoving earth! If Aristarchus was to convince the world, he needed a direct visual, persuasive proof of his idea. He had none. Nor did Copernicus, who some 2000 years later proposed the same idea. Foucault's pendulum (ca. 1850) was the first such simple, "*Show me!*" demonstration. Even children look in fascination at his brilliant demonstration of the earth's rotation, as a long, majestic pendulum tips over the next in a ring of blocks or bowling pins.

Attaching the Wire

In constructing a Foucault pendulum, the top of the long wire is usually attached to a gimbal, allowing the wire to turn freely with almost no torsion. Suppose that instead of wire, we use a long, fine link chain (think of a very strong necklace chain) not attached to a gimbal, but instead welded to the support. Assume that such a chain can twist a full 360° with no torsion. Would this make a satisfactory Foucault pendulum?

(a) Yes

(b) Only for a few days

(c) Not at all

(p. 259)

> Making a good Foucault pendulum is no cakewalk. For one thing, the air dampens the motion of the ball, so that within a few hours most of the swing is gone. To overcome this, a ring of magnets is usually installed near the top of the iron-containing wire or in the floor just below the ball. Periodically, the magnets give the pendulum a small radial tug, just enough to keep its amplitude constant. Even in starting the pendulum, one must be careful. To insure that its motion is truly radial, the ball is usually pulled back and tied in launch position with thread. When the ball is perfectly still, a candle is lit. The flame burns through the thread and the pendulum begins to swing.

> Léon Foucault did much more for science than most people realize. After constructing a 220-foot pendulum in the Panthéon in Paris, he gave a series of hugely popular demonstrations. During this period it occurred to him that a spinning mass might prove to be a yet more persuasive demonstration than a swinging one. To this end he developed the modern form of gyroscope, nicknaming it Greek for the "circle watcher"— *gyros* + *skopos*. Today, gyroscopes of all sizes and shapes have found a virtually unlimited variety of applications.
>
> Foucault also collaborated with his French compatriot Hippolyte Fizeau (1819–1896) in designing a now-famous apparatus for measuring the speed of light, c. It measured the time taken for light sent from the apparatus to return from a stationary mirror some 20 miles distant. Upon completing the round trip the light beam hit a very rapidly-rotating mirror in the apparatus, and during this instant the mirror had moved slightly, displacing the beam from its original position by a tiny amount. This was improved in the Michelson-Morley apparatus. The surprising conclusion that c measures the same, independent of the observer's speed, formed a basis of Einstein's special relativity.

A <u>Lunar</u> Foucault Pendulum?

Go to the moon's north pole and set up a Foucault pendulum, complete with a circle of bowling pins. How would it work there? It would knock down the circle of pins (choose all that apply)

(a) going clockwise.   (b) going counterclockwise.

(c) taking more than 24 hours.   (d) in 24 hours.   (e) taking less than 24 hours.

(f) It wouldn't knock down a whole circle of pins there.   (p. 259)

Retrograde Motion?

Mars is famous for its retrograde motion—it can appear to orbit one way, then slow down and go backwards for a few days, then turn around and continue in its original direction. (See **retrograde motion** in the Glossary.) This was mysterious and puzzling to ancient sky observers. The question: could an observer on Mars see the earth go retrograde, too?

(a) Yes, and up to delay for light travel, simultaneously with an earth observer seeing this motion

(b) Yes, but it might be delayed by a few days   (c) No   (p. 259)

Astronomy

Moonday

How long does one high-noon to high-noon lunar day last?

(a) 24 hours (b) One month (c) Forever (p. 259)

> In the first few hundred million years after our moon was formed, it was an awesome giant in the sky, far closer to earth than it is now. It also orbited the earth faster, and the more frequent tides were monumental—hundreds of feet between high and low several times in each of today's 24-hour day. These tides inundated a great proportion of existing land, greatly accelerating erosion. The tremendous energy loss through tidal friction slowed earth's spin rate. To conserve angular momentum, the moon's spin rate around the earth had to increase, and that boosted it into an ever-higher orbit. Our moon is currently receding from us about an inch and a half each year.

Full Moon

Suppose the moon looks like a disk in London:

A London skyline bathed in moonlight

Did it look like a disk to people in China seven or eight hours earlier?

(a) Yes (b) No (p. 260)

Full Moon, Again

When the sun and moon appear in the sky together, must the moon be a full disk?

(a) Yes (b) No (p. 260)

Moondial

Using our flagpole, suppose we try to make a moondial by marking off the hours on the ground on a clear night when the moon is shining brightly. Can we use this as a nighttime version of a sundial on other bright, clear nights?

(a) Yes (b) No (p. 260)

Moonrise

If you observe our moon rising at sunset, then the moon must be a

(a) thin crescent. (b) half moon.

(c) full moon. (d) Each of the above can occur. (p. 260)

Moonstruck?

Your friend tells you that because you weigh so much less on the moon, you can see the stars from anywhere there, whether you're standing on the lit or dark side. Is he crazy, or is there some actual connection?

(a) He's crazy.

(b) There really is a connection. (p. 261)

Earth's Phases

If you lived on the moon, you'd see the waxing and waning of the earth, from a new earth to a full earth and to waning earth. About how long would one full wax-wane cycle take?

(a) 12 hours (b) 24 hours (c) 2 weeks

(d) A month (e) 6 months (f) A year (p. 261)

Opening or Closing Parenthesis?

Suppose you lived at the center of the moon's disk, so you always get a full view of earth. Which way would the wax-wane cycle go? Would a new earth look like an opening parenthesis, or a closing one? It would look like

(a) an opening parenthesis.   (b) a closing parenthesis.   (p. 262)

Light Lumps, Massive Spheres

Asteroids usually have irregular shapes, while massive bodies like planets and stars are nearly spherical. This has the most to do with

(a) the greater chance for statistical averaging in forming a large body.

(b) gravity being stronger on a large body.

(c) the greater likelihood that a massive body has a hot interior.   (p. 262)

Dynamite Asteroids

Suppose an asteroid approaches the earth at 3 km/sec. By the end of its journey when it slams into the earth, the total energy released is about equal to the asteroid's weight in TNT. However, asteroids commonly travel ten times that fast. That speed translates to how many times the asteroid's weight in TNT?

(a) slightly over 3   (b) 10   (c) 30   (d) 100   (p. 262)

For essentially all its 4.5-billion-year life, everything from huge asteroids to fine interplanetary dust has rained down on our earth. Some 40,000 tons of space material fall on us each year. This, however, is nothing compared to the scene during earth's tumultuous growing-up period. In fact we really are just an enormous ball of asteroids and dust all melted down into a planet. Astronomers believe that when gravity compressed enough matter in a huge gas cloud to form our sun, it left an enormous "ring." The future solar system must have been striking—a freshly-ignited star with a halo. Earth began as gravity rounded up aggregates of cold halo material. We continued to grow as more matter slammed into us—now a nontrivial gravitational force in our own right. These violent impacts were so frequent that the released energy eventually melted earth into a molten ball. Iron sank to the center, and silicates—the material of future mountains, rocks and sand—floated to the top.

Impact Angle

Starting with the same speed a few hundred miles up, two asteroids of equal mass smash into the earth, but at different places and at different angles. Do they release almost the same amount of heat during their encounters with earth?

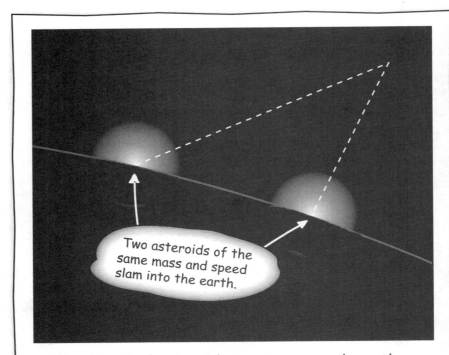

Two asteroids of the same mass and speed slam into the earth.

Two identical asteroids in outer space have the same speed. They both hit the earth, but at different angles. From start to finish, do they release equal amounts of heat?

(a) Yes   (b) No   (p. 263)

> Of the countless asteroids to hit us, it's thought that the largest ever, sometimes called *Theia*, was about the size of Mars, and it nearly tore us apart. The hypothesis goes that the splash flung up material, forming a halo around us. Somewhat like a fast-forward replay of the solar system's formation, the matter in this ring coalesced within a mere century to form a miniature planet—our moon. Its cratered surface is mute testimony to the thousands of asteroids that subsequently hit *it*.

Astronomy

A Manned "Deep Impact" Mission?

For centuries, the nature of comets has been shrouded in mystery. We now know that they are part of the original material that formed our solar system some 4.5 billion years ago. However, there remain many more questions than answers. Experiments like "Deep Impact" and the comet dust collector "Stardust" are helping to replace various conjectures with fact.

In NASA's Deep Impact Mission, its flyby spacecraft released a smaller impactor spacecraft and guided it into the path of the comet Tempel 1. This photo of the July 3, 2005 collision has become one of the most famous space pictures ever taken:

IMPACT!

Suppose you, an experienced fighter pilot, are chosen for a similar experiment, with the difference that this time the mission is manned—from the mother ship you are to manually guide the impactor into the comet's path. Your target comet is roughly spherical, two miles in diameter. Like Tempel 1, it's moving about 10 mi/sec. Safe in your mother ship, you can accelerate your released Deep Impact spacecraft as much as 32 ft/sec² in any direction. Your mother ship is equipped with a telescopic TV camera aimed at the comet. You know that an impact will occur if and only if the comet remains within a two-degree circle printed at the center of your monitor, so your mission is specific: keep the comet within the circle.

For some hours, the comet has remained exactly over the cross-hairs at the circle's center. When the comet is 400 miles away, the telescopic image of the comet has drifted off by about a tenth of a degree, and at 50 miles this has increased to about a quarter degree:

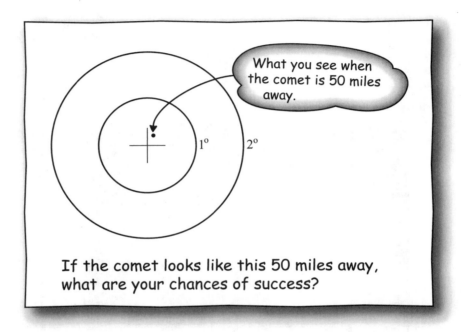

If the comet looks like this 50 miles away, what are your chances of success?

You fire the impactor's thrusters to correct. The chances of success are

(a) a certainty.　　(b) good, but gauging how long and hard to fire the thrusters is crucial.

(c) only fair—there's almost no room for error in firing the thrusters.

(d) hopeless.　　　　　　　　　　　　　　　　　　　　　　　　　　(p. 263)

Vaporizing Asteroid

In plowing through our atmosphere, an asteroid typically vaporizes completely. Occasionally, however, a chunk or chunks survive the trip and crash into the ground. Does the amount of energy released depend upon whether it completely vaporizes?

(a) Yes, more if completely vaporizes　　(b) Yes, less if it completely vaporizes

(c) No　　(d) Not enough information to say　　　　　　　　　　　　(p. 263)

Astronomy

> In summer, more than half a non-equatorial circle of latitude is always illuminated by the sun, while in winter, less than half is always illuminated. This is easy to read from a picture in this chapter's Introduction. A continuity argument shows that at two times in our yearly journey around the sun, just half of any such circle is illuminated. The time following winter is the *vernal equinox*, the one following summer is the *autumnal equinox*.
>
> "*Equinox*" comes from Latin "æquinoctium" which derives from "æquus" (equal), and "nox" (night—compare with *not, naught nocturnal*, etc.)— the times of the year when the length of the night equals the length of the day.

Vernal Equinox

Do the vernal equinoxes in London and in Sydney fall on the same day?

(a) Yes   (b) No   (p. 264)

Martian Equinox?

Does the Mars rover Spirit in a southern latitude of Mars, ever experience a Martian autumnal equinox?

(a) Yes   (b) No   (p. 264)

An Ancient "Aha!"

One of the most remarkable of the ancient Greek thinkers was Eratosthenes of Syene (276–194 BCE). An astronomer, mathematician and geographer, his contributions in any one of these areas would have earned him enduring fame. During the Dark Ages, many people believed the world was flat. But the ancient Greeks not only knew the earth was spherical, Eratosthenes determined its radius, accurate to within five percent of today's value. How did he do it?

He knew that at noon on any year's longest day (the summer solstice), the sun's light reached all the way to the bottom of a deep well located in Syene near the border of ancient Egypt. In modern-day units, it was 500 miles due south of Alexandria. Inspiration hit Eratosthenes: at the solstice noontime, the sun's rays would *not* reach the bottom of a similar well in Alexandria because the sun's rays travel in essentially the same direction there, but the well's axis would be different by some angle θ:

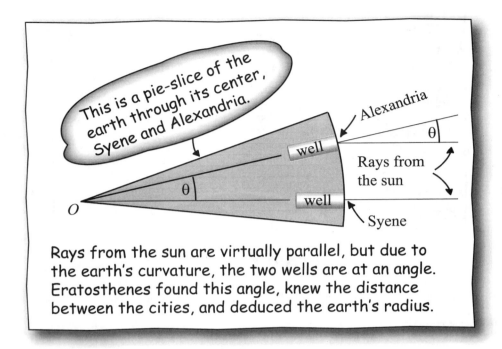

In Alexandria, position a straight pole so it exactly lines up with the sun's rays, the pole casting no shadow. Then θ is the angle between that pole and a plumb line. Of course, the plumb line lines up with an imaginary well in Alexandria. From θ and the known distance of 500 miles, Eratosthenes calculated the earth's radius. In today's units, he came up with approximately 3820 miles. Which of these is closest to the angle he found?

(a) 3°   (b) 4°   (c) 5°   (d) 6°   (e) 7°   (p. 264)

Smiley Face?

Is it possible to stand at just the right place on the earth so that the moon's crescent looks like a smiley face instead of being more vertical, like an opening or closing parenthesis?

(a) Yes   (b) No   (p. 264)

It seems reasonable to assume that unscattered sunrays hitting the earth are parallel. How good is this assumption? This picture will tell us:

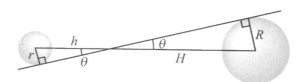

The earth is on the left and the sun is on the right. (Relative dimensions have been exaggerated.) The picture shows that the largest possible angle between rays hitting the Earth is 2θ. We know that the earth's radius $r \approx 3,950$ mi, that the sun's photosphere radius is $R \approx 431,200$ mi, and that the distance between their centers is $H + h \approx 93,000,000$ mi. Since θ is small, $\theta \approx \sin\theta = \dfrac{r}{h} = \dfrac{R}{H}$. It is easy to check that $\dfrac{r}{h} = \dfrac{R}{H} = \dfrac{r+R}{h+H}$. Since we know $r + R$ and $h + H$, we know $\dfrac{r}{R}$ and therefore $\sin\theta$. This leads to $\theta \approx \dfrac{435,150}{93,000,000} \approx 0.00468$ (in radians), which is close to $.268°$. That makes $2\theta \approx 0.536°$, about a half-degree. This is one reason why shadows have fuzzy edges even on a clear day.

Nature frequently replicates at different scales. The way conservation of energy and of angular momentum worked to form the solar system also works at far larger scales to form galaxies. Typically, a galaxy even has a high-energy central core such as a black hole that plays a role analogous to the sun. Here's the Sombrero Galaxy, photographed by the Hubble space telescope.

Archimedes' Principle

Introduction

Archimedes' principle is one of the earliest scientific principles, yet it has a surprisingly wide range of applications. This chapter illustrates some of them. The principle is

A body partly or completely immersed in a fluid is buoyed up by a force equal to the weight of the displaced fluid.

Not only was Archimedes the greatest mathematician of antiquity, he was something of a one-man Department of Defense. He was related to King Hiero II of Syracuse, a just, prudent, and generous man as well as a patron of the arts. The good ruler often called upon Archimedes for ideas, advice and for solutions to a wide range of problems. Once, Hiero commissioned a wreath-shaped crown of pure gold to be placed on the statue of a god or goddess. He began to suspect that he might have been defrauded—could the goldsmith have replaced some of the gold given to him by an equal weight of silver? The goldsmith knew that the wreath was a sacred object dedicated to the gods, which meant the King could not disturb it in any way. Archimedes' task: without harming the wreath, find out whether the goldsmith had cheated. Initially stumped, the story goes that Archimedes had a flash of inspiration as he felt his own buoyancy in a public bath. He was so excited by his insight that he hopped out, rushing naked into the street and yelling, "Eureka! Eureka!"—Greek for "I have found it! I have found it!"

A justification of the principle isn't hard. The basic assumption is that underwater pressure increases directly with depth. The picture at the top of the next page shows a white box, its top and bottom being at depths D_1 and D_2. Initially, suppose the box has no sides, so that just the top and bottom are sitting there, each surrounded by water. The force pressing down on the top is the weight W_1 of the column of water above it. There's an equal upward force under the top since the top isn't moving. The bottom isn't moving, either, so there's a similar downward weight W_2 at the box's bottom balanced by an equal upward force. The column of water over the bottom is longer, so W_2 is larger than W_1. Now add the sides of the box and remove the water from it, leaving an empty box. Then the only force on the top is the downward weight W_1 of the water above it, and the only force on the bottom is the upward force of size W_2. Their combined force is upward of size $W_2 - W_1$. But $W_2 - W_1$ is also precisely the weight of water that the box displaces. That observation is the heart of Archimedes' principle.

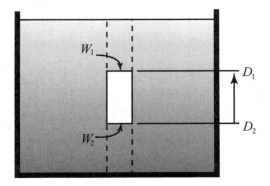

From this one-box justification of Archimedes' principle we generalize to arbitrary shapes. Consider a partly submerged rock and subdivide the part below the water level into many boxes. These boxes may be taller and thinner, and some may lie right above others, but the principle applies to each of them. The total buoyancy force is therefore the sum of the water-weights displaced by the columns. The total water weight displaced by the object is therefore the total buoying force on it. This picture shows a typical such box:

The statement of Archimedes' principle is slick, and says more than it would initially appear. It not only tells us something about bodies that float, it also tells us about ones that sink. Floating amounts to not sinking. A non-sinking body displaces a weight of water equal to the weight of the body—the upward buoyancy force equals the body's downward weight, so the body can float like an iceberg, or be suspended in the water like a submarine. What about a dense object resting at the bottom? *There's still an upward buoying force whose strength equals the weight of the displaced fluid.* Archimedes' principle continues to hold exactly as before, guaranteeing an upward force equal to the weight of displaced water. It's now simply that this buoying force isn't enough to make the dense object float. The object isn't moving, so where is the extra balancing force coming from to keep

Archimedes' Principle

the object stationary? The answer: it comes from the bottom of the tank or pond. That supplies the extra upward force.

Although we've used water, Archimedes' principle applies to any fluid, from liquids like alcohol or liquid mercury, to gasses like the atmosphere, which is really an ocean of air. A hot-air balloon, filled with warmed-up (and therefore less dense) atmosphere obeys Archimedes' principle just as much as a submarine, sunken gold, an iceberg or a floating bottle with a note inside. Archimedes' principle is not static—when conditions change, buoyancy changes. When the air in the balloon cools, air density increases and the balloon drifts downward; when the operator turns on the propane, air density decreases and the balloon rises.

Our justification of Archimedes' Principle is based on weight, and the weight of any mass depends on what gravitational force field it's in. Oil in liquid water rises faster here than on the moon, where gravity is weaker. At a nonaccelerating point in outer space where gravity is zero, oil wouldn't rise up at all—in fact, what direction is *up*, there?

Visual Demonstration?

A solution to the King's crown problem is sometimes presented in picture form. The two weights of the crown and a lump of gold are made to balance perfectly in air, but when carefully lowered into water, they come to rest at an angle because the goldsmith cheated, making the crown from material less dense than pure gold:

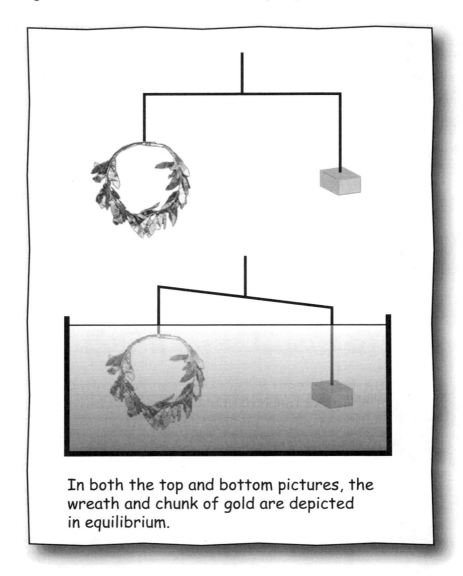

In both the top and bottom pictures, the wreath and chunk of gold are depicted in equilibrium.

(a) The picture is right.

(b) The picture is wrong. (p. 267)

Archimedes' Principle

Another Way?

Is Archimedes' principle equivalent to solving King Hiero's problem, or would it have been possible for Archimedes to solve it without using his principle?

(a) It was necessary.   (b) It wasn't necessary.

(p. 267)

Rain Gauge

A homemade rain gauge consists of a rectangular metal pan with ruler markings on a vertical side. An iron shot keeps the tin from blowing away.

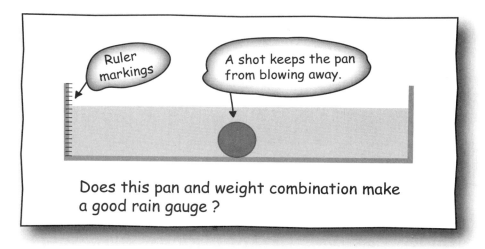

Does this pan and weight combination make a good rain gauge?

When reading the gauge, the shot

(a) should be removed

(b) should be left where it was when it rained

(p. 267)

Here is the first known rain gauge (Korea, AD 1441). This box-plus-tube was fine for rough readings of very heavy rain, but for accuracy the design is exactly backwards—the rain-catching opening should be large, and the readout container's cross-section small.

Beachball

A plastic inflatable beach ball with radius one foot floats in a swimming pool:

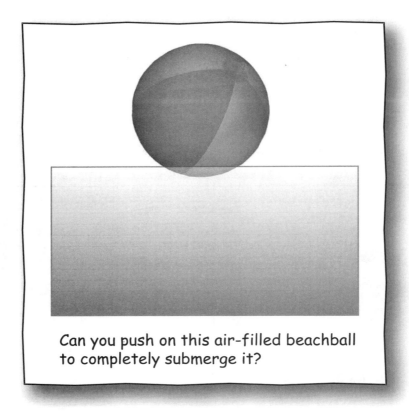

Can you push on this air-filled beachball to completely submerge it?

How likely is it that an adult of average strength, outside the pool, would be able to submerge the ball by pushing on it?

(a) Perfectly reasonable.

(b) A muscle-builder might have a chance at it—provided the plastic beach ball doesn't rupture first!

(c) Forget it. (p. 268)

> How heavy is water? Consider: one cc of water weighs a gram, less than half the weight of a dime. But doing the math shows that a cubic meter of water must therefore weigh a metric ton—100^3 = one million grams, or a thousand kilograms. This is 2204.6 lb, more than the short ton of 2000 lb, and 17% more than a Honda Insight whose curb weight is "only" 1880 lb.

Archimedes' Principle

Top versus Bottom

Does it take more force to keep a plastic inflatable beach ball submerged at the bottom of a six-foot-deep pool or at the top?

Is the beachball more buoyant when it's near the water surface?

(a) It takes more at the top. (b) It takes more at the bottom.
(c) It's the same either way.

(p. 268)

> Water is hard to compress. But under extreme conditions, compress it does. If water were truly incompressible, then the world's oceans, with an average depth of some 12,000 ft, would be about 105 ft higher. At 12,000 ft, water pressure is around 6,000 pounds per square inch, and this compresses water volume 1.75%. Compression varies nearly linearly with depth. Even taking into account the slight decrease in land mass from the higher water level, the average compression from sea level to ocean bottom is about half of 1.75%. This lowers the sea level $12,000 \cdot \frac{1}{2} \cdot 0.0175 = 105$ ft.

Pouring Buckshot

A block of ice with a cavity floats in a vat of water. Some buckshot is poured into either the cavity—as illustrated in the figure—or into the water itself.

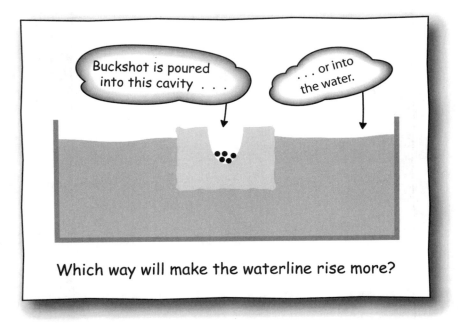

Which way will make the waterline rise more?

The waterline rises

(a) more if the buckshot is poured into the cavity.

(b) more if the buckshot is poured into the water.

(c) the same either way.

(d) Depends on whether the cavity's bottom is below or above the waterline.   (p. 268)

Wooden Spheres

Suppose that in the last problem, the buckshot is replaced with little balsa-wood spheres. They float, so now the waterline rises

(a) more if the wooden spheres are poured into the cavity.

(b) more if they are poured into the water.

(c) the same either way.

(d) Depends on whether the cavity's bottom is below or above the waterline.   (p. 269)

Archimedes' Principle   75

Hot Buckshot

Suppose the buckshot in the cavity is hot, so that the cavity's sides and bottom begin to melt. The buckshot does not break through the bottom:

As this happens, the vat's water level

(a) goes up.   (b) goes down.   (c) stays the same.   (p. 269)

> You might think that hot buckshot would have little trouble melting ice. Actually, most metals aren't very good at holding heat. As an example, an ice cube can cool down an equal weight of hot iron by more 150 Fahrenheit degrees! A cylinder's base can model heat-holding capacity, with a large base representing greater heat-holding capacity. The cylinder's height models temperature, and volume corresponds to the actual heat content. Below is such a representation of water in liquid and solid forms. As ice, its heat-holding capacity is less than as a liquid. Transiting between water's solid and liquid phases requires gaining or losing a *lot* of heat. Iron's base is smaller than any of these.
>
>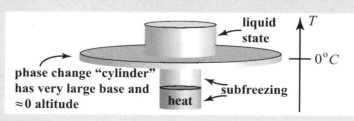

Hot Buckshot Melts Through

At some moment after the hot buckshot has made the cavity's bottom precariously thin, the bottom finally gives way, and the buckshot drops to the vat's bottom:

Hot buckshot has eaten through the cavity's bottom.

When this happens, the vat's water level

(a) quickly rises.   (b) quickly falls.   (c) Nothing happens suddenly.   (p. 269)

> Archimedes' principle doesn't care about the size of the buckshot or wooden spheres. So instead of an everyday number of everyday-sized objects, one could have millions of very tiny particles such as oil spherules in water. Their lower density means that such submerged particles feel a buoying force greater than their weight. They rise when surrounded by water, achieving a steady state when all the oil is floating. Go small enough, however, and the gravity driving Archimedes' principle meets a competitor: molecular bombardment. We see this in emulsions and colloidal suspensions, where Brownian motion drowns out gravity. Smoke, ink, paint, homogenized milk and mayonnaise are examples. Centrifuge the suspension, however, and the resulting centrifugal force serves as a very strong artificial gravity, once again making Archimedes' principle dominant. The newly-created buoyancy usually forces the less dense components to separate and rise.

Air Pocket

Suppose a 50-lb block of ice has an air pocket in it:

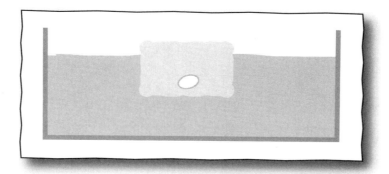

Compared to a 50-lb block of ice with no air pocket, the waterline on the container will be different if

(a) the air pocket is above the waterline.   (b) the air pocket is below the waterline.

(c) It will be different either way.   (d) It will be the same either way.   (p. 269)

> Even the layered structure of our earth reflects Archimedes' principle. When very young, our earth was a molten ball, and iron sank toward the center. Today, iron forms the earth's core. Lighter silicates rose, forming the mantle and crust. A yet lighter compound—liquid water—now floats atop that, with air floating above the liquid water. The lightest gasses, hydrogen and helium, slowly migrate toward the top of our atmosphere, a small proportion of them annually escaping into outer space.
>
> In heavenly bodies, Archimedes' principle isn't the only mechanism leading to layers of decreasing density. Our sun, for example, consists mostly of hydrogen and helium, but atomic motion drowns out any separating power that Archimedes' principle might have on gasses there. Gravity, on the other hand, has crushed the sun's gas core to a density of about ten times that of lead. This very heavy material didn't sink to the center—gravity created high density there. It turns out that the sun's density decreases nearly exponentially with distance from its center, so the sun actually has no distinct radius. While pictures of the sun show a sharp-looking disk, the disk is just the photosphere, the part so hot and dense that it radiates lots of light. The sun's corona extends millions of miles beyond that and, though much hotter, is only about a trillionth as dense as the photosphere and radiates about a millionth as much light.

Melting Ice Cube

Here's a glass of water with an ice cube in it:

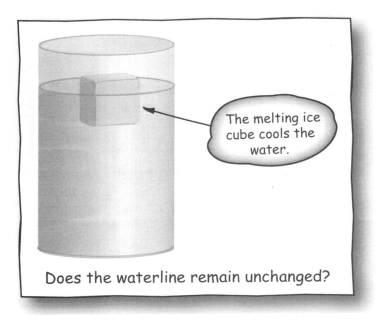

The water starts at room temperature. As the ice cube melts, the water cools, so its density changes slightly. As this happens, the fraction of the ice cube's volume above the waterline

(a) increases slightly. (b) stays the same.

(c) decreases slightly. (d) Depends on the shape of the ice cube. (p. 269)

> At room temperature, water expands approximately in proportion to temperature rise. Galileo combined this fact with Archimedes' principle to create the first modern thermometer. This is still commonly seen as colorful and functional decor: several glass spheres are partly filled with liquid, all inside a sealed column of water. Hanging from each sphere is a small metal weight, calibrated so the one marked 72°, say, matches the density of water at $72°F$, making that sphere freely floating. Less dense spheres (marked with higher-temperature tags) rise to the top, and denser ones sink to the bottom. If none of them is freely floating in the middle, take the average of the lowest-marked sphere at the top and the highest-marked sphere at the bottom. The sphere liquids are usually variously colored for easy temperature identification and for decoration.

Float Capacity

Here's a cylindrical glass of water with a floating, cylindrical ice-berglet:

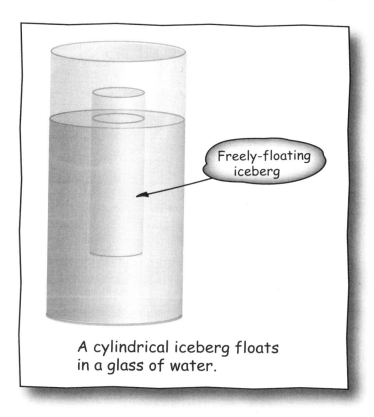

A cylindrical iceberg floats in a glass of water.

Assume that 10% of the volume of any freely floating iceberg is above the waterline. What's the greatest percentage of the water in the glass that can freeze into a cylindrical shape, yet be a freely floating iceberg? The answer is closest to

(a) 90%.   (b) 100%.

(c) Neither, because the glass might initially be over 90% full.   (p. 270)

Carbon Monoxide Detector

If you know some chemistry, use Archimedes' principle (and common sense) to decide where a CO detector should go.

(a) It should go near the ceiling.   (b) It should go near the floor.

(c) It doesn't matter.   (d) There's not enough information to say.   (p. 270)

Arctic Blast

A tube delivers a stream of arctic-cold air to the water's surface, freezing some water and forming a little berg:

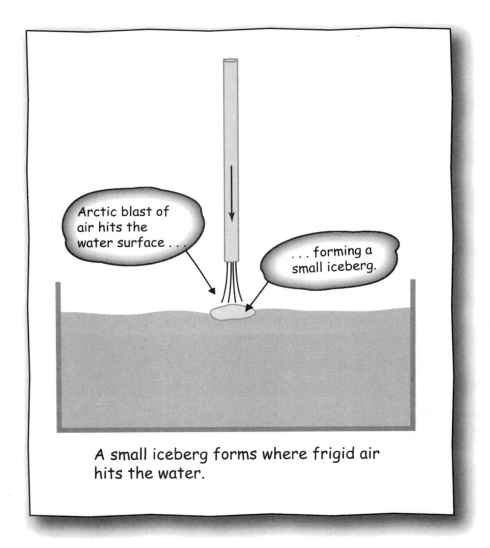

What happens to the waterline?

(a) It rises.

(b) It falls.

(c) It stays the same.

(p. 270)

Liquid Nitrogen

A small steel capsule containing some liquid nitrogen has been dropped into a glass of water:

Ice continues to form around the capsule containing liquid nitrogen.

Ice keeps forming around the capsule; let's assume the ball never gets stuck to the bottom. Eventually the ball rises to the top, forming a floating iceberg. During this process, the water level

(a) only rises.

(b) only falls.

(c) first rises, then falls.

(d) first falls, then rises.

(e) Nothing happens to the water level.   (p. 270)

Archimedes' Principle?

A soup of $\frac{1}{8}$-inch diameter steel ball bearings is poured into a vat. Lubricant ensures that they're sliding around each other with almost no friction. Does Archimedes' principle apply to this tub of ball bearings? Toss a one-inch cube of aluminum into it. Does it float? What about a one-inch cube of tin? Assume that steel weighs 7.85 gm/cc, aluminum weighs 2.7 gm/cc, and that tin weighs 7.28 gm/cc.

(a) Both sink. (b) Tin sinks, aluminum floats. (c) Both float. (p. 271)

Speaking of $\frac{1}{8}$-inch diameter ball bearings, what's the greatest number you could fit in a barrel? Kepler famously—and correctly—conjectured that the densest possible packing of spheres in space just continues the familiar way oranges are stacked in a grocery store. It turns out the spheres stacked this way occupy $\frac{\pi}{\sqrt{18}} \approx 74\%$ of space. That is about 10% higher than if you simply poured them into a barrel.

A Tale of Two Boats

Two boxed-shaped toy boats are open at the top, both boats being cut out from the same large piece of sheet metal. The outside surface area of each boat is one hundred square inches. One of them is two inches high and the other is six inches high, and both have square bottoms. Which can hold more evenly-distributed weight before sinking?

(a) The two-inch high boat

(b) The six-inch high boat

(c) It's a tie.   (p. 271)

Once, a submarine was exploring the ocean floor off the Gulf of Mexico and at one spot was unable to reach the floor, even though just a few minutes earlier it had been at a greater ocean depth. Operators saw the submarine making ripples on an invisible surface, as if it were an ordinary boat on the surface of the water. How to explain this unexpected phenomenon?

It turned out that concentrated brine was seeping up through a crack in the ocean floor. Brine is denser than water, and the brine was heavier than ordinary ocean water. It collected into a large pool, its weight keeping it at the ocean bottom. The submarine's average density matched that of ordinary ocean water and therefore could freely maneuver in it. However, the submarine wasn't dense enough to sink into that even denser underwater pool. Archimedes' principle pops up in the most unexpected places.

In the microgravity of space, the effects of Archimedes' Principle are greatly reduced, and this can have applications. For example in space, it's possible to create more uniform metal alloys. On earth, Archimedes' Principle constantly tries to separate the components of differing density as the molten mixture cools.

Probability

Introduction

The roots of probability go back to prehistoric games of chance and gambling. Today's dice are descendants of knucklebones from various animals such as sheep, deer, oxen and llamas. These early dice had four sides that were often roughly tetrahedral, and on specimens unearthed from prehistoric American Indian burial mounds we see marked sides and a polish resulting from extensive use. There's also evidence of extensive gambling by the ancient Greeks and Romans. Sundry gaming tables were scratched or carved into stone corridors of the Roman Coliseum, and even into steps leading to the temple of Venus. Gambling was an amusement that was popular from soldiers in barracks to the highest strata of society. No upscale banquet was complete without games of knucklebone dice.

With the aim of winning, people must have thought about strategies. It is safe to say that people believed dice could sometimes be hot, the day lucky or the planets fortuitously aligned, and not surprisingly, there are examples of loaded dice from those far-off days. A few even tried to gain an edge by using rudimentary probability—this we know from recorded attempts by ancient Romans to understand gambling odds. But probability theory is notorious for leading even keen thinkers astray. For example . . .

Gerolamo Cardano (1501–1576) was a highly-regarded doctor, not to mention that he earned mathematical immortality through his contributions to solving the general cubic equation. But as a happy-go-lucky college student, he devoted considerably more time to merry-making and gambling than he did to hitting the books. In time, he became addicted to gambling and consorted with an unseemly crowd, even carrying a knife for protection. As a 25-year-old, he wrote a book on games of chance (*Liber de Ludo Alea*). In it, he argues that if you throw a die three times, the odds of getting a six are 50-50, and if you roll a pair of dice 18 times, the odds of getting a double-six are better than 50-50. On the surface, his arguments seemed plausible enough.

Fast forward about a century, and enter one Chevalier de Méré (1607–1684). A familiar figure at the court of Louis XIV, he was the apotheosis of courtly charm. He had risen to serve the highest reaches of nobility as a respected arbiter of arguments ranging from trifling to consequential. A well-regarded author as well as philosopher, the good man would roll over in his grave were he to discover his true, enduring legacy: gambler. But gamble he did, and through many years of it he arrived at some shrewd guesses about winning as well as non-winning strategies.

He learned the hard way that Cardano's strategy of rolling a die only three times to get a six was a losing one. Rolling it four times, he found, won him money over the long term. And getting a double-six? Once again, losses taught him that Cardano's 18-roll strategy just didn't work. De Méré re-argued Cardano's logic, concluding that 24 should be a money-maker—after all, in four rolls of a die you're more likely than not to get a six, and in rolling a pair, that second die has to show *some* value from one to six. It seemed to him that in a sequence of six four-roll experiments, there ought to be one time the second die would be a six, thus forming a double-six. But reality intruded: when he placed even odds on getting a double-six in $6 \times 4 = 24$ rolls, he continued to lose money. It was a 25-roll strategy that paid off. Could logic and mathematical reasoning be trusted to work in the mysterious world of gambling? He needed wisdom, and went to the top for it.

In 1654, de Méré challenged his French compatriot, friend and brilliant mathematician Blaise Pascal (1623–1662). Might the genius be able to solve this and various other conundrums about odds? De Méré's challenges initiated a now-famous series of letters between Pascal and Pierre de Fermat (1602–1665), and their correspondence eventually led to basic concepts such as experiment, sample space, events, expectations and the binomial distribution, thereby establishing the foundations of today's probability theory. We next sketch these definitions and ideas. It will then be easy to see why Cardano was off on both counts, and why the Chevalier de Méré's gambling intuition was correct on both.

Some basic notions in probability and statistics. At the very root of probability theory is the notion of **experiment**. Part of this idea is that it is a repeatable activity having recordable results. The result actually recorded is called an **outcome**. One performance of the activity, together with recording the outcome, is a **trial** of the experiment. An unordered list of all possible outcomes of an experiment is called the experiment's **sample space**, which we'll denote by S. The elements of S are called **sample points** and often (including in this book) they are called **elementary events**. Here are some examples of experiments:

- Repeatedly toss a coin (fair or not) and record the outcomes of heads or tails. One toss-plus-recording is a trial of this experiment. The sample space of the experiment is $S = \{H, T\}$.

- Repeatedly roll a die and record the face value as the outcome. The sample space is $S = \{1, 2, 3, 4, 5, 6\}$.

- Repeatedly randomly draw a card from a deck of 52 and record the card. S has 52 points.

Experiments can involve multiple objects: tossing twenty coins, rolling two dice, randomly picking three cards from a deck of 52. But repeated physical acts of tossing or random selection doesn't itself define the experiment. The recording is an essential part and can lead to very different experiments and different sample spaces. To illustrate, here are four distinct experiments, all involving rolling the same pair of dice. To allow tracking outcomes exactly, assume the dice distinguishable in some way—say, one die has red dots, and that the other has green dots.

Probability

- The two dice are rolled, and the number of red dots and the number of green dots are each recorded, such as "5 red points, 2 green points." In a natural way, the sample space S consists of the 36 ordered pairs (i, j), $(1 \leq i, j, \leq 6)$. In rolling two dice, it's commonly assumed that we don't lose track of which die is which, though just how this is done is not often spelled out. It could be by keeping an eye on them, or marking on them, or coloring their dots as we're doing. Keeping track of the dice leads to a 36-point sample space.

- Let's roll the same two dice, but now record only the number of dots on each, not their color. This is a different experiment, because now we're not keeping track of which die is which. Correspondingly, this experiment's sample space, instead of consisting of ordered pairs, consists of only 21 unordered pairs.

- A third experiment: repeatedly roll the same two dice, but record only the total number of red and green dots. Now the sample space $\{2, 3, 4, \ldots, 12\}$ consists of only 11 points.

- Roll the same two dice and record the nonnegative difference between the number of red and green dots. This leads to the 6-point sample space $\{0, 1, 2, 3, 4, 5\}$.

The notion of **experiment** is completed by adding to it two other notions. The first notion is **event** (sometimes called a **compound event**). Events are subsets of S, sometimes restricted in some way to form a class of subsets of S. For us, when S is finite, the events are arbitrary subsets of S. In any case, the collection of events is assumed to be closed under union, intersection and complementation, and includes the empty set $\emptyset$ and S itself. An event is often defined by some verbal condition. For example, in rolling a die one event could be "the die comes up even," a verbal way of defining the subset $\{2, 4, 6\}$.

The second notion is that any experiment is endowed with a **probability function** P that assigns to each event E a probability $P(E)$ taking values in the unit interval $[0, 1]$. P is assumed to satisfy the intuitive requirements that $P(\emptyset) = 0$, $P(S) = 1$, and if events A and B are disjoint—that is, $A \cap B = \emptyset$—then $P(A \cup B) = P(A) + P(B)$.

Examples:

- Conditions like "the die comes up even" and "it comes up less than 5" can be combined to define the event such as "the die comes up even *and* less than 5"—that is, the subset $\{2, 4\}$. Simple events are always disjoint because a die can't come up showing two different numbers on the top face. So the definition of P implies that if the die is fair, every simple event has probability $\frac{1}{6}$, and that on simple events P is additive: the probability of $\{2, 4\}$ is $\frac{1}{6} + \frac{1}{6} = \frac{1}{3}$.

- Fancier experiments generate fancier sample spaces. In rolling our pair of dice with red and green dots, the sample space of the experiment can be represented as a 6×6 array of ordered pairs, say with (1,1) at the top left and (6,6) at the bottom right. For fair dice, all 36 outcomes are equally likely, meaning each sample-space point has a probability of $\frac{1}{36}$. Here is this 6×6 matrix:

(1, 1)	(1, 2)	(1, 3)	(1, 4)	(1, 5)	(1, 6)
(2, 1)	(2, 2)	(2, 3)	(2, 4)	(2, 5)	(2, 6)
(3, 1)	(3, 2)	(3, 3)	(3, 4)	(3, 5)	(3, 6)
(4, 1)	(4, 2)	(4, 3)	(4, 4)	(4, 5)	(4, 6)
(5, 1)	(5, 2)	(5, 3)	(5, 4)	(5, 5)	(5, 6)
(6, 1)	(6, 2)	(6, 3)	(6, 4)	(6, 5)	(6, 6)

In rolling a pair of dice, the array shows there are six points in the event of getting a seven, since the event's points form the long diagonal of six squares going from bottom left to top right. So the probability of getting a seven is $\frac{6}{36} = \frac{1}{6}$. By counting squares, we see there are five ways to get an eight, so the probability of getting an eight is $\frac{5}{36}$. Similarly, the probability of getting at least one odd die is $\frac{27}{36} = \frac{3}{4}$.

- The language of set theory is useful in probability theory. For example let's denote the above probability of getting a seven by A, of getting an eight by B, and of getting at least one odd die by C. Set-theoretic operations can then make wordy statements or phrases compact. Using an overbar to denote set-theoretic complement, $\overline{A}$ stands for rolling a pair and not getting a seven; $A \cup B$ corresponds to getting a seven or an eight, and $B \cap C$ comes down to getting either $(3, 5)$ or $(5, 3)$. It's then natural to write the corresponding probabilities as $P(A)$, $P(B)$, $P(C)$, $P(\overline{A})$, $P(A \cup B)$, and $P(B \cap C)$.

In each of our examples so far, the sample space contains only finitely many points. A point has dimension zero, as does any set of finitely many points. But many probability problems often lead in a natural way to sample spaces having higher dimensions. The sample space might be an interval, or a bounded area of the plane, or a bounded n-dimensional region in $\mathbb{R}^n$. Each of these represents a problem in geometric probability. Here's an example:

- *Pick a point within a square of side one. What's the probability that the point is less than $\frac{1}{4}$ from the square's boundary?* Looking at a sketch shows that the answer is $\frac{3}{4}$ since the area is within the square but outside a smaller one of side $\frac{1}{2}$.

In general, the probability of an event in n-dimensional sample space is the n-volume of the event divided by the n-volume of the sample space. We will always assume the sample space is bounded. All the definitions given above extend in a natural way to geometric probability theory.

No matter what the sample space S, events are always subsets of S. It is natural to ask whether there are laws for evaluating probabilities of various combinations of events A, B, C, ... in terms of $P(A)$, $P(B)$, $P(C)$, ... The answer is yes. Here are some general laws for complementation, union and intersection.

Complementation. $P(\overline{A}) = 1 - P(A)$. In rolling a die, let A be the event of getting a seven. Then the probability of not getting a seven is $1 - \frac{6}{36} = \frac{5}{6}$, or about 83.3%.

Union. Rolling a pair of dice motivates the essential idea. Let D denote the event of getting a double (meaning that both dice show the same value). What is the probability of getting an eight or a double? That is, what is $P(B \cup D)$? In the 36-point array, B contains 5 points, D contains 6 points, and $B \cup D$ forms an X-shape with 10 points. Directly adding the points in B and D counts the cross-point of the "X" twice, giving 11 instead of 10 points. If we subtract the number of double counts, then everything is counted once, as it should be. In symbols, the general formula is $P(B \cup D) = P(B) + P(D) - P(B \cap D)$. In our example, this correctly gives

$$\frac{5}{36} + \frac{6}{36} - \frac{1}{36} = \frac{5}{18} \approx 27.8\%.$$

Intersection. Roll two dice, let E denote the event of the first die coming up even, and let F denote the event of the second die showing a value greater than 2. Neither die knows what the other is doing—what one die does is independent of what the other does. Counting in the 6×6 array shows that $P(E) = \frac{18}{36} = \frac{1}{2}$ and $P(F) = \frac{24}{36} = \frac{2}{3}$. An outcome in $E \cap F$ means the that first die is even and that the second die shows more than 2. There are just 12 such outcomes, so $P(E \cap F) = \frac{12}{36} = \frac{1}{3}$. For these independent events the probabilities multiply: $P(E \cap F) = P(E) \cdot P(F)$. That is, $\frac{1}{3} = \frac{1}{2} \cdot \frac{2}{3}$. More generally, two events A, B are called *independent* if and only if $P(A \cap B) = P(A) \cdot P(B)$. In the experiment of throwing a dart at a U.S. map, the events of landing in California and landing in San Francisco are not independent. It can't happen that the dart misses California yet lands in San Francisco.

These three laws are general. For any events A, B,

- $P(\overline{A}) = 1 - P(A)$;
- $P(A \cup B) = P(A) + P(B) - P(A \cap B)$;
- $P(A \cap B) = P(A) \cdot P(B)$ is the defining equation for A and B to be independent.

It has been said that it is in probability and statistics that seasoned mathematicians are most likely to make mistakes. One reason is that it is often crucial to realize just how much is known when an experiment is performed. For example, if you know nothing about the outcome of rolling a pair of dice, and someone asks you what the probability is that a total of eight was rolled, the answer might well be just what we saw a moment ago: $\frac{5}{36} \approx 13.9\%$. However, if you happen to notice that one die came up even, then instead of five ways to get an eight, there are now only three—$(2, 6)$, $(4, 4)$ and $(6, 2)$. However, the sample space has changed, too. Knowing that one die is even means that all the odd-odd entries in the original 6×6 array are now eliminated. The sample space shrinks to 27 points, so the actual probability, with your new knowledge, is

$$\frac{3}{27} = \frac{1}{9} \approx 11.1\%.$$

This new result was conditioned by what you learned about the outcome when you happened to notice something about one die.

Here's a more general set-theoretic picture of the situation:

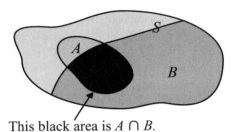

This black area is $A \cap B$.

The entire area is the original sample space S (all 36 points in the dice example), and A is the central region (getting an eight in the dice example). B is the smaller sample space after you learn something—some points of S are no longer possible outcomes. The black region is $A \cap B$, the possibilities for A remaining after your knowledge increases. The probability of getting A knowing B is denoted by $P(A|B)$. This probability is $P(A \cap B)$ normalized to make the new sample space B play the role of the original one, which we do by dividing $P(A \cap B)$ by $P(B)$. That is,

- $P(A|B) = \dfrac{P(A \cap B)}{P(B)}.$

Multiplying both sides of this formula by $P(B)$ turns it into $P(A \cap B) = P(A|B) \cdot P(B)$, a formula for $P(A \cap B)$ that holds whether or not A and B are independent. When A and B are independent, knowing B tells us nothing additional about A, which is another way of saying that $P(A|B)$ is no different from $P(A)$. Then $P(A \cap B) = P(A|B) \cdot P(B)$ reduces to $P(A \cap B) = P(A) \cdot P(B)$.

When A and B are independent, then compared to the last picture, S can be drawn somewhat more informatively as a rectangle, the product $S = S_1 \times S_2$ of two sample spaces. The horizontal side represents the sample space S_1 of the first event and the vertical side the sample space S_2 of the second event. A then consists of full rows in S, and B consists of full columns in S:

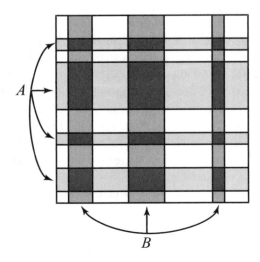

Probability

In a roll of two dice, if A is the event that the first die turns up odd and B is the event that the second die turns up 2, 5 or 6, then there are three full A-rows and three full B-columns. In this picture, the A-rows are lightly shaded, the B-rows more heavily shaded, and their intersection is shaded the darkest:

(1, 1)	(1, 2)	(1, 3)	(1, 4)	(1, 5)	(1, 6)
(2, 1)	(2, 2)	(2, 3)	(2, 4)	(2, 5)	(2, 6)
(3, 1)	(3, 2)	(3, 3)	(3, 4)	(3, 5)	(3, 6)
(4, 1)	(4, 2)	(4, 3)	(4, 4)	(4, 5)	(4, 6)
(5, 1)	(5, 2)	(5, 3)	(5, 4)	(5, 5)	(5, 6)
(6, 1)	(6, 2)	(6, 3)	(6, 4)	(6, 5)	(6, 6)

A and B are independent events because how the first die lands has nothing to do with how the second one lands. Two events being independent says that one event consists of entire rows and the other consists of entire columns. Using the first die's sample space S_1 of six points, the probability that it turns up odd is $\frac{3}{6} = \frac{1}{2}$. Now look at this event within the entire 36-point sample space $S = S_1 \times S_2$. The probability is the same, $\frac{18}{36} = \frac{1}{2}$. The vertical bars multiply numerator and denominator of $\frac{3}{6}$ by 6, leaving the quotient unchanged.

One can even use this picture to illustrate geometric probability. Regard the rectangle as a 1×1 square so that the area of any region is that region's probability. Denote the areas of the four A-rows by a_1, a_2, a_3, a_4, and of the three B-columns by b_1, b_2, b_3. Then $A \cap B$ consists of the 12 black rectangles, so $P(A \cap B)$ is their total area $a_1 b_1 + \cdots + a_4 b_3$. Now comes the magic: by the distributive law, this sum of $4 \cdot 3$ terms factors into the product $(a_1 + a_2 + a_3 + a_4) \cdot (b_1 + b_2 + b_3)$. This gives an intuitive justification for the formal definition of two independent events: A and B are independent if and only if $P(A \cap B) = P(A) \cdot P(B)$.

Another experiment modeling a wide variety of everyday situations is the coin-toss, sometimes called a binomial experiment because a coin has two sides. With dice and coins alike, there's no need to restrict to only one or two tosses. We can flip a penny a thousand times or drop a thousand identical pennies on the floor—probabilistically, the experiments are the same. Coin tosses are independent, and the distributive law works the same magic as above. With a thousand tosses, instead of distributing a product of two sums we distribute a product of a thousand identical sums $p + q$, where p is the probability the penny lands on heads, and $q = 1 - p$ is the probability of showing tails. For a fair penny, $p = q = \frac{1}{2}$. In distributing this product, one can go from left to right, picking either the left term p or right term q in each of the thousand factors $(p + q)$. After all is done, for each r collect terms of the form $p^r \cdot q^{1000-r}$. In $(p + q)^n$, the number of terms $p^r q^{n-r}$ is $\frac{n!}{r!(n-r)!}$. This can be made more symmetric by writing $n = r + s$, giving $\frac{(r+s)!}{r!s!}$. This form suggests how to extend this to products of, say, three terms, such as $(p_1 + p_2 + p_3)^n$, which corresponds to performing a three-outcome experiment n times.

Our coin-flipping experiment has a nice physical analogue, a kind of pinball game:

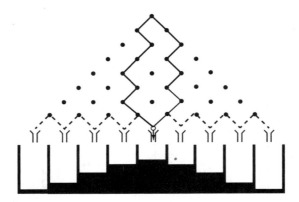

Small ball bearings start a downward journey from the top, moving downward along $\pm 45°$ paths and finally dropping into a bin at the bottom. Two possible paths to the same bin are shown. This models flipping a fair penny eight times. Each flip corresponds to the ball bearing going to the left (for tails, say) or right (heads). If the ball bearings are very small and the experiment is repeated many times, the bins slowly fill up, and standing at some distance, we'll see better and better approximations to a symmetric bell curve. If the penny is flipped a thousand or a million times (or that many fair coins are dropped on the floor), the picture would have many more rows of pins and many more bins. Spacing the pins closer together to make the bins have a common small base Δx produces a smoother curve. Notice that there are more ball bearings in the central area because there are more paths leading to bins near the center than to bins at the ends. In fact, there's only one way for a ball bearing to get to the leftmost bin: it must always go to the left. Similarly for the rightmost bin. The curve approximates a Gauss bell curve better as the number of flips increases. If the coin is unfair, the curve is skewed. If it favors heads, the peak of the curve would move to the right.

Most of the problems in this chapter are about probability, but there are a few on statistics. How are probability and statistics related? Probability is an area of pure mathematics: it starts from hypotheses—realistic or not—and deduces consequences. Probability might assert that in flipping a fair coin twice, the probability of getting two heads is precisely $\frac{1}{4}$. When you physically test this assertion, it is seldom borne out: in ten, a thousand, a million double flips, you usually get something other than $\frac{1}{4}$. Statistics is applied probability. In flipping coins, for example, statistics makes a statement like this: in some particular number of double flips, 95% of the time you'll get double heads within the range 24% − 26%. Probability and statistics converge in the limit, in that probability's assertion becomes true in the limit of infinitely many flips, still assuming the coin remains fair and doesn't wear away to nothing. With increasingly many flips, statistics' buffer zones tighten up, 95% heading toward 100%, and 24% − 26% heading toward 25% − 25%. Probability is deductive, while statistics often has an inductive side. The Chevalier de Méré felt that placing even odds on getting a six in three die tosses lost money, while in four tosses, it made money. He also suspected that placing even odds on getting a double-six in 24 rolls was a losing strategy, while in 25 rolls it was a winner. In these examples he had in effect mined statistical data derived from his past experience, and inferred trends. Pascal proved that de Méré's intuitions were right. In a great many trials of each experiment, statistical observation and probabilistic prediction converge.

Toward the end of this chapter, there are some statistics questions that come down to randomly throwing darts at a target and counting the proportion landing in some specified subset of the target. This is the Monte Carlo method and can be used to estimate areas or volumes that may be difficult to find with calculus. Many other statistics problems can be massaged into looking like this, which means determining the proportion of a sample space occupied by some subset of interest. That proportion is approximated by having a computer randomly choose perhaps hundreds of thousands of points in the sample space, always keeping track of successes versus total tries. The computer uses a random-number generator to create dart-tosses.

It's now time to make good on a promise—an easy probabilistic argument showing why de Méré's intuitions were on target. Here it is. In the fair one-die game, the probability of getting a six on one toss is $\frac{1}{6}$, making the probability of not getting a six $\frac{5}{6}$. The question then becomes, how many times can you can roll before the probability of *not* getting a six falls below .5? In this Russian-roulette question, the die tosses are independent so the probabilities multiply. The question then becomes, how large can n get before $\left(\frac{5}{6}\right)^n$ goes below .5? One can check that $\left(\frac{5}{6}\right)^3 \approx .579$ while $\left(\frac{5}{6}\right)^4 \approx .482$—four rolls is a money-maker. In the two-dice counterpart, the chance (that is, the probability) of getting a double-six is $\frac{1}{36}$, making the chance of not getting a double-six $\frac{35}{36}$. Then: how many times must we roll before the chances of missing a double-six fall below .5? In other words, how large can n get before $\left(\frac{35}{36}\right)^n$ goes below .5? One can verify that $\left(\frac{35}{36}\right)^{24} \approx .509$ while $\left(\frac{5}{6}\right)^{25} \approx .494$, making 25 rolls a winning strategy.

Red and White

Suppose a bin contains 1000 red and 1000 white ping-pong balls, and assume they are thoroughly mixed. After being blindfolded, you draw six balls. The probability that you get equal numbers of each color is closest to

(a) 50%   (b) 40%   (c) 30%   (d) 20%   (e) 10%   (p. 273)

Red, White and Blue

Suppose 1000 blue ping pong balls are added to the above bin with red and white balls, and the bin's contents are thoroughly mixed. Once again, you are blindfolded and you draw six balls. The probability that you get equal numbers of each color is closest to

(a) 33%   (b) 30%   (c) 20%   (d) 10%   (e) 5%   (p. 273)

Dropped Penny

Assume that the diameter of a penny is $\frac{3}{4}$ inch. Drop one on a floor tiled with square linoleum tiles twelve inches on a side:

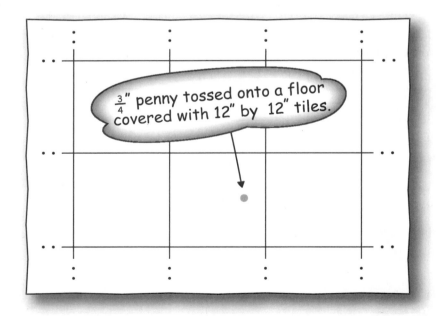

The penny might touch just one tile. Or it could touch two. What about three? four? (Choose all that apply.)

(a) 3   (b) 4   (p. 273)

What are the Chances?

Assume that the diameter of a penny is $\frac{3}{4}$ inch. What are the chances it will land on exactly one of the above linoleum tiles? The answer is closest to

(a) 95%   (b) 90%   (c) 85%   (d) 80%   (e) 75%   (p. 274)

Teachers versus Students

As part of a term project, a group of sociology students sent questionnaires to all teachers and students at their university. One of the items: "For each of your courses, write down how many students are in your class." In their report, the group calculated the university's average class size using responses from the teachers, and again using responses from the students. Assuming everyone is honest, which calculation should be more trustworthy?

(a) Data from the teachers   (b) Data from the students   (p. 274)

(c) The calculations should be about the same.

Exit Polls

In the 2006 U.S. congressional election, exit polls reported that 8 in 10 voters who were for the war in Iraq voted Republican, and 8 in 10 who were against the war voted Democrat. Using this information, one can conclude that the number of voters who were neither for nor against the war is closest to

(a) 1 in 10   (b) 2 in 10

(c) 4 in 10   (d) 6 in 10

(e) Insufficient information to say   (p. 274)

Ideal Tetrahedron

Suppose a fair regular tetrahedron has one face marked 1, another marked 2, a third marked 3, and one face blank. When you toss the tetrahedron, it comes up (actually lands face down on) either 1, 2, 3 or blank. As you repeatedly toss the tetrahedron, you record only the outcomes 1, 2 or 3. Is this experiment equivalent to repeatedly rolling a fair cube whose six faces are marked 1, 1, 2, 2, 3, 3 ?

(a) Yes   (b) No   (p. 274)

Sex and Money

• A couple has two children, and you're told that at least one of them is a girl. Knowing that, what's the probability that both children turn out to be girls?

• You flip a fair coin twice. The first time, you see that it comes up heads. Knowing that, what's the probability that both flips turn out to be heads?

The respective answers are:

(a) $\frac{1}{3}$, $\frac{1}{3}$ (b) $\frac{1}{3}$, $\frac{1}{2}$ (c) $\frac{1}{2}$, $\frac{1}{3}$ (d) $\frac{1}{2}$, $\frac{1}{2}$
(e) $\frac{1}{2}$, $\frac{2}{3}$ (f) $\frac{2}{3}$, $\frac{1}{2}$ (g) $\frac{2}{3}$, $\frac{2}{3}$ (h) none of these. (p. 275)

Three Children

A couple has three children, and you're told that at least two of them are girls. After seeing two of the children but before seeing the third, the probability that the third child is a girl is

(a) $\frac{1}{4}$ (b) $\frac{1}{3}$ (c) $\frac{1}{2}$ (d) $\frac{2}{3}$ (e) $\frac{3}{4}$ (p. 275)

Average Distance

What's the average distance between two randomly-chosen points in the interval [0, 1]?

(a) $\frac{1}{6}$ (b) $\frac{1}{4}$ (c) $\frac{1}{3}$ (d) $\frac{1}{2}$ (e) $\frac{2}{3}$ (f) $\frac{3}{4}$ (g) $\frac{5}{6}$ (p. 276)

Short Sticks

Randomly choose two points in the interval [0, 1] and let them be the endpoints of an interval or stick. What's the probability that the stick is less than $\frac{1}{2}$ long?

(a) $\frac{1}{2}$ (b) $\frac{1}{3}$ (c) $\frac{2}{3}$ (d) $\frac{1}{4}$ (e) $\frac{3}{4}$ (f) none of these (p. 276)

The word *statistics* is related to words like *state*, *estate* and *static* and, in contrast with man's early nomadic life, connotes something fixed, settled and permanent. During the 1700s, the German term "Statistik" meant political science, and some of the earliest collections of data to be statistically analyzed were extensive birth and death records—affairs of the state.

Average Distance-Squared

What's the average square of the distance between two randomly-chosen points in a square of side 1?

(a) $\frac{1}{6}$ (b) $\frac{1}{4}$ (c) $\frac{1}{3}$ (d) $\frac{1}{2}$ (e) $\frac{2}{3}$ (f) $\frac{3}{4}$ (g) $\frac{5}{6}$ (h) None of these   (p. 276)

High-Noon Shadows

A one-foot ruler is tossed into the air 1000 times. Each time, while it's still airborne, a camera snaps a life-sized, vertical-projection ("high-noon") photo of its shadow. Assuming the ground is flat and horizontal, the average of these 1000 shadow lengths is closest to

(a) .1 ft   (b) .2 ft   (c) .3 ft   (d) .4 ft   (e) .5 ft

(f) .6 ft   (g) .7 ft   (h) .8 ft   (i) .9 ft   (p. 277)

Seeing Two Sides

A regular hexagon of side 1 is centered in a disk of radius 2:

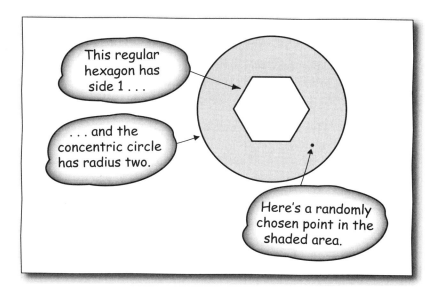

If you randomly pick a point in the shaded region, what's the probability the point can see exactly two sides of the hexagon? The answer is closest to

(a) 80%   (b) 75%   (c) 70%   (d) 65%   (e) 60%   (p. 278)

How Grim the Reaper?

It all started innocently enough—he was shopping in a mall and saw that a local hospital had a tent with a big sign announcing FREE MEDICAL SCREENING. He asked himself, "Why not?" A few days later, he opened up a letter saying that his test proved positive for a life-shortening disease. The letter informed him that 1 in 1000 of the general population has the disease, but went on to offer a bit of consolation, telling him that of all the tests given, 5% of them falsely report positive, indicating presence of the disease when it's not actually there.

What is the probability that he has the disease? Choose all that apply:

(a) About 95%   (b) Over 50%   (c) Under 50%

(d) About 5%   (e) About 2%   (f) About 0.1%   (p. 278)

Can You Trust This Headline?

A baseball fan is extremely confident of his hero Big Al's batting ability and makes a substantial bet that Al will end up with the team's highest batting average for the season. Halfway through the season, Al doesn't disappoint—he leads with .300, followed by his rival's average of .280. In the second half of the season Al keeps his magic, making the team's top average of .250, his rival coming in second with .200. At the end of the season a sports story appears, its headline announcing a new hero: Al's rival. Towards the end, the article notes that Al came in second this year.

The fan sees the story and is livid, fuming "This guy is nuts!" He immediately phones the article's writer, who insists that he is correct. Our fan, not surprisingly, counters with the obvious—simple math shows Big Al had the higher batting average.

What do *you* think? Should the fan win the bet?

(a) Assuming both betting parties are honest, yes, he clearly should win.

(b) The writer shows the fan some less simple math, and the fan gives up. (p. 279)

Common Birthdays

In a room of $365 \times 2 = 730$ people, the chances that each person shares a birthday with exactly one other person in the room is small, but just *how* small? Assuming no February 29th birthdays, the probability is

(a) between 10^{-10} and 10^{-30} (b) between 10^{-30} and 10^{-60}

(c) between 10^{-60} and 10^{-90} (d) less than 10^{-90} (p. 279)

Probability

A Serious Waiting Game

You're looking at an atom of material X that you are told is radioactive. Day after day it doesn't split, and the weeks and months go by. You check on your atom after 100 million years, and yet again after another 100 million years, and find that nothing has happened. When you ask if X is actually radioactive, you're assured that it is—just check again after another half-billion years! By then, you're told, the odds that it will have split are 50-50.

(a) Material X is a lousy candidate for nuclear fuel.

(b) By looking at a billion atoms, one might split in only a year! Notwithstanding, X remains a poor choice.

(c) X could be a great choice for nuclear fuel. (p. 279)

A Random-Integer Game

Randomly choose an integer between 0 and 1000. Do this again and again, until the sum of your choices equals or exceeds 1000, then quit. After playing this game hundreds of times, you are most likely to find that the most frequently-encountered number of draws making the sum equal or greater than 1000 is

(a) 2   (b) 3   (c) 4   (d) 5   (e) over 5   (p. 280)

Skewed Left or Right?

Take a symmetric bell-shaped curve and pretend that it is rubbery. When you pull its left tail leftward, the resulting curve is skewed to the left. If instead you pull the right tail rightward, the curve becomes skewed to the right. In either case the relative location of the peak changes—if curve is skewed to the left, the peak seems to have moved to the right.

Is a histogram of the heights of all professional basketball players skewed to the left or to the right?

(a) It is skewed to the left.   (b) It is not skewed.

(c) It is skewed to the right. (p. 281)

HARPER COLLEGE LIBRARY
PALATINE, ILLINOIS 60067

Bottles versus Boxes

Bottles of beer coming off a conveyor belt at a filling plant are individually weighed for consistency. Suppose the distribution of a thousand such weighings is flat, like this:

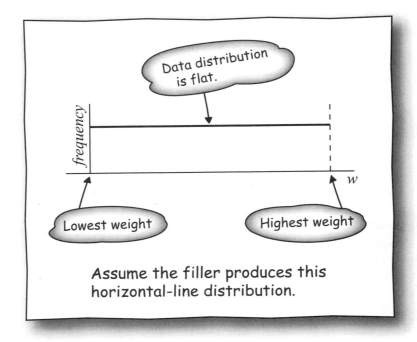

Instead of weighing individual bottles, the company decides to weigh a thousand cases of four six-packs. The distribution of these larger weights will be

(a) another horizontal-line distribution.

(b) bell-shaped. (p. 281)

The Gauss bell curve illustrated in this chapter's Introduction is arguably the most important curve in statistics. Its equation is

$$y = \frac{1}{\sigma\sqrt{2\pi}} e^{-\frac{(x-\mu)^2}{2\sigma^2}}.$$

There's exactly one point where the first derivative of y is zero. It is at the curve's center of symmetry, the mean μ of the data. There are exactly two points where the second derivative of y is zero, and these are at the curve's two inflection points, each a distance of one standard deviation σ from μ; σ measures the data's spread.

Throwing Darts

In the (x_1, x_2)-plane, the disk of unit radius centered at the origin consists of points satisfying $x_1^2 + x_2^2 \leq 1$, and the smallest coordinate square containing it consists of those points with $-1 \leq x_i \leq 1$ ($i = 1, 2$). Suppose we randomly choose a large number of points in the square, counting a dart-throw (a_1, a_2) a success if it lands in the unit disk— that is, if $a_1^2 + a_2^2 \leq 1$. For a large enough number of throws, the proportion of successes to the total number of tries, times the area of the square, is likely to approximate the area of the disk to within any specified degree of accuracy. The actual area is, of course, $\pi \approx 3.14$.

This idea generalizes to higher dimensions. In dimension $n = 3$, count a dart-throw (a_1, a_2, a_3) in the cube $-1 \leq x_i \leq +1$ ($i = 1, 2, 3$) a success if it lands in the unit ball—that is, if $a_1^2 + a_2^2 + a_3^2 \leq 1$. As the number of tries increases, the proportion of successful tries times the box volume approaches $\frac{4\pi}{3} \approx 4.19$. In a similar way, by randomly choosing points in the n-cube $-1 \leq x_i \leq 1$ ($i = 1, 2, \ldots, n$) in R^n, and calling a point $(a_1, \ldots, a_n)$ a success whenever it lands in the unit ball $B(n)$—that is, when $a_1^2 + \cdots + a_n^2 \leq 1$—we can approximate the volume of $B(n)$. In dimension 4, we get a 4-dimensional volume of about 4.93, and in dimension 5, a volume close to 5.26.

But what about for really large dimensions? The volume of our n-cube is 2^n, which approaches ∞ with n. What about the volume of $B(n)$ as $n \to \infty$? It

(a) approaches ∞, too.

(b) always increases, but stays finite.

(c) turns around and decreases to 0.

(p. 282)

Monte Carlo

Dunk a donut into a cup of coffee. Assume that the donut is a solid torus of known dimensions, that you know how far down you dunk it, and at what angle. How much of the donut stays dry? Sans the coffee cup, an example of this impossible-looking problem appears at the top of the next page.

A challenging integration problem like this can sometimes be turned into a straightforward statistical, dart-throwing problem. Here's the general idea: assume you know the equations of the torus and of liquid's surface, which is a plane. Fit an imaginary box around the torus. Randomly pick a point within the box, then another, then another..., each time counting as a success any point landing above the plane and within the torus. After thousands of such dart-throws, compute the proportion of total tries to successful ones. This approximates the proportion of the box occupied by the dry part of the torus. This proportion times the box volume is then our approximation to the volume of the dry torus region.

The equation of the tilted torus does not promise to be neat or simple. Fortunately, the equation of a torus in standard position is well known, and so is the equation of a tilted

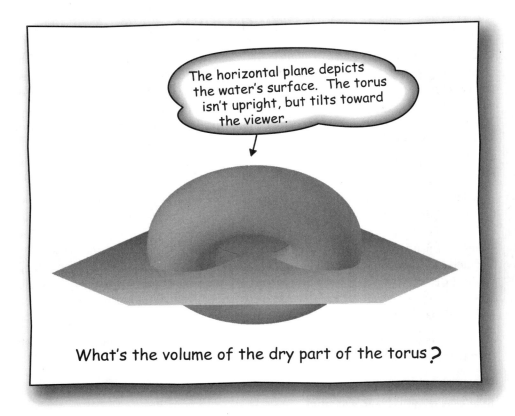

What's the volume of the dry part of the torus?

plane. Therefore, let's simply tilt the whole picture to make the torus equation simple, letting the torus a rest on a tabletop. As for shape, assume the torus has radius 2 from its center point to the circle running along the middle of the tube, and that the tube has radius 1. An equation of such a torus is

$$\left(2 - \sqrt{x^2 + y^2}\right)^2 + z^2 = 1.$$

Of course we need to tilt the plane, but that's easy: one plane tilted symmetrically with respect to the x-, y- and z-axes has equation

$$x + y + z = 2.$$

With these two equations it's now surprisingly easy to test for success—a successful point must satisfy both equations, with = changed to ≤. In practice, it's best to choose a box not much larger than the object whose volume we want. If the box is far oversized, many more random points would have to be picked to obtain a significant and representative number of them within the object. In this case, we're lucky, because the torus fits perfectly in a $6 \times 6 \times 2$ box, just touching all six sides.

Maple or Mathematica code can generate random numbers for the x-, y- and z-coordinates of a point in the box. For each success, increment an integer variable SUM. By suppressing screen printing to speed things up, cycling through a million throws takes less than 5 minutes and usually yields an answer accurate to within two-tenths of a percent. Repeating this a few times narrows the likely error even further.

The volume sticking up above the water is closest to

(a) 29.6 (b) 29.8 (c) 30.0 (d) 30.2 (e) 30.4 (p. 283)

Monte Carlo and Average Distance

A question on p. 97 asks for the average distance-squared between two randomly chosen points in a unit square. But how about the average distance itself? Use the Monte Carlo method to estimate the average distance between two randomly chosen points in a unit square. The answer is closest to

(a) .5 (b) .52 (c) .54 (d) .56 (e) .58 (p. 283)

Monte Carlo and the Needle Toss

A needle is repeatedly tossed on a plane in which parallel lines are drawn one unit apart. If the needle is very short, it will seldom land touching any line. At the other extreme, if the needle is very long, it almost always will touch some line. For some intermediate length, the odds are 50-50 that it lands touching some line. What is that length? The answer is closest to

(a) .65 (b) .7 (c) .75 (d) .8 (e) .85 (p. 283)

Short Sticks Revisited

In the problem Short Sticks on p. 96, we randomly chose two points in the interval $[0, 1]$, letting them be the endpoints of an interval or stick. Now randomly choose two points in the unit square $[0, 1] \times [0, 1]$ and let them be endpoints of a stick. What is the probability that the stick is less than $\frac{1}{2}$ long? The answer is closest to

(a) 50% (b) 75% (c) 80%
(d) 85% (e) 90% (f) 95% (p. 284)

Classical Mechanics

Introduction

Newton's mechanics fundamentally altered man's vision of the universe, affecting the way we conceive of the physical world. Others before him, such as Copernicus (1473–1543), Kepler and Galileo (1564–1642), had helped clear a long trail of false starts and half truths that *they* had inherited. Some of the earliest physical "laws" turned out to be utterly wrong. Aristotle, for example, made pronouncements coming from pure thought that were largely untested by physical experiment, and so did Plato (ca. 428–348 BCE), who incorrectly concluded that planets travel in perfect circles. Even during the early decades of the Renaissance such teachings were considered gospel. Plato's ancient dictum "Save the circles" became so much a part of Kepler's motivating inspiration that when he finally did discover the truth, that planets travel along ellipses, the disillusioned scientist regarded it as a sad commentary about our universe.

We've become much wiser through the ages, and now know that science has two sides, strategic and tactical. That is, science requires both the armchair for thought and the laboratory for data-collecting. Nicolaus Copernicus realized this and made careful observations of the heavens, culminating in his life's great opus, *De revolutionibus* (*On the Revolutions*). Galileo eventually embraced the Copernican model and then openly supported it. This created great consternation in the Church, for it was angered yet had to tread carefully: Galileo was articulate, brilliant and immensely popular. Hundreds had looked through his new eye on the universe, the telescope, an instrument many church officials refused to peek through, fearing what they might see. Galileo was one of the first to realize the fundamental role of observation in understanding the physical world and even as a young instructor, set up a home physics lab and ran tests to check Aristotle's pronouncements. He found that dropped objects did *not* fall at a constant speed as Aristotle taught, but instead accelerated, and Galileo formulated the right form of the law. It was in his later years that he did some of his finest theoretical work on terrestrial physics. This included establishing the parabolic trajectory of a projectile.

Taking its place alongside terrestrial physics was celestial physics. Kepler's dream was to understand precisely how the planets moved. He could become deeply inspired by geometrically beautiful explanations of the heavens, but he also realized that any law had to square with observation. In pre-telescope days, the preeminent astronomical observer was the Danish nobleman, Tycho Brahe (1546–1601). Through many years of highly professional observation, Tycho had acquired the world's most extensive and accurate set of astronomical data. But he was egocentric and carefully guarded the most important of his observational treasures. Both he and his most gifted assistant Kepler passionately wanted to crack the secret of planetary motion. But it was only after Tycho died that Kepler was finally able to squeeze, drop by slow drop, the observational nectar from the nobleman's highly protective and suspicious family.

At last, fate decreed it to be Kepler's turn. Finally able to sift through data on Mercury's orbit, he filled hundreds upon hundreds of pages with cramped mathematics and philosophical ramblings. But Plato's circles obstinately refused to cooperate. In frustration, Kepler at last turned to an ancient Greek curiosity, the ellipse. To Kepler it was a bent-out-of-shape, distinctly second-rate circle with but one advantage—it worked. As years passed, Kepler distilled his discoveries into what are now known as Kepler's Three Laws:

    I Each planet moves in an elliptical orbit with the sun at one focus.

    II The line segment from the sun to a planet sweeps out equal areas in equal times.

    III The period of any planet is proportional to $a^{3/2}$, where a is the length of the semi-major axis of the planet's elliptical orbit.

Kepler's celestial physics and Galileo's terrestrial physics were each signal accomplishments, but left the worlds of the heavens and of the land unconnected. Few people even suspected there might be a single theory encompassing both. In hindsight, however, each of the two theories had disturbing limitations. Kepler's Laws spoke only to actual orbits but never explained why they should be elliptical. Absent a broader explanation, Kepler never realized that his "second rate" ellipses were simply part of a larger picture—that heavenly bodies can travel along parabolas or hyperbolas, too. As for Galileo's parabolic cannonball path, he never realized that gravitational strength varies with altitude. That variation in fact implies that his cannonball's parabolic path is really an arc of an ellipse—or, if the speed is great enough, of a hyperbola.

What was needed was a more thoroughgoing, basic approach that led to truly universal laws instead of ones dealing with only the celestial or terrestrial. The year Galileo died, Newton (1642–1727) was born. By the time young Isaac began to brood upon a philosophy of the universe, there was greater acceptance of the heliocentric system, especially in England. Church persecution was one burden Newton never had to bear. After much contemplation over what was truly basic and fundamental, he proposed certain quantities that he thought to be utterly bedrock, such as length, time and mass. In translating his ruminations into written form, he was strongly influenced by studying Euclid as an undergraduate. He also read Descartes and didn't like it. In the spirit of Euclid, length, time and mass were taken as undefined terms, akin to Euclid's point, line, circle and angle. Just as Euclid defined common geometric constructs (polygons, parallel lines, areas, ...) in terms of the undefined irreducibles, Newton (and others after him) similarly constructed from length, time and mass other quantities that commonly appear in the physical world: force, momentum, energy, ... Newton then proposed three universal laws that play a role

Mechanics

analogous to Euclid's five postulates. They all involve forces. In the first law, the forces are balanced. The second, unbalanced. The third considers a pair of forces. In modern terminology, his Three Laws of Motion are:

> I An object at rest will remain at rest unless acted upon by an external and unbalanced force. An object in motion continues to move at a constant velocity unless acted upon by an external and unbalanced force.
>
> II A force acting on a body gives it an acceleration that is in the direction of the force. The magnitude of the acceleration is inversely proportional to the mass of the body.
>
> III For every action force there is an equal but opposite reaction force.

In the first two books of the three-book *Principia*, Newton derives a multitude of consequences from his Definitions and Three Laws.

The third book is devoted to his law of universal gravitation, stating that the force between two point masses m_1, m_2 located a distance r apart is $G\frac{m_1 m_2}{r^2}$. G is the universal gravitational constant; its value depends on the units used.

One can modernize the scalar distance to the vector quantity displacement **r** from one point to another. A cluster of classical quantities use the zeroth, first and second time derivative of **r**, together with their products with mass:

- **v** = velocity = $\frac{d\mathbf{r}}{dt}$
- **a** = acceleration = $\frac{d\mathbf{v}}{dt} = \frac{d^2\mathbf{r}}{dt^2}$
- $m\mathbf{r}$ = moment of mass positioned at the tip of the vector **r** about the vector base
- $m\mathbf{v}$ = momentum
- $m\mathbf{a}$ = force

Although each of Newton's Three Laws involves force, today it is momentum that is often regarded as the more fundamental. There is a law of conservation of momentum, and also of angular momentum, but not of force. Knowing momentum tells us exactly the force, since force is the time derivative of momentum. Force does not uniquely determine momentum.

With great wisdom, Newton said "Never frame hypotheses." We now realize that basic conservation laws of momentum, angular momentum and energy are more fundamental than his Three Laws: the conservation laws apply to all physics—modern, old, quantum, everyday-sized, or of astronomical dimensions. It took another two centuries after Newton before physicists realized there was such a thing as conservation of energy. Intuition suggested that energy was *not* conserved—after all, a rolling ball eventually comes to a stop. But as it rolls, it creates noise and slightly warms itself, the surface it rolls along as well as the air it moves through. Its initial kinetic energy simply radiates away from it. Add up every bit of escaped energy, and the total remains constant. Only later was it realized that energy takes a variety of other forms too, including chemical energy and energy in electromagnetic radiation.

Even before relativity and quantum mechanics, Newton's Laws were transformed into more easily used forms. Solving a mechanics problem using his laws necessitates detailing all forces. Some "forces" were only apparent, like the so-called Coriolis forces, and could cause confusion and disagreement. Later mathematicians like Lagrange (1736–1813) and Hamilton (1805–1865) supplied a powerful alternative to solving problems: instead of finding all forces acting on all particles, one need only find each particle's total energy. This was a tremendous simplification, in part because energy is scalar while force is vector. Lagrange's and Hamilton's equations involve taking partial derivatives, and this is mostly a matter of being careful, or simply letting modern software like Mathematica or Maple do the job. Such software can numerically solve the resulting equations of motion, and then create a real-time animation. Such animations can give students and professionals alike a deeper appreciation of the nature of various physical systems.

Here's a brief list of some classical forms of energy:

- Kinetic Energy: $KE = \frac{1}{2}mv^2$
- Gravitational potential energy at everyday scale: $U = mgh$
- Potential energy stored in a spring of stiffness k: $U = \frac{1}{2}kx^2$
- Energy as work performed as force $\mathbf{F}$ through directed length $\mathbf{r}$: $W = \mathbf{F} \cdot \mathbf{r}$

Throughout his life Galileo was both observant and thought about what he observed. As a young man, this potent combination collided head-on with Aristotle's theory about falling objects, accorded blind veneration for centuries. Aristotle taught that objects fall at a constant rate and that heavier objects fall proportionally faster. Galileo found that neglecting air resistance, falling objects *accelerate*, while weight does *not* matter, differing from Aristotle on both counts. Many years after Galileo's "weight doesn't matter" hypothesis had become famous, he said he had first thought about the question during a hailstorm, when he noticed that both large and small hailstones hit the ground at the same time.

If Aristotle were right, he reasoned, then large and small hailstones would hit the ground simultaneously only if the larger ones all began their fall from higher up in the sky than the lighter ones—assuming they all start falling around the same time. Either that, or all the lighter ones had to start falling earlier than all the heavier ones. Neither seemed to make much sense to Galileo. Instead, the simplest explanation was that heavy or light, all hailstones fall with the same speed.

How Much Work?

One often hears that "Work equals force through distance: $W = Fd$." Is this actually true? Let's apply it to a tetherball swinging around its pole:

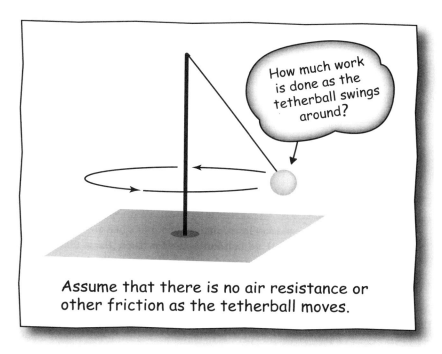

Assume that there is no air resistance or other friction as the tetherball moves.

As the ball rotates, the tether is under tension, so $F \neq 0$. Since the ball is moving, $d \neq 0$. Therefore $Fd \neq 0$, so work is performed. Is there anything wrong with this argument?

(a) The argument is OK.

(b) The argument has a flaw in it. (p. 285)

Toy Car

On a frictionless horizontal plane, a resting 1-lb toy car is pushed straight ahead with a constant force. In the course of covering 10 feet, it sped up to 10 ft/sec. How much work was done in accelerating the car in that 10 feet? The answer is closest to

(a) .5 ft-lb.   (b) 1 ft-lb.   (c) 1.5 ft-lb.   (d) 3.2 ft-lb.

(e) 10 ft-lb.   (f) Not enough information to say   (p. 285)

Grandfather Clock

One type of grandfather clock is powered by a weight suspended from a chain whose links mesh with a drive gear in the clock. The energy given the clock when raising the weight keeps the pendulum going by replacing kinetic energy lost in friction between the gears, through air resistance against the pendulum, and so on. As the drive gear turns, the weight descends at a slow, constant speed.

Suppose we take the clock to the moon. There, the work required to raise the weight is only about a sixth of that needed on earth. Assuming it actually runs there, the grandfather clock on the moon (choose all that apply)

(a) runs slower.

(b) keeps good earth time.

(c) needs winding more often.

(d) needs winding less often.

(e) needs winding just as often as on earth.   (p. 285)

Gravity's Horsepower

When you drop a brick, gravity causes it to speed up as it falls. Increasing the brick's speed requires work, so gravity must be performing work during each small interval of time. Now power—including horsepower—is work done per unit time. Since the brick is speeding up, the horsepower of gravity

(a) is increasing.   (b) is decreasing.   (c) remains constant.   (p. 286)

Spinning Turntable

A penny rests on a phono turntable's platter, which we assume is frictionless. Around the circumference of the platter there's a half-inch vertical wall to keep the penny from falling off, but in our fanciful design this little wall can quickly disappear, sliding down at the flick of a switch. With the penny resting on the platter and very near the wall, the platter is sped up to ten revolutions per second. Then the switch is flipped. What's the fate of the penny? It

(a) flies off tangentially to the disk's edge.

(b) flies off in a radial direction.

(c) flies off at some other angle.

(d) travels in a parabolic path.

(e) does none of these.   (p. 286)

A Running Start

Except for their color, two blocks Red and Green are identical, and each slides down a frictionless inclined plane. Green starts from rest, and when it reaches the bottom it's traveling 5 ft/sec. Red is already going 5 ft/sec at the top:

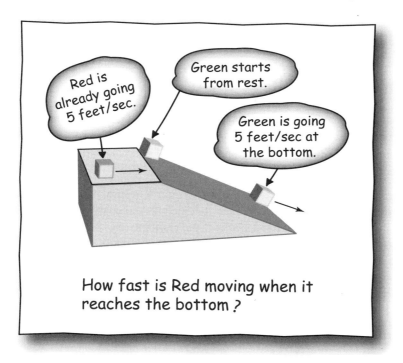

How fast is Red moving when it reaches the bottom?

How fast is Red going when it reaches the bottom? The answer is closest to

(a) 7 ft/sec.   (b) 10 ft/sec.   (c) 13 ft/sec.   (d) 15 ft/sec.

(e) It depends on the plane's slope. (p. 286)

Squeezing the Earth

Suppose the earth were compressed to half its present diameter. Standing on the smaller but equally-massive earth, a 200-lb person would weigh

(a) 50 lbs.   (b) 100 lbs.   (c) 200 lbs.

(d) 400 lbs.   (e) 800 lbs. (p. 287)

Paring Down the Earth

In the last problem, let's be unrealistic and assume that the composition of the uncompressed earth is uniform throughout. Suppose material is pared away and transported far away, leaving a ball of half the earth's present diameter. Then a 200-lb person standing on this smaller earth would weigh

(a) 50 lbs (b) 100 lbs (c) 200 lbs (d) 400 lbs (e) 800 lbs (p. 287)

Galileo's Cannonball

Toward the end of his illustrious and colorful life, Galileo faced the Inquisition on suspicion of heresy. He was found guilty and sentenced by the Church to house arrest for the remainder of his life. His offense? He openly preached that the sun was the center of the solar system, and that the earth moved around it.

Turning from such dangerous celestial physics to safer terrestrial physics, the now-isolated giant wrote some of his greatest work, and that included an analysis of projectile motion. His famous Law of Projectiles states that if air resistance is ignored, then a mass shot from a gun or cannon traces out an arc of a parabola. Exactly *which* parabola depends on several things: the speed v_o with which the projectile leaves the cannon, the angle θ of the muzzle's inclination above the horizontal, and the earth's gravitational strength g. Taking the origin as the starting point of the projectile's path and the x-axis as the earth's surface, the equation of the parabola is, in modern notation,

$$y = (\tan\theta)x - \frac{g}{2v_o^2 \cos^2\theta}x^2.$$

Without gravity—that is, if $g = 0$—the coefficient of x^2 disappears, and the equation is $y = (\tan\theta)x = mx$. That is, the projectile travels along a line in whatever direction it is launched. Add a bit of gravity, and the negative term grows with the square of x, the curve forming a downward-turning arch. The larger the coefficient of x^2, the faster the downturn and therefore the shorter the projectile's range. This coefficient $\frac{g}{2v_o^2 \cos^2\theta}$ says that strong gravity or small initial speed each contribute to a shorter range. What about θ? It appears two places in the displayed formula and leads algebraically to a short range in two ways: a shallow angle of inclination or a steep one.

Differentiating this formula with respect to θ and setting the derivative equal to zero shows that for fixed g and v_o, the range is greatest when θ is 45°. But what about equal deviations above or below 45°—that is, $45° + \alpha$ and $45° - \alpha$? Assuming $0° < \alpha < 45°$, then

(a) the range is greater for $45° + \alpha$ than for $45° - \alpha$.

(b) the range is greater for $45° - \alpha$ than for $45° + \alpha$.

(c) the range is the same either way. (d) The answer depends on α. (p. 287)

Muzzle-Speed

Other things being equal, what effect does doubling the muzzle-speed of a projectile have on the range? (Neglect air resistance.)

(a) It doubles it.

(b) It makes it four times greater.

(c) It makes it eight times greater.

(d) None of these   (p. 288)

Muzzle-Speed, Again

Assuming other things equal, what does doubling the muzzle-speed of a projectile do to the flight time needed to cover the range? (Neglect air resistance.)

(a) It doubles it.

(b) It makes it four times greater.

(c) It makes it eight times greater.

(d) None of these   (p. 288)

Muzzle-Speed, One More Time

Other things being equal, how does doubling the muzzle-speed of a projectile affect how high it goes? (Neglect air resistance.)

(a) It multiplies it by $\sqrt{2}$.

(b) It doubles it.

(c) It makes it four times greater.

(d) It makes it eight times greater.

(e) None of these.   (p. 288)

Galileo's reasoning in deriving his law of projectiles was flawless, but he had no way of knowing that gravity's force weakens slightly as the cannonball rises. This may seem to make almost no difference, but the ball actually moves along the path of an ellipse, not a parabola. In fact if the projectile weren't interrupted by landing on the earth—say, if earth were somehow suddenly compressed to the size of a basketball before the cannonball landed—then the cannonball would find itself in an elliptical orbit around the tiny earth.

What would this orbit look like? Although the cannonball may be traveling fast in terrestrial terms, in astronomical terms it's virtually crawling, almost as if the ball were just dropped. Just dropping it means it would travel in a straight line to the basketball-earth. The actual cannonball, with its small but nonzero horizontal speed, would travel in an orbit so thin and elongated that to our eyes it would in fact seem to be slowly oscillating back and forth along a straight line.

Greatest Range

Suppose that the greatest possible range for a cannon is 1000 ft. Now suppose the earth's mass is squeezed down to half its present volume. What's the maximum possible range for the cannon on this denser earth?

(a) 1000 ft   (b) 500 ft   (c) 250 ft   (d) 125 ft   (e) None of these   (p. 288)

How Much Faster?

A weightless spring is placed between, but not attached to, two identical blocks on a frictionless plane:

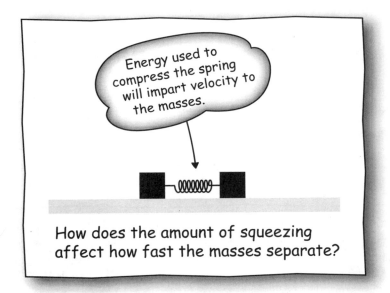

You squeeze the blocks together to compress the spring one inch. When you let go the spring decompresses, accelerating the blocks in opposite directions. Let's suppose each block ends up traveling 10 ft/sec.

Now rerun this experiment, this time squeezing the blocks together so the spring is compressed *two* inches. Which is closest to the final speed of each block?

(a) 10 ft/sec   (b) 14 ft/sec   (c) 20 ft/sec

(d) 28 ft/sec   (e) 40 ft/sec   (p. 289)

Mechanics

Pulling Strings

Two springs are stretched between fixed walls and hooked together to create a yard-long variant of an archer's bow:

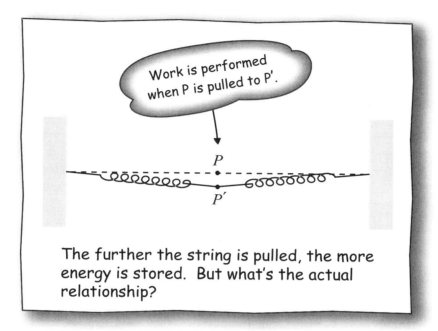

Point P is pulled one inch to position P'. (This inch is exaggerated in the picture.) Doing that requires work, and the work adds an amount E of potential energy to the system.

Pulling the point an additional one inch increases the amount of stored energy by how much?

(a) $3E$ (b) $4E$ (c) $7E$ (d) $8E$ (e) $15E$ (p. 289)

Where Does the Arrow Land?

In Galileo's parabolic projectile path, one can look at an archer's bow as a cannon and the arrow as the cannonball. Neglecting air resistance, suppose that in the last question, when P' is at the one-inch position, the arrow has a maximum range of 100 yards. When P' is at the two-inch position, what's the maximum range of that arrow?

(a) 200 yards   (b) 280 yards   (c) 400 yards

(d) 800 yards   (e) A mile   (p. 290)

A Buoyancy Machine

A remarkable phenomenon leads to an interesting theoretical machine based on it.

First, the phenomenon: Take a glass filled with water and slowly push into the water a closely-fitting lightweight piston that is as tall as the glass. As you push the piston to the bottom, almost all the water overflows—only a little bit of water remains between the piston and the inside of the glass. When the piston is at the bottom, you're pushing with the entire weight of the displaced water—almost all the water originally in the glass!

Now the machine: Assume the piston fits the glass so well that when the piston is at the bottom, there is only one drop of water remaining in the glass. Because there's that substantial buoyancy force on the piston, we can use that force to do some work—replace the downward force you're applying by a smaller weight, and the piston would lift up that smaller weight. After letting the piston do this work, open a little valve at the bottom of the glass and let the drop of water drain out. The piston falls back to the bottom of the glass. Now let a vessel of water above the piston supply another drop. The pressure goes up, and you do some more work. Sounds like a free lunch, doesn't it?

What's wrong with this "near perpetual machine"? (Choose all that apply.)

(a) Despite the substantial pressure, almost no work is done by the piston.

(b) Just adding a drop of water cannot in reality produce all that pressure.

(c) You did more work in raising the vessel of supply water above the apparatus than you'd ever get out of the apparatus.

(d) Surface tension from the close fit ruins the scheme. (p. 290)

For centuries, the promise of perpetual motion lured cranks and serious scientists alike. Even the "father of chemistry" Robert Boyle (1627–1691) was constantly on the lookout for off-the-beaten-path phenomena that might possibly lead to fulfilling the dream. He was one of the first to observe oscillating chemical reactions, but these eventually fade out. Probably his most famous theoretical argument was Boyle's fountain. As the picture shows, this vessel of water had a small capillary tube draining from the bottom and feeding back at the top. The capillary action was supposed to keep the water flowing.

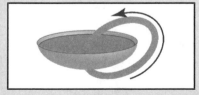

It was more than 150 years after Boyle's death before the first two laws of thermodynamics (*ca.* 1850) proved the futility of all such attempts.

Mechanics

A Pendulum's Critical Speed

Suppose an ideal pendulum consists of a 3-lb point mass at the end of a weightless 3-ft rod free to spin without friction all the way around its horizontal suspension pin. If you lightly tap the bob with a hammer, the pendulum will start swinging, while a true wallop could make it buzz around like a one-bladed propeller. There's an intermediate wallop speed bridging these two behaviors—a speed too small to make it go over the top, yet too great for the bob to stop and begin a return swing. At this theoretical, critical speed the bob will inch up to the vertical position without ever quite getting there. It literally spends an eternity just climbing.

Suppose the critical speed for the pendulum is 15 ft/sec. If the 3-ft rod is replaced by a 6-ft rod, what's the critical speed for this longer-rod pendulum? The answer is closest to

(a) 20 ft/sec   (b) 30 ft/sec   (c) 40 ft/sec

(d) 60 ft/sec   (e) 90 ft/sec

(p. 290)

On the Moon

Take the last problem's pendulum up to the moon. Assume that gravity there is a sixth of what it is here. What is the pendulum's critical speed there? The answer is closest to

(a) 2.5 in/sec   (b) 5 ft/sec   (c) 6 ft/sec

(d) 11 ft/sec   (e) 15 ft/sec

(p. 290)

Have You Heard of <u>This</u>?

Force fields are basic in the physical world. Suppose you pick a point a few thousand miles away from the earth and let your space ship serve as a test particle there. It feels gravity from the earth, moon, sun, as well as from other planets, their moons, asteroids, and so on. The net effect is a single force on the test particle space ship. It has both magnitude and direction, so it's a vector. If you draw the force vector at each point in a region of space, you obtain a gravitational force field in the region. Similarly, electric charges in space generate an electric force field. This idea generalizes in many ways. For example, electric charges in a flat copper sheet generate a force field in a plane.

The picture on the next page depicts a continuous force field **F** in the plane, together with a smooth path Γ from point A to point B. To calculate the work done in traveling along the path from A to B, integrate: if **t** is the unit tangent vector to the path at P, then the work done is

$$W = \int_{P \in \Gamma} \mathbf{F} \cdot \mathbf{t} \, ds.$$

For certain force fields **F** there's a potential function $\phi(x, y)$ so that the work done going from A to B is $\phi(B) - \phi(A)$, independent of the particular path from A to B.

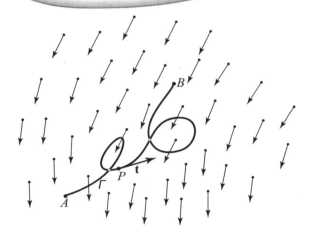

This could depict a force field, a velocity field, or even a strain field, telling how points in a material get displaced when the material is slightly deformed.

If this is a force field, then to find the work done as the point P moves along the path Γ from A to B, multiply each little path segment length ds by the component of the vector field **F** along the unit tangent vector **t** to the curve at P, and sum these along the path.

All these are in two- or three-space, but what about a one-dimensional analogue of this cluster of ideas? If there is such a thing, one never hears much about it. The reason(s):

(a) The analogous potential function ϕ is always trivial.

(b) Such a potential function ϕ doesn't exist in a line.

(c) The paths are degenerate.

(d) **F** and ϕ are talked about, but in a language so different that you don't recognize it.

(p. 291)

Mechanics

Now What Will Happen?

Let's reconsider the chain problem on p. 4 in the Sampler chapter. There, the angle between the inclined planes is 90°. But suppose the angle between the planes is different from 90°, for example:

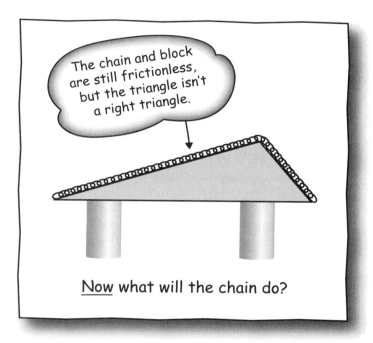

The angle is slightly larger than 90°. Does your answer to the 90°-question still hold?

(a) Yes (b) No (p. 291)

> In the hands of Galileo (1564–1642), the inclined plane played a pivotal role in the birth of physics. It seemed to him that things sped up as they fell. That contradicted the Aristotelean physics that Galileo's fellow teachers taught, but Galileo decided to pursue his suspicions. A problem was that things fall fast, and in 1600 the closest thing to a stopwatch was just a water-clock, the ancient Greek *clepsydra* (pronounced *klep-suh-druh*). Galileo would open and close the water tube at the beginning and end of an experiment and weigh the collected water. But could one open and close it fast enough and precisely enough? His solution was to use an inclined plane to slow the rate of fall, in effect slowing down time and thereby making it easier to measure . . .

Twanging Rod

Here's a problem involving rotational dynamics (torque, moment of inertia and the like). The picture is of a conservative system. Tightly clamped in a lathe chuck is a uniform steel dowel that is one yard long and a sixteenth of an inch in diameter. The lathe gears are frictionless, so when a small torque is briefly applied to the chuck, the chuck and rod will continue slowly rotating. The rod is initially given a good bump so it twangs up and down:

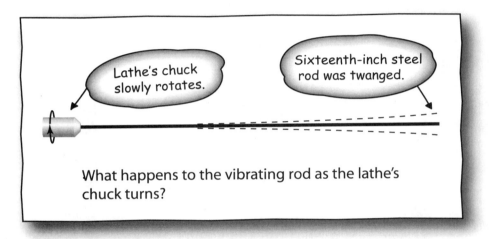

Though the initial twang is up and down, the chuck is rotating. As you watch the rod vibrate, you'll see that the rod's to-and-fro motion mainly

(a) stays vertical.

(b) rotates with the chuck. (p. 291)

Galileo's inclined plane was about 20 feet long, and down its middle ran a groove, polished to minimize friction. Galileo marked equal lengths beside it, and timed how long it took a bronze ball to get to the first mark. He took that as his unit of time. If the fall rate was constant as Aristotle said, then the lengths covered between successive time-units would be 1, 1, 1, 1, 1 ... Galileo found that the lengths were 1, 3, 5, 7, 9 ...—the ball sped up, as he'd suspected. Measuring from the origin, the sums of distances are 1, 1 + 3 = 4, 1 + 3 + 5 = 9 ..., so the speed increases as the square of elapsed time. The inclined plane was in a sense the grandfather of the formula $y = 16t^2$ for the number of feet a dropped object falls in t seconds.

Total Torque

An L-shaped exercise bar is bolted to a wall. A young gymnast hangs from the long, horizontal side of the bar, each hand equally sharing her 100-lb weight. One hand is 1.5 ft from the L's bend; the other hand is 3 ft from the bend:

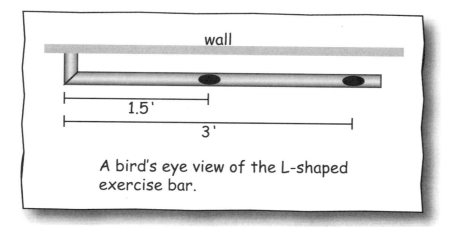

A bird's eye view of the L-shaped exercise bar.

In ft-lbs, what's the total torque on the bar?

(a) 300   (b) 225   (c) 150   (d) ≈ 262   (e) None of these   (p. 292)

The spinning earth is an interesting mechanical system. Its daily rotation makes it an immense gyroscope. Just as in a toy gyroscope, its spin gives it angular momentum so that its axis of spin resists change. The earth, with its tremendous mass, has great axial inertia, and that rotation gives it stability, keeping the north-south axis from randomly tumbling around. As a gyroscope, the earth's axis (tilted 23.5°), slowly precesses, taking 25,800 years to sweep out a cone.

It was only a matter of time before some inventive geophysicist would build a model of the spinning earth to demonstrate its physics. That happened in 1817: Johann von Bohnenberger (1765–1831), one of the pioneers of German geodesy, built what he simply called "the machine," and it beautifully demonstrated the earth's precession. He used it in his physics and astronomy lectures at the University of Tübingen, and word about it spread quickly. The famous mathematical giant Laplace began talking it up as a great teaching aid, and a highly-impressed Napoleon recommended introducing it into the French school system. One student in the system, Léon Foucault, later modified the model by replacing the globe by a disk to make what is now the gyroscope's modern form.

Saturn's Rings

Saturn's rings are mostly composed of billions of ice chunks, ranging in size from houses to bits of dust. Although the rings are virtually invisible when viewed edge on, at least one of them is over a half mile thick.

Seen by Cassini's camera from 1.3 million miles away, this is Saturn eclipsing the sun. The backlighting revealed two new rings. The tiny dot inside the white box to the left is the earth.

In theory, would it be possible to compress the half mile thickness so that the ice crushes together to form a solid ring?

(a) No   (b) Yes   (c) This will occur within 1.5–2 billion years.   (p. 293)

> Most people think that as tides rise and fall, it is the water that rises and falls. Not so. The moon pulls our earth's water mass into a slightly oval shape and the earth spins within that fixed shape, reminiscent of a brake's disk rotating between the brake's calipers. Land masses slowly plow through the oval shape and in a rising tide, land sticking above the ocean experiences the same thing as you do in wading from the shallow end of a pool to the deep.

Electricity and Magnetism

Introduction

Practical needs stimulated the growth of ancient counting, geometry and astronomy. Not so with electricity and magnetism. Various curious properties uncovered by the ancient Greeks seemed to offer no solutions to the practical problems of the day, and instead became fodder for myth, legend and misconception.

The ancients had some familiarity with both magnetism and static electricity, although in those early days no distinction was made between the two. Awareness of magnetism came first. The Greek historian Pliny the Elder (23–79 CE) wrote that when the Greek shepherd Magnes chanced upon a lodestone, the iron tip of his staff as well as his iron hobnail boots were attracted to it. The shepherd's name has become immortalized in *magnet*, as well as the region's name Magnesia and the element magnesium. Pliny put the date at around 900 BCE, but there are Chinese writings going back to 4000 BCE that also mention naturally magnetic iron.

It's thought that around 300 BCE the Chinese general Huang-ti fashioned a compass, possibly by putting a small lodestone in a bowl floating in a pan of water. Lodestone compasses were used by Chinese military commanders during the ensuing century, and after the Dark Ages the effect was rediscovered by the Italian scholar Petrus Peregrinus (1220–?). He wrote the first important study of magnetism, *Epistola de Magnete*, which includes a description of magnetic attraction and repulsion.

What about static electricity? By 600 BCE it could be reliably created: amber, the fossilized resin of trees, had become a prized material for artisans. They undoubtedly polished some of their creations and if fur was used for this, considerable static would build up, making the fur stand up on end. The amber itself, just after buffing, would attract feathers and pieces of straw. The Greek mathematician Thales (ca. 624–ca. 546 BCE) experimented with such materials and recorded his findings. The root of the word *electricity* comes from the Greek *elektron*, meaning amber. Genuine amber is anywhere from 30 to 90 million years old.

It wasn't until the sixteenth century that magnetism and electricity received serious scientific attention. The prominent London physician William Gilbert (1544–1602) had among his clientele many high-profile aristocrats, in no small part because they believed in a fundamental relation between magnetism and health. Gilbert was at the forefront in exploring possible connections. He had a strong interest in science, much of it centered around magnetism, and performed numerous careful studies of it. He was eventually appointed personal physician to Queen Elizabeth I with a stipend large enough to give him ample time to carry out his investigations. Among his many contributions:

- He was the first to describe the magnetic field of the earth, and built scale models—spheres with appropriate lodestones to mimic the earth's field—to demonstrate how a compass worked. (Before him, people thought a compass was somehow attracted to the Big Dipper.)

- He introduced the word *electricity*.

- He found that an iron rod could be magnetized by repeatedly stroking it with a natural lodestone, and that an unmagnetized iron rod aligned with the earth's field gradually became magnetized.

- He discovered that sufficiently prolonged and intense heating would demagnetize any magnet.

- He wrote up his discoveries in a book *On the Magnet, Magnetic Bodies, and the Great Magnet of the Earth* published in 1600. For generations, this was Europe's standard reference work on magnetic and electric phenomena.

- He carried out experiments to debunk myths and the idea that magnetism was magic.

In the 1700s, progress in electricity picked up. The Leyden jar was invented around 1746 by Pieter van Muschenbroek (1692–1761), and independently by Ewald Georg von Kleist (1700–1748) about a year before that. It is essentially a capacitor that can store enough static electricity to create a sizable spark. It represented a quantum leap from the tiny sparks created by rubbing amber. The ever-curious and inventive Benjamin Franklin (1706–1790) experimented with them and noticed a number of similarities with lightning: a cracking sound, bluish light, and even the same peculiar smell. (He really knew storms—one of his passions was to chase them on horseback.) His 1752 kite experiment confirmed his hunch: lightning is a giant electric discharge—easily *two* quantum leaps beyond those Leyden jar sparks! He had united phenomena that had been assumed separate and unrelated, a hallmark of scientific progress. His discovery led to another of his inventions—the lightning rod. He also introduced several words into the vocabulary of electricity: battery, conductor, condenser, charge, discharge, negative, plus, minus, electrician.

Our story continues with Luigi Galvani (1737–1798), a professor of medicine at the University of Bologna. In 1782, he was dissecting a frog on a tin laboratory dish and noticed that touching the frog's leg with a knife made the leg twitch as if shocked. It seemed to him that frog muscles had to contain what he eventually termed "animal electricity." Word of this curiosity got around, and Alessandro Volta (1745–1827), a physics professor at the University of Pavia, repeated the experiment some six years later. Volta suspected that the frog wasn't the central player, that role going instead to dissimilar metals—the aluminum dish and the iron in the knife. And the frog? It was wet, salty and sandwiched between two different metals! Electricity flowed through the frog, and that current made the legs jump. It was not much different from the jerking reflex we humans have when shocked.

In 1794, Volta decided to try metals alone, using no animal tissue. He filled bowls with brine and connected them with metal arcs dipping from one bowl into the next, one end being copper and the other tin or zinc. This non-animal setup did indeed generate a small, continuous electric current, which Volta called "metallic electricity." This string of bowls was the world's first electric battery. Volta made it more compact by stacking alternating copper and zinc disks with brine-soaked cardboard disks sandwiched between them. The wet, salty disks played the role of the bowls of brine—the frog.

Volta had made scientific history by capturing the genie of continuously flowing electricity, and it had far-reaching implications for laboratory science. The Leyden jar, which supplied sudden jolts of energy, was now supplemented by a controlled flow. A continuous electric current was basic to extending our understanding of electricity, and what he termed an "artificial electrical organ" powered several historic experiments. Furthermore, the size of the flow could be varied since more sandwiched elements in a stack meant higher voltage and therefore greater flow. Pushing to an extreme, he generated enough voltage to melt steel. This caught the attention of Napoleon, who in 1801 sent for Volta to demonstrate his stack at the French Academy of Science. An impressed Napoleon eventually made him a count of Lombardy. Numerous medals and decorations followed, including the Legion of Honor. Eventually both *volt* and the *voltaic* (volta-ic) *pile* were named after him.

Poor Galvani! The great gentleman couldn't give up his animal electricity, however impressive the voltaic pile might have been. The two scientists remained friends, but supporters gathered around animal versus metallic, the two camps arguing incessantly for years.

Except for Franklin's lightning rod, which spared many structures from damage, electricity and magnetism hadn't yet found any serious real-world applications. And the two phenomena *still* seemed mostly unrelated. Although small continuous electric currents were available, it turned out the world needed big, not small. The British chemist Humphrey Davy (1778–1829) took the bull by the horns and assembled a whole roomful of batteries built on the voltaic pile idea—2000 cells in all. The incredible monster produced enough current to electrolytically produce a few grams of the metal potassium in its pure form. But practical? Hardly. The huge battery was expensive, non-portable, messy. And it stank. Earlier, Davy had concluded that the voltaic pile was chemical in nature (something Volta had missed), and the heart of Davy's Rube Goldberg current generator was indeed chemical. But progress, it seemed, required a different mind set.

One April evening in 1820, a crucial observation fell into the lap of a Danish scientist, Hans Christian Oersted (1777–1851). He was giving a science demonstration in his home for friends and students. The two highlights of the evening were demonstrating how a current heats the wire it flows through, and showcasing various properties of magnetism. For current he used a voltaic pile, and for properties of magnetism, a compass. He noticed that every time he switched on the current, the compass needle jumped. He had connected the wire to a voltaic pile, and that wire happened to rest over the compass. Oersted said nothing at the time and finished his demonstration, but in the following weeks he kept trying to make sense of the weird phenomenon. It seemed that when the compass needle moved, it tried to make a right angle with the wire instead of pointing North on the compass. He never could figure out what was going on, so he ended up publishing a description of the odd phenomenon with no explanation. On the following September 11, the brilliant André Ampère (1775–1836) had a major flash of inspiration and in a single week he wrote a report with his explanation of the phenomenon: electric current created a magnetic field around the wire, and that field competed with the earth's magnetic field, causing the compass needle to move away from North. Here is a pictorial paraphrase of what Oersted noticed on that April evening:

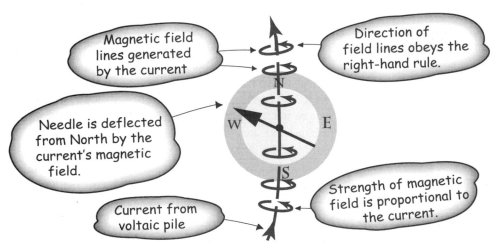

The circles around the wire are magnetic lines of force later invented by Michael Faraday (1791–1867).

Oersted, explanation or not, had stumbled upon a colossus: electricity and magnetism are intimately related. He'd connected two major areas of science, and that April night saw the birth of electromagnetism, a major player in the industrial revolution that lay ahead. Those were eventful times, for within five years William Sturgeon (1783–1850) had invented the electromagnet, a year later Georg Simon Ohm (1789–1854) established Ohm's Law and four years after that, Joseph Henry (1797–1878) constructed an electromagnet an order of magnitude more powerful than Sturgeon's.

Enter Michael Faraday. The son of a blacksmith, he had but a few years of formal schooling, and as an apprentice to a bookbinder the conscientious youngster would read books brought in for rebinding to fill in his education. One of these was a volume of the *Encyclopedia Britannica* that contained an article on electricity. Reading it filled his youthful soul with inspiration and excitement. As fate would have it, at that time Britain's foremost chemist, Humphrey Davy, was giving a series of popular lectures. Faraday attended, drinking in the information and taking meticulous notes. Some time later, Davy needed an assistant. Faraday applied, and when he showed Davy the complete and careful notes he'd taken, the conscientious lad was soon hired. Faraday had just begun one of the world's most far-reaching scientific careers. Today, many historians regard him as the best experimentalist in the history of science.

After experimenting with electromagnets, it seemed to Faraday that if an electric current could produce a magnet, then a magnet ought be able to produce an electric current.

He had invented magnetic field lines, calling them "lines of force," and in 1831 got the crucial insight: moving a magnet through a coil of wire induced current to flow through the wire. This picture gives the idea:

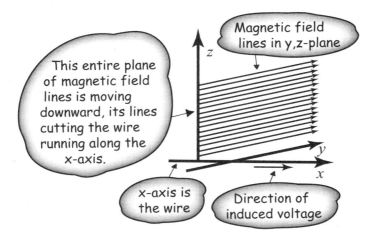

This genius knew almost no mathematics. Instead, he thought in pictures and used them to predict and corroborate experimental results. Based on his observation of induced current, Faraday was able to build the world's first electric motor. It was crude and weak, but the idea was all there. Over the next few years Faraday ran hundreds of experiments to fill out his understanding of how moving magnetic fields create electric current. In 1833, Emil Lenz (1804–1865) added to Faraday's work by discovering Lenz's law, which says that current in a loop has inertia, somewhat like mass does. Its magnetic field lines will change so as to oppose any change in current flow:

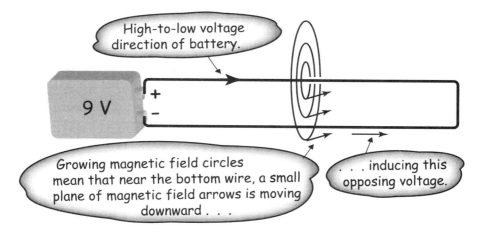

Faraday did not live to see this brainchild blossom. Mankind had to wait another forty years before Thomas Edison (1847–1931) constructed a practical direct-current generator. It lit his large laboratory at Menlo Park in New Jersey, and later powered electric lamps on an entire street. Edison always tended to think big, and in this case he envisioned lighting a full square mile of Manhattan. For this he built six "Jumbo" generators, each a 27-ton Goliath with a capacity of 100 kilowatts—enough for 1200 light bulbs. These were powered by steam engines, which James Watt (1736–1819) did so much to improve. The most common unit of electric power is the watt, and the name is certainly appropriate, for it was by coupling Edison's generators to Watt's steam engine that enough wattage was produced to meet the unprecedented demand.

Lenz never realized it, but his law has turned out to be a milestone in the history of electricity. Lenz's law fills out a cluster of ideas revealing an astonishing brotherhood with Newtonian mechanics. Newton's fundamental quantities of length, time and mass, various other quantities constructed from them, the three Laws of Motion as well as many consequences derived from them, are all mathematically mirrored in the world of electricity. This linkage is all the more astonishing because the actual physical mechanisms in the two worlds are quite different. Yet these two worlds are linked through equations having precisely the same mathematical form.

The star exhibits here are two differential equations, one describing mechanical oscillations in a mass-spring system with damping, the other describing electrical oscillations in a LC-circuit with resistance.

The basic differential equation for a damped mass-spring system:

$$m\frac{d^2x}{dt^2} + D\frac{dx}{dt} + kx = 0.$$

The basic differential equation for an LCR loop:

$$L\frac{d^2q}{dt^2} + R\frac{dq}{dt} + \frac{1}{C}q = 0.$$

This remarkable pair suggests symbolic correspondences such as

$x \sim q$ between spring displacement x and capacitor plate charge q;

$M \sim L$ between mass M and electrical inductance L;

$D \sim R$ between mechanical damping and electrical resistance.

Time t is common to both differential equations.

As for $x \sim q$ and $M \sim L$:

- With signed distance, points on the positive side of the real number line correspond to positive distance, and points on the negative side to negative distance. Analogously, positive and negative charges correspond respectively to points on the positive and negative side of the real number line, with zero charge corresponding to the origin.

- Mass is a measure of inertia—it resists any change in velocity. The picture illustrating Lenz's law depicts the electrical analogue. When a wire is connected to a battery to

Electricity

form a loop, the wire generates (or induces) a voltage opposing that of the battery, giving the loop "electrical inertia" or *inductance*, denoted L. This opposing voltage ("electrical force") is analogous to a pushed mass pushing back with mechanical force. A loopy circuit consisting of a coil with many turns has greater inductance; all else being equal, the inductance of a coil is proportional to the number of turns.

Newton's three Laws of Motion, which appear on p. 107, have electric analogues. These analogues can be paraphrased this way:

I A loop with no current running through it will continue that way unless an external, unbalanced electromotive force is applied to the loop. A loop with current running through it continues to have that same current running through it unless an external, unbalanced electromotive force is applied to the loop.

II An electromotive force applied to a loop steadily increases the loop's current in the direction of the electromotive force. The magnitude of this increase is inversely proportional to the inductance of the loop.

III For every electromotive force applied to a loop there is an equal but opposite induced electromotive force applied to that loop.

The box on the next page lists some of the natural mathematical correspondences between classical mechanics and electricity.

Classical Mechanics vs. Electricity

$x =$ distance $\sim q =$ charge

$\frac{dx}{dt} = v =$ speed $\sim \frac{dq}{dt} = i =$ current

$\frac{d^2x}{dt^2} = \frac{dv}{dt} = a =$ acceleration $\sim \frac{d^2q}{dt^2} = \frac{di}{dt} =$ rate of current increase

$m =$ mass $\sim L =$ inductance

$F =$ force $\sim V =$ voltage $=$ electromotive force

$F = ma \sim V = L\frac{di}{dt}$

$D =$ viscous damping $\sim R =$ electrical resistance

$v_o = \frac{\text{pressure}}{\text{viscosity}} =$ terminal speed in viscous medium $\sim i = \frac{V}{R} =$ Ohm's law

$k =$ spring stiffness $\sim \frac{1}{C} = \frac{1}{\text{capacitance}}$

$F = -kx =$ Hooke's law $\sim V = \frac{1}{C} \cdot q$

Basic differential equation for damped mass-spring system:

$$m\frac{d^2x}{dt^2} + D\frac{dx}{dt} + kx = 0$$

Basic differential equation for LCR loop:

$$L\frac{d^2q}{dt^2} + R\frac{dq}{dt} + \frac{1}{C}q = 0$$

Frequency of an undamped mass-spring varies as $\sqrt{\frac{k}{m}}$.

Frequency of a resistanceless inductance-capacitance (an "LC") loop varies as $\sqrt{\frac{\frac{1}{C}}{L}} = \sqrt{\frac{1}{LC}}$.

Staying Power

Two identical flashlight bulbs are connected by resistanceless wires to two identical, fresh D-size batteries. In the top picture, the bulbs are wired in series, and in the bottom, in parallel. Which battery lasts longer?

(a) The two batteries will last equally long.

(b) The battery in the top figure will last longer.

(c) The battery in the bottom figure will last longer.   (p. 295)

> Oersted chanced upon the seed of electromagnetism, but it was Ampère who singlehandedly initiated its serious development. He created a theory (encapsulated in "Ampère's law") describing the magnetic force between two electric currents. It not only accounted for electromagnetic phenomena already seen, it also predicted many new ones. This genius—shy, awkward, impossibly absent-minded—also invented the galvanometer for precisely measuring current. Appropriately, the SI unit of electric current *ampere* is named after him. Incidentally, Ampère's talents extended to chemistry. In 1811 he coined the name *fluorine*, from the Latin *fluere*, meaning "to flow".

Survivors

Repeat the last experiment, this time comparing the light bulbs. As the batteries run down, each light bulb starts to dim, eventually going out.

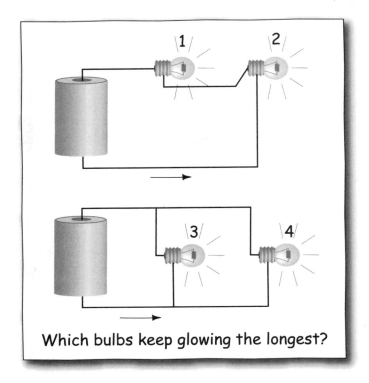

Which bulbs keep glowing the longest?

The arrows indicate the direction of actual electron flow. Theoretically, (choose all that apply)

(a) bulbs 1 and 2 go out simultaneously.

(b) bulb 1 goes out before bulb 2.

(c) bulb 2 goes out before bulb 1.

(d) bulbs 3 and 4 go out simultaneously.

(e) bulb 3 goes out before bulb 4.

(f) bulb 4 goes out before bulb 3.

(g) bulb 2 goes out before bulb 3.

(h) bulb 3 goes out before bulb 2.

(p. 295)

Electricity

A Current Difference?

In the figure on p. 132, what happens at the very beginning of the experiment? How many times greater is the current at the arrow in the top circuit compared to the arrow in the bottom circuit?

(a) Four (b) Two (c) They're the same. (d) One half

(e) One quarter (f) There's not enough information to tell. (p. 295)

Later

What's the answer to the last question after the experiment has nearly run its course and the batteries are almost dead? Compared to the current in the top circuit on p. 132, in the bottom circuit

(a) the current is greater. (b) the current is less.

(c) the current is the same as in the top circuit.

(d) There's not enough information to tell. (p. 296)

Voltage Change

In a simple loop circuit consisting of copper wire (of low but nonzero resistance) and a single flashlight battery lighting one bulb, the greatest voltage change is in the

(a) battery.

(b) wire.

(c) filament. (p. 296)

> James Watt, of steam engine fame, never guessed that one day his name would appear on tens of billions of funny little things called "light bulbs." In fact, he knew little about electricity. What he did was merely jumpstart the Industrial Revolution—he reinvented the Newcomen steam engine, making it so efficient that it soon mushroomed into the primary power source of the coming age. Watt tied first place with Thomas Edison in a survey of 229 giants in the history of technology, and Edison himself used Watt's steam engine to power his ambitious street-lighting projects. In 1882, 63 years after Watt died, the British Association designated the term *watt* as the SI unit of electrical power.

Current Difference, Again

Assume in the figure on p. 132 that the batteries are fresh. How many times greater is the current through bulb 3 compared to bulb 1?

(a) Four   (b) Two   (c) They're the same.   (d) One half

(e) One quarter   (f) There's not enough information to tell.   (p. 296)

> Discovering that things once considered separate are in fact related can be enormously powerful in science or mathematics. Examples: Descartes related algebra and geometry in analytic geometry; Einstein found that mass and energy are related via $E = mc^2$; Oersted saw that electric current creates a magnetic field; Faraday demonstrated that a moving magnetic field induces an electrical current; Maxwell predicted a grand, united electromagnetic spectrum from long-wave radio waves through visible light to ultraviolet light, X-rays and shorter—it's a profitible exercise to extend the list as far as you can . . .
>
> Before Ohm's now-famous 1827 book *The Galvanic Circuit Investigated Mathematically*, voltage and current were thought to be entirely separate. Ohm's cornerstone law $i = \frac{V}{R}$ shows how they're related. Ohm was perennially strapped for cash, and even had to make his own metal wire for experimentation. He was fanatically precise with all details of his experiments, but his mathematically-oriented writings were rambling and got a very frosty reception from his fellow countrymen—they weren't used to using much mathematics in physics. More than a quarter-century after his death, the Electrical Congress in Paris appropriately designated the *ohm* as the SI unit of electrical resistance. The reciprocal of this, "conductance," is his name spelled backwards: *mho*.

Type AA

Replace each D battery in the figure on p. 131 with an AA battery, and run the experiment again. Assume both batteries are fresh, from the same manufacturer and of the same kind—say alkaline. Choose all that apply:

(a) Initially, by just looking at the bulbs, you couldn't tell any difference.

(b) The D experiment will finish before the AA one.

(c) The AA experiment will finish before the D one.

(d) They'll finish at the same time.   (p. 296)

> In a larger sense, Ohm taught us that the primary driver of current is a difference in voltage, or "electrical pressure," bringing it into what is today a larger fold: *fluids flow from locations of high pressure to low*. So just as electric current flows from higher voltage to lower, the pressure-difference concept describes wind (which arises from a difference in barometric pressure), water flow (coming from a difference in hydrostatic pressure) and heat flow (from a difference in temperature, or "thermal pressure").

Most Massive

Assuming an imaginary scale so sensitive you could detect any difference, when does a rechargeable battery have the most mass?

(a) When fully charged   (b) At some point when it's partially discharged

(c) When fully discharged   (d) It's exactly the same in all cases.   (p. 296)

> Charles Coulomb (1736–1806) left a rich legacy of scientific accomplishments. Of particular importance, he invented the torsion balance. It could measure very small forces, and helped confirm Coulomb's Law, his electrical analogue of Newton's Universal Law of Gravitation. This law gives the strength of force F between two charges q_1, q_2 a distance r apart: $F = k \frac{q_1 q_2}{r^2}$. Coulomb was a major force on transforming electricity from a qualitative exploration into a quantitative science.

A Question of Polarity

In each of these four pictures, a 9-volt battery is connected to an insulated copper coil wound around a soft iron core:

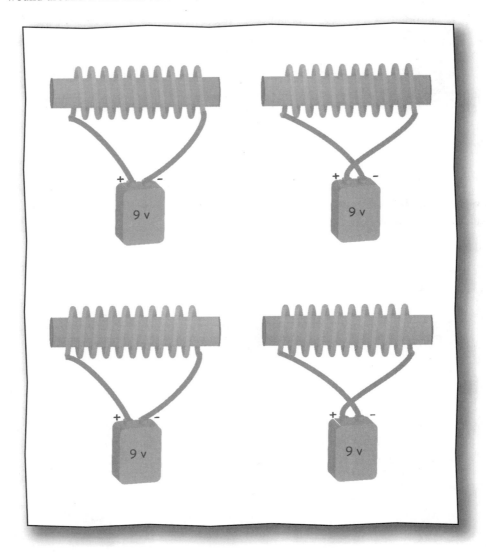

In which picture(s) is the magnet's North-South polarity the same as the one at the top left?

(a) Top right

(b) Bottom left

(c) Bottom right

(p. 297)

A Coil Around a Coil

Suppose that to an iron core plus coil, we introduce a second coil surrounding the first:

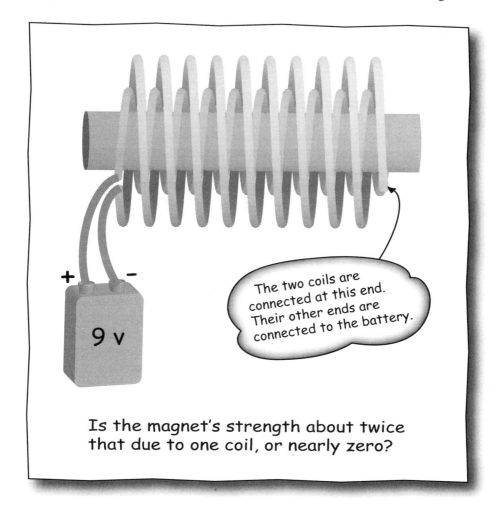

The two coils are connected at this end. Their other ends are connected to the battery.

Is the magnet's strength about twice that due to one coil, or nearly zero?

The larger coil has the same number of turns, but the winding goes the opposite way, as shown.

What does adding this second coil do to the strength of the iron-core magnet?

(a) It nearly doubles it.

(b) It exactly doubles it.

(b) The effects of the two coils nearly cancel out.

(d) The effects of the two coils exactly cancel out.

(p. 297)

A Fatter Wire

Suppose that we make two electromagnets. Their iron cores and batteries are identical, they both use insulated pure copper wire, and both have same number of turns. However in one of them, the copper wire is thicker. Will the thicker wire change the electromagnet's strength? The one with the thicker wire

(a) will be stronger.

(b) will have the same strength.

(c) will be weaker. (p. 297)

Forgetting to Turn It Off

Suppose we turn on an electromagnet consisting of an iron core with turns of wire around it, but then we forget to disconnect the battery. Assume that the wire and battery are ideal, meaning neither has any resistance, and that the battery never wears out. Which of these apply?

(a) Since there's no resistance, the current will immediately become infinite.

(b) Up to relativity effects, the current will keep increasing at a steady rate.

(c) The current increase is fastest at the beginning, but long before there are any measurable relativity effects, the current approaches a finite value that depends on the battery's voltage.

(p. 297)

Building a Large Magnet

Here's a totally off-the-wall way to make a large magnet: You've got thousands of tiny copper coils, each a small version of the ones shown on p. 136, and super-miniature batteries that never wear out, making each little coil into a small magnet. You drop them into a large cylinder containing melted paraffin and maintain a strong magnetic field along the length of the cylinder as the paraffin hardens. You've created a portable magnet this way. How does this creation compare to an ordinary permanent magnet? The image of a multitude of these super-tiny magnets lining up to make a much larger magnet is

(a) a pretty good description.

(b) fanciful, but basically baloney. (p. 298)

A Larger Capacitor

A capacitor and a coil are connected by resistanceless wire to form a loop:

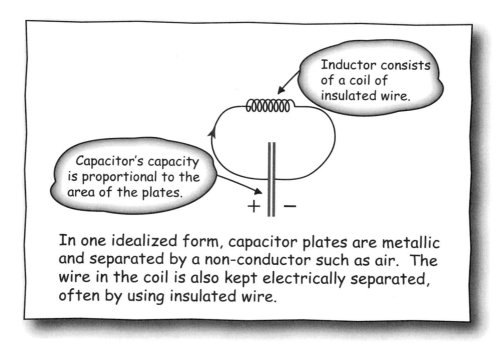

Suppose the capacitor plates are two closely-spaced metal disks. By analogy with a one-mass, one-spring system, an initial charge on one of the disks will flow back and forth between the plates, making the charges on the plates oscillate sinusoidally.

If you double the diameter of the disks, the oscillation frequency will be

(a) doubled.   (b) multiplied by $\sqrt{2}$.   (c) unchanged.

(d) divided by $\sqrt{2}$.   (e) halved.   (p. 298)

More Turns

A capacitor and a coil are connected as in the last problem, so an initial charge on either capacitor plate will cause the plate charges to oscillate sinusoidally.

If you double the number of turns in the coil, the oscillation frequency will be

(a) doubled.   (b) multiplied by $\sqrt{2}$.   (c) unchanged.

(d) divided by $\sqrt{2}$.   (e) halved.   (p. 298)

Warm Resistors

Two identical resistors are connected to batteries, but the resistor on the right has twice the voltage applied to it:

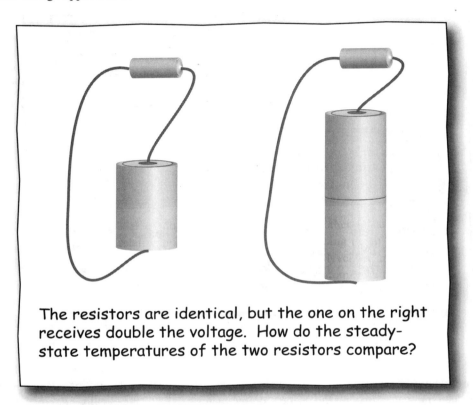

The resistors are identical, but the one on the right receives double the voltage. How do the steady-state temperatures of the two resistors compare?

Assume the batteries never run down. Suppose that after a long time the resistor on the left is 1° above a fixed room temperature. After a long time, about how much above room temperature will the right resistor be?

(a) 1.414° (b) 2° (c) 2.828° (d) 4° (e) 6° (p. 298)

> James Watt knew very well that a steam engine was far more efficient than a team of horses. He measured the power that a healthy, strong horse could deliver, coined the term *horsepower*, and expressed the power of his steam engines as, say, a "20 horsepower engine." For each engine he sold, he calculated how much the company would save, and charged a third of that savings, payable yearly for 25 years. He died a very wealthy man.

Four Charges

Place four equal electric charges on a wire circle, as shown in the left picture:

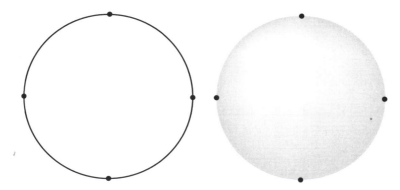

Since like charges mutually repel, they arrange themselves on the circle so that, relative to the circle's center, neighboring charges are 90° apart. They are in an arrangement that is stable in the sense that if you perturb one, they'll rearrange themselves so that neighboring charges are still 90° apart.

Now suppose these four charges are similarly arranged on a sphere. Are they stable *there*?

(a) Yes

(b) No—moving one of them slightly will make them jump to some other arrangement.

(p. 299)

Energy Loss

In the capacitor-coil setup on p. 139, if there's no charge on the plates then it is natural to call this the zero-energy state—it takes energy to separate the charges to make a positive charge on one plate and a negative one on the other.

Suppose a separation of charge has been accomplished and now a small resistance is put into the loop. As the charge flows through the loop, it has to work to get through the resistor, and we see the effects of that spent work—the resistor warms up slightly. As resistance increases, the charge must work harder to make it through. Claim: higher resistance means faster energy drain.

(a) This is always true.   (b) It's sometimes true.   (c) It's never true.   (p. 299)

The Shrinking Capacitor

In a loop containing a capacitor and a coil, suppose that the plates have equal and opposite charges, and that each plate is a metallic iris from a camera. Assume there's no friction in opening or closing either iris, and that when charges are initially put on them, both irises are closed—that is, each allows just a small amount of daylight through. After they're charged, will it require work to open them up?

(a) Yes

(b) Each iris neither requires nor donates energy as it opens.

(c) No, they both donate energy as they open.

(d) One iris gives energy, the other takes energy.

(p. 299)

Heat and Wave Phenomena

Introduction

You're in your living room on a winter night holding a cup of hot chocolate and enjoying a crackling fire. Soft strains of music mingle with the slow tick-tock of an old grandfather clock. The heat from the cup of chocolate and the crackling fire may seem to have little connection with the music's sound waves or the clock's stately pendulum motion, but viewed through the prism of mathematics the two phenomena reveal a brotherhood through the equations governing them.

Early forays into heat flow and wave phenomena progressed along mostly separate paths. Here's a brief look.

Heat. It was Newton who first gave us the germ of today's theory in his law of cooling. He never wrote it as a formula, but described it in words. It says:

> **The rate of change of a small object's temperature is proportional to the difference between its temperature and the surrounding temperature.**

This is easy to translate into a differential equation. If t is time, T is how much hotter the object is than the surrounding temperature, then $\frac{dT}{dt}$ is how fast the body is warming up. Since a constant of proportionality K relating physical quantities is typically taken to be positive, and because the body is actually cooling, Newton's law can be written as

$$\frac{dT}{dt} = -KT, \quad K > 0.$$

The solution is a decreasing exponential $T = T_0\, e^{-Kt}$, where T_0 is the initial temperature difference—that is, T at time $t = 0$. Physically, K measures how easily heat moves between the object and its surroundings. The reciprocal of K is then how difficult it is for heat to move. (Think "R-value.") A paradigm for Newton's law of cooling has become part of mathematical folklore: a cup of coffee exponentially cooling down towards room temperature.

Waves. Galileo observed during a church sermon that a pendulum's frequency seemed independent of the amplitude of swing. It was a milestone in the study of wave phenomena and suggested an application to time-keeping. However, translating it into a decent clock met a host of engineering problems. Galileo never succeeded in overcoming them, and only in 1656, 14 years after Galileo's death, did the Dutch mathematician Christiaan Huygens (1629–1695) construct the first accurate pendulum clock. But accuracy was needed on ships, too, since one must know the time to figure out a ship's longitude, and Huygens' clock was designed to work on *terra firma*, not on rolling, pitching and yawing ships. The need for a good marine timekeeper remained as critical as ever. Enter one Robert Hooke (1635–1703), a polymath with a breathtaking range of scientific interests. He helped open up the field of microbiology and coined the term *cell*. He improved upon the microscope, telescope, barometer and air pump, and invented the familiar spirit level that every carpenter and mason uses. Not only was he a highly-regarded architect, he also worked on a universal theory of gravitation. That big plum, of course, went to Newton.

As if all that was not enough, Hooke was also an expert clockmaker. He realized that when it came to marine use, the pendulum approach to timekeeping was fatally flawed because ships roll, pitch and toss, while a pendulum clock needs to stay steady. (Today we know a gyroscope can serve as a stabilizer, but Foucault's 1852 invention of the modern gyroscope was nearly two centuries in the future.) Hooke saw that a marine clock must be unaffected by the ship's chaotic motion—a requirement at odds with the constant, downward force of gravity. A *spring*-driven clock, however, seemed to offer hope, because the spring, as an integral part of the clock, would move right along with it. So Hooke turned himself into an expert on elasticity. For a time he even believed that gravity acted on a planet like an enormous invisible spring, holding it in orbit.

From empirical observations on springs, he induced his spring law:

A small deflection of a spring is proportional to the force producing it.

As with Newton's law of cooling, this easily translates to a formula. Imagine a helical spring with its long dimension lying along the x-axis, its left end fixed. If you stretch the other end to the right a small amount Δx, the spring pulls left with a small force which we'll denote by $-\Delta F$. Writing k for the (positive) constant of proportionality, we get

$$\Delta F = -k \Delta x.$$

This relation holds equally well when Δx is negative. That is, pushing the right end left compresses the spring, and the spring pushes back to the right. Physically, k is the stiffness of the spring. A watch hairspring has a small k while a garage-door spring has a large one. To ensure that Hooke's law closely approximates the physical situation, Δx is always assumed to be small relative to the dimensions of the spring.

Hooke's law holds for all sorts of spring shapes: the familiar coil (helix-shaped), spiral (like a watch hairspring), leaf (something like a musical reed, with one fixed end and the other vibrating) or even a straight rod that's stretched or compressed. Whatever the shape, if we choose a variable that parameterizes a small range of deflections (the Δx), there is then an appropriate proportionality constant k associated with the spring, the variable and the choice of units. The picture at the top of the next page shows a few examples.

Hooke's marine-clock idea was to replace a pendulum driven by gravity by a balance wheel driven by a spring. Think of a watch with a hairspring: one hairspring end is anchored

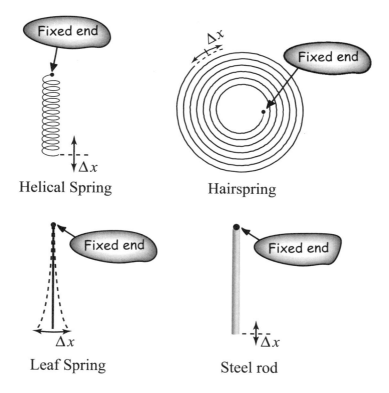

Helical Spring

Hairspring

Leaf Spring

Steel rod

to the watch body, the other is attached to the rotating balance wheel. The restoring force in Hooke's law makes the wheel oscillate with all the regularity of a pendulum. In fact, it's even better than a pendulum, for the period of a spring-driven balance wheel stays constant while the period of an ideal pendulum increases slightly with amplitude. Hooke originally used a tiny leaf spring. Others later improved this to a spiral hairspring with several turns, so it was really a much longer leaf spring curled up to save space, and in addition, the relative deflections are smaller, making the spring's behavior closer to ideal. As part of his redesign, Hooke also resolved the many problems with the traditional 300-year-old escapement design by inventing the far superior weighted anchor escapement still used today.

Although Hooke's law applies to a wide variety of spring shapes, we get even more because his general law powerfully and effortlessly dovetails with Newton's $F = ma$. The spring force F is just $-kx$, so this leads to $ma = -kx$, or $m\ddot{x} = -kx$, where dot indicates differentiation with respect to time. When we solve this, we learn how all sorts of spring shapes vibrate. And solving is easy. For example, when m and k are 1, $\ddot{x} = -x$ is satisfied by $x(t) = \sin t$ and $x(t) = \cos t$, since double differentiation just changes the sign of sine or cosine. For $m\ddot{x} = -kx$, one can check that $x(t) = \sin\sqrt{\frac{k}{m}}\, t$ and $x(t) = \cos\sqrt{\frac{k}{m}}\, t$ work.

Just as there's a paradigm for Newton's law of cooling (the cooling cup of coffee), there's also one for Hooke's spring law: an oscillating conservative mass-spring illustrated by the picture in the box on p. 153. The connection between heat and wave phenomena is especially simple with these two paradigms. Here's the idea:

In the box on p. 130, we met the differential equation $m\ddot{x} + D\dot{x} + kx = 0$ describing damped mass-spring oscillation. In this equation, if only m is zero, then the differential equation describes the cooling cup of coffee. If only D is zero, then the differential equation describes our wave motion when there's no energy loss. Now these two extremes can morph into each other, and to illustrate, look at

$$r\ddot{x} + (1-r)\dot{x} + x = 0.$$

As r increases from 0 to 1, the differential equation goes from describing one paradigm, the cooling cup of coffee, to describing the other, the motion of the undamped mass spring on p. 153.

The extreme cases corresponding to $r = 0$ and $r = 1$ cast long shadows in that they govern a variety of physical situations. Here are some physical setups obeying an $r = 0$ differential equation. In each case, "something leaks," and the rate is slow. This slowness minimizes the role of m in $m\ddot{x} + D\dot{x} + kx = 0$.

- **Mechanical.** The differential equation describes a massless dashpot. One can create a laboratory version this way: starting with the picture on p. 162, make the mass closely fit the casing, like a piston. Use a very light piston and drill a tiny vertical hole through it. As the cylinder tries to reach equilibrium, oil must make its way through the small-diameter hole. The rate at which the cylinder moves is proportional to its displacement from equilibrium. Many shock absorbers are based on this dashpot idea, as are door closers that keep the door from slamming shut.

- **Electrical.** The differential equation also describes capacitor leaking its charge through a resistor. (See the Glossary for the definition of capacitor.)

- **Liquid:** It can also describe the leaking tank that we will meet in the next chapter. The water in a cylindrical tank leaks through a tiny resistive hole at a rate proportional to the amount of water in the tank, in analogy with charge leaking though a resistor at a rate proportional to the amount of charge. The picture on p. 192 illustrates this.

- **Heat.** The differential equation can describe a cooling cup of coffee (although liquid in a thermos is a better and more general example). Heat slowly leaks through the casing—effectively a heat resistor—at a rate proportional to the difference in temperature between the contents and its surroundings.

Here are two classical examples governed by the $r = 1$ differential equation:

- **Mechanical.** This corresponds to the undamped mass-spring setup depicted in the picture on p. 153.

- **Electrical.** The differential equation describes an electrical loop with a capacitor and an inductor, or coil.

In the mass-spring paradigm, the spring is *ideal:* it satisfies Hooke's law and is massless.

Heat and Wave Phenomena

The idea here is that a spring is able to store only potential energy, no kinetic energy. Dually, the point-mass is ideal: it has no springiness, so that it can store no potential energy of this form. As the mass oscillates, energy flows back and forth between the mass and spring, illustrated in the boxes on pp. 153 and 154. We make it official:

In this book, both springs and masses are assumed to be ideal.

In all the examples, the setup can be miniaturized, and many identical copies can then be strung together in a line. One could line up one hundred tiny leaking tanks on a tabletop and connect successive tanks by little pipes so that they drain into each other, with the two endmost pipes draining to the environment. In a lineup of one hundred tiny equal masses, one could attach neighboring masses by tiny springs and attach the two end springs to fixed supports. In the first case, the tanks drain into each other according to the same law that heat flows, providing a kind of discrete heat-conducting rod. Mathematically, the endmost pipes drain water to the environment like the endpoints of a horizontally-heat-conducting rod conduct heat to ice blocks kept at each end. In the second case, the hundred connected masses give an approximation to a vibrating string—that is, a discrete version of a string.

The above $r=0$ and $r=1$ examples generalize to a line of many of them, leading in spirit from a zero-dimensional setup to a one-dimensional setup. The idea is encapsulated in this very simple picture that works for either heat flow or wave motion:

○●●●○

As a model of heat flow, the picture approximates an insulated rod with the empty circles depicting fixed temperatures at the ends. The black dots are heat-containing point masses, and the little lines connecting them, heat conductors of constant K. At any instant, heat flows from any point mass to each of its neighboring points according to Newton's cooling law $\frac{dT}{dt} = -KT$. Then, an instant later, the scenario is repeated. Heat transfer continues to progress in this way.

The picture can also depict an approximation to an elastic segment such as a violin string or the air column in a flute. The two empty circles then depict the segments's two fixed ends, the black dots represent point-masses, and the little lines connecting them are the minuscule springs. If the dots vibrate in line—in the direction of the line of dots—the vibrations are called *longitudinal*. The elastic column of air in a flute vibrates like that. If the dots vibrate perpendicularly to the line, the vibration is called *transverse*; that's the way strings vibrate, as on a piano, violin or guitar. The mathematics is similar in each case, but transverse vibrations make for easier visualization, so that is what we'll now assume. At any instant, the string's motion at any point essentially takes initial conditions of position and velocity, feeds these through a local version of $m\ddot{x} = -kx$ to get the output at an instant later. This scenario repeats over and over, wave propagation evolving in this way.

In the above one-dimensional heat and wave setups, what does the evolution look like? An appreciation can be gained by looking at the discrete case using, say, a thousand elements.

For heat conduction, take a simple case of a temperature spike. Imagine an ice-cold rod with a concentration of heat at the rod's midpoint. The concentration can be likened to a tight bundle of tiny jumping beans there, each ready to jump with equal likelihood to the right or to the left during each time interval Δt. With a large number of miniature tanks for holding jumping beans, this turns out to be approximated by the bell-shaped curve encountered in statistics. As time goes on, the curve spreads out ever more, increasing its standard deviation. The area under these curves corresponds to the total amount of heat and remains constant until the jumping beans jump beyond either end point, when each is irretrievably absorbed by the ice there. As the bell curve approaches the horizontal axis, it flattens. The area under the curve approaches zero, corresponding to all the original heat that created the spike leaking out the ends of the rod.

A more general temperature distribution can be approximated by the union of a thousand temperature spikes whose heights agree with the temperature distribution on the thousand points. One can imaginatively give all the beans in each particular bean-holding tank a common, distinctive color that each bean keeps as it jumps about. As the experiment begins, each jumping bean executes its own random, right-left succession of jumps, unconcerned about what any other bean may do. If you look at just one color, you'll see the approximate bell curve of that color gradually spreading. The actual temperature at any point corresponds to the total number of beans—of any and all colors—in that point's tank. In tanks that hold many beans, more will jump than from a tank with just a few beans, a discrete version of Newton's Law of Cooling.

Actually, a temperature spikes can be regarded as the particles in particle-wave duality. What sort of temperature distributions correspond to waves? The answer: sine-shaped ones! That is, in a conducting rod coinciding with the interval $[0, \pi]$, they are $T_n(x) = \sin nx$, for $n = 1, 2, \cdots$. Performing the jumping-bean game on these distributions leads to a result of astonishing simplicity: the amplitude of the sine curve decreases exponentially through time. Just as you can put together a bunch of single spikes to create a more general temperature distribution, it happens that an initial temperature distribution also breaks down into a sum of $\sin nx$'s of various amplitudes. Once again, these sine-shaped distributions act independently, doing their own thing. Once you know the rate at which each $T_n(x) = \sin nx$ melts down, the rod's temperature can be followed through time.

What about wave motion in a string? Once again, look at a straight string with a single spike. We may regard the spike as the sum of two spikes of half that height, and it happens that when the string is released from rest, one half-height spike travels at a constant speed to the left, the other travels at that same speed to the right. When a traveling spike encounters a fixed end, the spike reflects off that wall, changing its direction. If the spike pointed up just before the reflection, it points downward just afterward, and vice-versa. (Think of a tennis ball bouncing off a wall.) This is a special case of a remarkable fact, proved by d'Alembert (1717–1783), which we just state: starting at $t = 0$ from any initial shape $y = f(x)$, if the string is released from rest, then the string's shape at time t is $y(t) = \frac{1}{2}[f(x + ct) + f(x - ct)]$, where, in appropriate units, c is $\sqrt{\frac{T}{\rho}}$, T being the string tension and ρ, the string's mass per unit length.

As with temperature distribution, an arbitrary initial string shape can be thought of as the union of spikes. When released from rest, each spike does it own thing, and the overall shape of the string at any instant is then the sum of all the many traveling spikes at that instant. As before, on a stretched string coinciding with $[0, \pi]$, spikes have particle-wave duals—the string shapes $T_n(x) = \sin nx$, for $n = 1, 2, \cdots$, and it can be shown that these wave shapes oscillate, their amplitudes varying sinusoidally with time. The solution of tracking the history of a wave-shape is one of the great chapters of science and mathematics.

If you are unfamiliar with Fourier solutions to wave and temperature problems, then for some of the problems in this chapter you should read the Glossary's **Fourier solutions**. It gives the gist of the story and provides enough information so you can solve this chapter's problems. The fuller story is worth learning, and texts such as [1] or [8] do an admirable job. For a popular text pitched at a high school level, see [9].

A Cooling Rod Problem

Here's a picture of a cooling rod problem:

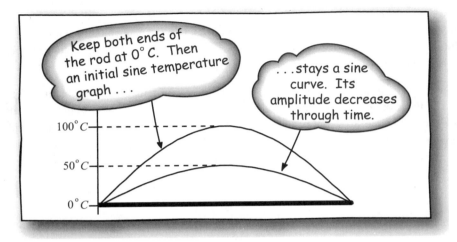

The picture suggests a remarkable property of sine temperature distributions. But the remarkableness doesn't stop there—there's a temperature half life. If it takes one hour for the center to cool from 100°C to 50°C (that is, half-way to 0°C), then after another hour, the center has dropped from 50°C to 25°C, and after another hour, to 12.5°C, and so on. Furthermore, every point of the rod has the same half-life, and this is what maintains the sine-shape through time. With a sine temperature distribution, all points of the cooling rod change in lockstep, so that knowing what any one of them is doing allows us to deduce what all the others are simultaneously doing.

The half life depends on the rod's length. For example, suppose that our rod is five feet long with a half-life of an hour. If the sine-arch were spread out over a much longer rod, say, to 50 feet, then the temperature differences between nearby points would be less and the heat flow would be slower. It makes sense that in this case the half-life would be greater.

Suppose our rod is five feet long with a one-hour half-life. Slice off a one-foot section, put blocks of ice at each end of this one-foot section, and give it an initial one-arch sine temperature, with maximum temperature of 100°C at the midpoint. In this short rod, how long does it take for the midpoint to cool from 100°C to 50°C? The answer is closest to

(a) 1 hour.

(b) 10 minutes.

(c) 2 minutes.

(d) 20 seconds.

(e) a few millionths of a second.

(p. 301)

Heat and Wave Phenomena

How Hot is Hottest?

In the one-foot section of the last problem, which of these is closet to the temperature at the midpoint after an hour?

(a) 20°C (b) 4°C (c) 1°C (d) 0.1°C

(e) A few millionths of a degree

(p. 301)

Does the Peak Walk?

The last problem has a dual: replace the one-arch sine temperature distribution by a single hot spot, the particle end of wave-particle duality. The picture on the left depicts the temperature due to heat concentrated at a point. The rest of the rod starts at 0°, and the ends are maintained at 0°. We can think of the hot-spot as a tight bundle of tiny jumping beans, each ready to do a left-right random walk in the rod. As the experiment begins, they venture out, the temperature graph soon approximating a tight bell-shaped curve that, with time, spreads out more and more. This random walk is reminiscent of the pinball game pictured on p. 92.

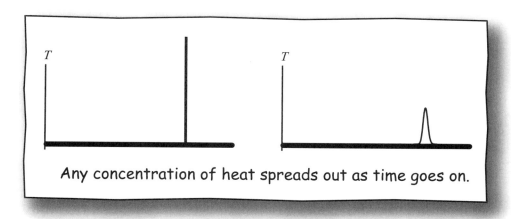

Any concentration of heat spreads out as time goes on.

What about the peak—that is, the hottest point? As time goes on, does it stay put, or does it move?

(a) The peak remains stationary.

(b) It drifts to the right.

(c) It drifts to the left.

(p. 302)

Going to Zero

Suppose the rod is insulated except at its endpoints, which is kept at 0°C by blocks of ice held at there. As the rod cools toward 0°C, which of these four initial temperature functions $f(x)$ keeps its shape during the cooling process, so that the temperature function is $cf(x)$, with $c \to 0$? (Choose all that apply.)

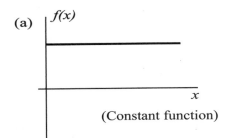

(a) (Constant function)

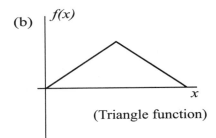

(b) (Triangle function)

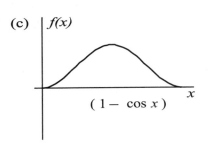

(c) $(1 - \cos x)$

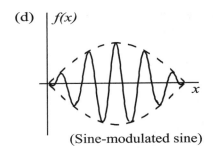

(d) (Sine-modulated sine)

(e) None of these

(p. 303)

How Far?

Suppose a 1-lb weight, when suspended from an ideal spring, makes the spring stretch four inches. Attaching an identical spring-plus-weight to it will make the top weight drop down further. How many inches further?

(a) 2   (b) $2\sqrt{2}$ (c) 4   (d) 6   (e) 8   (p. 303)

Hefty Block, Wimpy Spring

An certain ideal spring stretches one inch when pulled with a force of one ounce ($\frac{1}{16}$ lb). Suppose one end of this weak spring is attached to a one-ton cement block, and the other end is pulled up to a sky hook so high that the block dangles with its bottom just one foot above the ground.

If a two-year-old sits down on this block, the final resting place of the block's bottom would be

(a) on the ground.

(b) very slightly less than a foot above the ground.

(c) depending on the weight of the two-year-old, some distance between 1 in and 11 in above the ground. (p. 303)

A paradigm for much of classical vibration theory is an ideal, conservative mass-spring setup, such as this mass on a frictionless horizontal plane:

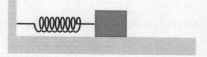

Its oscillation can be described simply by $x = \sin t$, giving position versus time. The mass serves as an "energy container"—a vessel or cup holding kinetic energy. The spring is also like a cup; it holds potential energy, and energy itself can be thought of as a liquid. As the mass oscillates, "liquid energy" gets poured back and forth between the two cups:

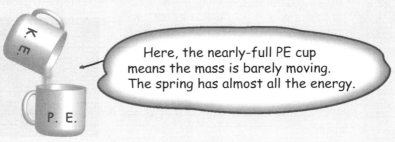

Here, the nearly-full PE cup means the mass is barely moving. The spring has almost all the energy.

The assumptions "ideal, conservative" tell us a lot . . .

Spring Stiffness

An ideal uniform foot-long coil spring hangs from a ceiling. Suppose that when you hook a 1-lb weight to the free end, the spring stretches 1 in. You cut the spring into two 6-in pieces and cut the mass into two $\frac{1}{2}$-lb lumps. How far does each 6-in spring with its $\frac{1}{2}$-lb weight stretch?

(a) 2 in   (b) 1 in   (c) about 0.7 in   (d) $\frac{1}{2}$ in   (e) $\frac{1}{4}$ in   (p. 303)

Spring Stiffness, Again

Two helical springs having Hooke constants 2 and 3 are joined end to end. Which of these is closest to the Hooke constant of this longer spring?

(a) .2   (b) 1   (c) 1.2   (d) 2   (e) 2.5   (f) 3   (g) 5   (p. 304)

... *Conservative* says that no liquid is ever lost throughout all the pouring. And *ideal* says something about both the spring and the mass.

Spring. An ideal spring satisfies Hooke's law: the amount of force needed to stretch the spring is proportional to the stretching. That is, $F = -kx$. The spring is assumed to be massless. Because of this, the spring can't store any kinetic energy.

Mass. In a dual way, our ideal mass satisfies Newton's Second Law, $F = m\ddot{x}$. The mass is assumed "springless," in that it is not able to store potential energy through being deformed (such as sqeezing or stretching).

The sinusoidal nature of the vibration means that each cup must be empty before it receives liquid from the other. Therefore when the spring is unstretched (corresponding to the PE-cup being empty), the mass is traveling the fastest (all the energy being in the KE-cup)—that's because $\dot{x} = \cos t$ is a maximum when $x = \sin t$ is zero. Similarly, when the spring is stretched or compressed the most (all the energy being in the PE-cup), the spring is at rest (the KE-cup is empty).

These two energies are like the two sides of a coin. See how similar their equations are:

$$KE = \frac{1}{2}m\dot{x}^2, \qquad PE = \frac{1}{2}kx^2.$$

The second of these follows from integrating force kx through a distance x.

Switching Off Gravity

Suppose this oscillating weight

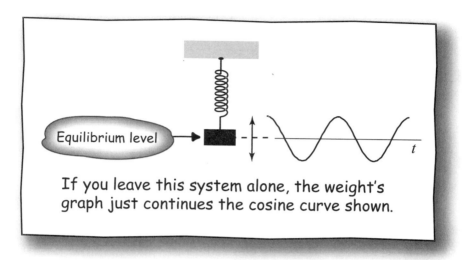

loses no energy through time, so the graph repeats endlessly.

What would happen if, at one of the times the mass reaches its lowest point, gravity is suddenly switched off? Which of these could depict the graph if that happened?

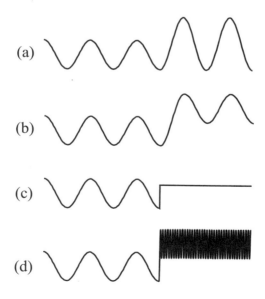

(p. 304)

Method versus Madness

In this mass-spring setup, pull the weight a little to the right or left and release it:

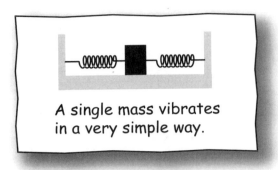

A single mass vibrates in a very simple way.

If there's no energy loss, you'll see a predictable, simple motion: the mass oscillates sinusoidally with constant frequency and amplitude. But what if there are *two* masses, like this?

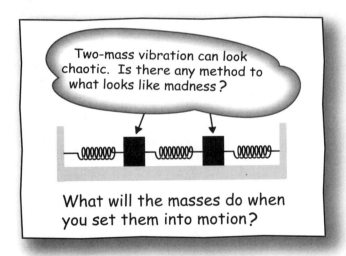

Two-mass vibration can look chaotic. Is there any method to what looks like madness?

What will the masses do when you set them into motion?

When you pull a little on one or both masses, you'll generally see jumbled movement, with no discernable order. This occurs even if we make nice assumptions, such as equal masses, all identical springs and no friction.

For this phenomenon, our eyes are not very enlightening instruments of observation. However, behind this apparent madness there lies an inherent simplicity. Here's the idea: there are two special ways to position the masses so that when you release them from rest, they will move just as simply and predictably as the mass in the first figure. Even better, these two nice cases actually form a basis for motion in general, in the sense that any motion is a linear combination of the two nice ones.

Heat and Wave Phenomena

Let's assume identical masses, identical springs and no friction. Here are the two special positionings.

- First way: move both masses the same distance and both in the same direction. Then release them from rest.

- Second way: move them equal distances but in opposite directions, then release them from rest.

In the first case, the two masses will move sinusoidally and with identical frequencies. In the second, they'll also move sinusoidally and with identical frequencies. However, the frequency in the first case is different from that in the second case. Which gives the higher frequency?

(a) The first method   (b) The second method

How much greater is the faster frequency, compared to the slower one?

(a) Less than double   (b) Exactly double   (c) More than double   (p. 305)

Suppose you had a tremendous number of masses—say, a horizontal row of 1000 small, equal masses connected by 1001 tiny, identical springs. By slightly displacing each mass horizontally and releasing all 1000 at once, we get very complicated motion. To more vividly see the apparent chaos we can just as well use transverse displacements—think of a grand piano string vibrating up and down. Is there still some method to the madness?

The big truth: the string of masses tries to emulate a musical string. Complicated phenomena are often combinations of simple basis objects, and for strings they are sine-shapes: if the string's length is π, then arbitrary shapes obtained by vertical displacements are linear combinations of $\sin(x), \sin(2x), \sin(3x)\ldots$, and these modes vibrate at relative frequencies of $1, 2, 3, \ldots$. Each component wave $\sin(nx)$ independently vibrates at frequency n, with its own amplitude. The entire motion, which may appear chaotic, is actually just the superposition of all the independent motions.

In trying to act like its grown-up cousin, our 1000-mass string has exactly 1000 fundamental modes of vibration. In any one of these, the motion looks just like the musical string's $\sin(nx)$-motion in that every mass in the lumped string mimics what the ideal musical string does, only at a different rate. In our two-mass problem, the two fundamental modes are those described in the bullets. The corresponding musical-string modes are $\sin(1x)$ and $\sin(2x)$. At the one-third and two-third points of the string, the values of $\sin(1x)$ are equal, corresponding to the first bullet. The values of $\sin(2x)$ at those two points are equal but of opposite sign, just the situation in the second bullet.

A Ceiling Under Stress

In this picture, a 1-lb weight is suspended from a spring and is resting at equilibrium:

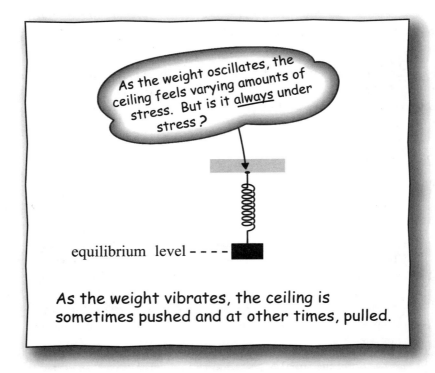

Suppose the weight is pulled down 2 in and released. It oscillates, and during this oscillation the ceiling undergoes varying stress at the point where the spring is attached. We assume that the spring itself is weightless, and that pulling the spring with 1 lb of force makes the spring stretch 1 in. The ceiling experiences the least amount of force when the weight is

(a) 2 in above the equilibrium level.

(b) 1 in above the equilibrium level.

(c) at the equilibrium level.

(d) 1 in below the equilibrium level.

(e) 2 in below the equilibrium level.

(p. 305)

Adding Waves

Graphs of sine curves can be used as building blocks for more general graphs. Sine waves of various amplitudes and frequencies can be added, their graphs summed pointwise over all the points on the x-axis, just as one sums the graphs of $y = x$ and $y = 2x$ pointwise to obtain the graph of $y = 3x$.

Here are two sine curves:

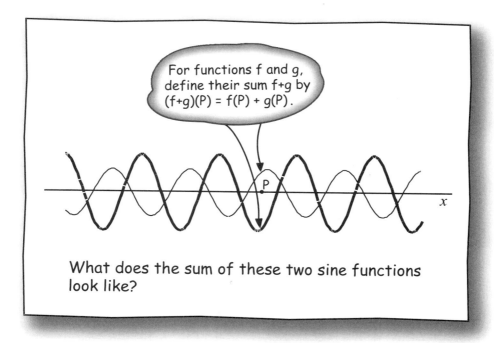

The more heavily-drawn sine curve was manufactured from the other one by stretching vertically to increase the amplitude and then by translating to the right. When you add these two sine curves pointwise, the resulting curve is

(a) a jumble and isn't periodic (that is, doesn't repeat itself).

(b) a jumble, but is periodic.

(c) smooth, periodic and not too complicated, but not sinusoidal.

(d) sinusoidal.

(p. 306)

Losing Energy

In this vibrating mass, suppose the only friction comes from air resistance:

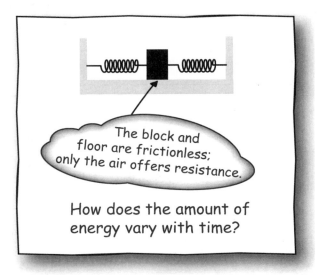

Assuming the mass is oscillating, the amplitude of the vibration decreases as the system runs down. Which of these plots of the total energy TE against time could describe the amount of energy remaining in the system?

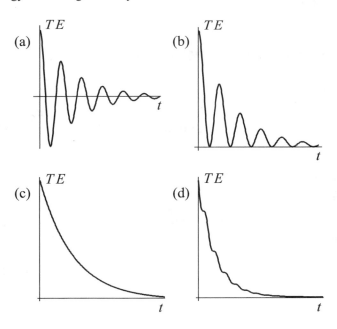

(p. 307)

Losing Energy, Again

In the last problem, we assumed that air resistance was the sole source of friction. Now let's suppose that in addition, we recognize that some small horizontal force on the block is required just to get it moving. This represents static friction. With these two types of friction, which one(s) of these could be the graph of the block's motion?

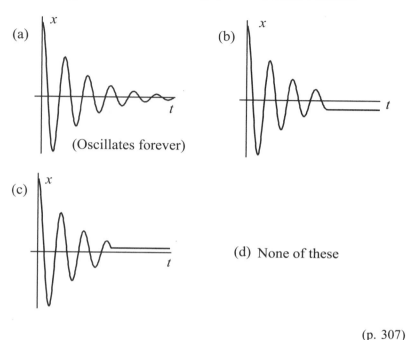

(d) None of these

(p. 307)

Friction and Frequency

Suppose the mass-spring is mounted inside a container filled with oil, as depicted on p. 162. The mass now has to plow through the oil as it does its thing, and this decreases the oscillation rate. For speeds that aren't too great, we can assume that at any instant the oil resistance is proportional to the speed of the mass. This means the setup can be described by the differential equation $m\ddot{x} + D\dot{x} + kx = 0$, where m, D and k are all positive constants. What happens as the experiment progresses and the amplitude of oscillation and the maximum speed of the weight have both decreased? Theoretically, is the slower rate of oscillation maintained? Or, as more and more energy leaves the system, does the rate of oscillation change?

(a) The rate never varies.   (b) The oscillation rate continues to slow down.

(c) As the amplitude approaches zero, the resistive force decreases, making the frequency increase slightly.

(p. 308)

Going, Going, Gone!

This picture represents a mass-spring system mounted inside a container filled with fluid:

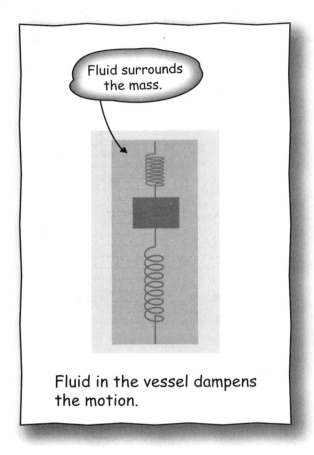

Fluid surrounds the mass.

Fluid in the vessel dampens the motion.

The mass has been pulled up, and the system initially has all the energy that went into thus cocking it. When the mass is released from rest, the fluid damps the motion, and damping means that energy drains out of the system. The fluid used obviously affects the amount of damping—water damps the motion faster than air does. If we run the experiment many times, slightly increasing the viscosity of the fluid between successive runs, we'll see that the amplitude of any oscillation decreases more quickly, a consequence of energy leaving the system faster. In short, the thicker the fluid, the faster the energy drain.

The question is, do you buy this last sentence? Is it true in general that the thicker the fluid, the faster the energy drain?

(a) Yes   (b) No

(p. 308)

Heat and Wave Phenomena

Change of Scene

Suppose this mass-spring system is sent to the moon, where gravity is one-sixth of earth's:

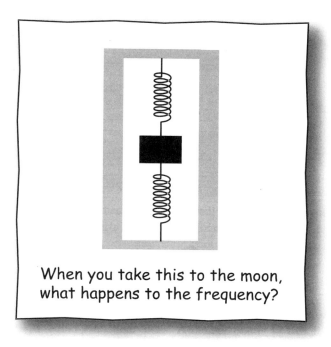

When you take this to the moon, what happens to the frequency?

Assume that the two springs are identical, and that there's no friction or other energy loss to the system. How would the change of scene affect the oscillation rate?

(a) It would make it faster.   (b) It would make it slower.

(c) There would be no change.

(p. 308)

Change of Scene, Again

In the last problem, suppose the apparatus is modified by using two springs of different strengths. How would its oscillation rate on the moon compare to that on the earth? On the moon it would be

(a) faster.   (b) slower.

(c) There would be no change.

(d) There's not enough information to say.

(p. 309)

Pendulum versus Mass-Spring

Suppose that here on earth, this pendulum oscillates twice as fast as the mass-spring:

On the moon, where gravity is one-sixth of earth's, the pendulum oscillates

(a) more than twice as fast as the mass-spring.

(b) twice as fast as the mass-spring.

(c) less than twice as fast as the mass-spring, but still faster than the mass-spring.

(d) at very nearly the same rate as the mass-spring.

(e) more slowly than the mass-spring.

(p. 309)

"Fixing" Our Moonbound Pendulum

Before taking the last question's apparatus to the moon, which has a sixth the gravity of earth's, suppose we modify it by replacing the pendulum bob with one weighing six times more. Once on the moon, we put our modified pendulum in competition. Compared to the unchanged mass-spring, the pendulum will oscillate

(a) more than twice as fast.

(b) twice as fast.

(c) less than twice as fast, but still faster than the mass-spring.

(d) at the same rate.   (e) more slowly.   (p. 309)

The pendulum has played a central role in the development of science. To test any physical law involving time, an accurate clock is needed, and for almost three centuries the best clocks available were pendulum-based. The genesis of the pendulum clock started with an observation by Galileo in his first year of college. The story goes that he was bored by a sermon in the cathedral of Pisa, and began counting how many of his heart beats it took for the cathedral's huge bronze chandelier to make a swing. Drafts in the cathedral sometimes made the chandelier swing more, but that didn't seem to affect the time of swing.

Later, Galileo discovered experimentally that quadrupling the pendulum length doubled the length of time it takes for the bob to make one full to-and-fro motion—the period of the pendulum. Further experimentation corroborated that the period varies as the square root of the length L. Eventually it was realized that gravitational strength g also enters into the picture, everything finally crystallizing into this famous formula for the pendulum's period P:

$$P = 2\pi \sqrt{\frac{L}{g}}.$$

It is reasonably accurate only for small pendulum swings, not wide ones.

Experience taught Galileo that transforming his pendulum ideas into an accurate timepiece was anything but trivial, and he had to make do with water-clocks for most of his time-based experiments. Only some 15 years after Galileo died did the world see the first successful pendulum clock, the brainchild of the brilliant Dutch mathematician Christiaan Huygens. It was astonishingly accurate relative to standards back then, to within a minute a day.

A Shorter Period? Or Longer?

For an ideal pendulum having a point mass, weightless rod and no friction, the period depends on the angle of swing. A large angle such as $90°$ means that the bob initially receives the full force of gravity, so the acceleration is greater, a vote for a shorter period. On the other hand, a bigger angle means the bob must travel further in one round trip, a vote for a longer period. Who wins? Assuming that initially raising the bob higher always either shortens or lengthens the period, which do you vote for? A large swing means

(a) a longer period.   (b) a shorter period.   (p. 309)

> Even in its early days, the pendulum taught us a surprising lesson about the shape of the earth. In 1671 the French government sent the distinguished scientist Jean Richer (1630–1696) to Cayenne, the capital of French Guiana, to measure the parallax of Mars. His measurements were to be used with measurements from other points of the earth to find the distance between us and Mars. On that trip, Richer also tested a pendulum clock that had been calibrated in Paris and discovered that it lost time in Cayenne. Using Galileo's formula $2\pi\sqrt{\frac{L}{g}}$ for a pendulum's period, Richer correctly concluded that because the pendulum rod's length L hadn't changed, gravitational strength g must have. So although Cayenne isn't mountainous, it must somehow be further from the center of the earth than Paris.
>
> Cayenne (after which Cayenne hot peppers are named) is only 5° north of the equator, and the earth's radius is in fact greater near the equator than at the poles. Newton and Huygens used Richer's data to make this precise, showing that the earth is very nearly an ellipse rotated about its minor axis—an *oblate spheroid*. The bulge comes from the earth's rotation, and we know today that the average equatorial radius is about thirteen miles greater than the polar radius. That may sound substantial, but it's nothing the eye would spot from space. Our earth, despite its rotation, is more perfectly round than a regulation billiard ball!

Which Runs Faster?

Let's revisit the grandfather clock described on p. 110. Take two identical such clocks, but in one, double the weight suspended from the chain. Which runs faster?

(a) The clock with the normal driving weight runs faster.

(b) The clock with double the driving weight runs faster.

(c) They run at the same rate.   (p. 310)

Inclined Mass-Spring

Here's a mass-spring system mounted on a frictionless inclined plane hinged at the top:

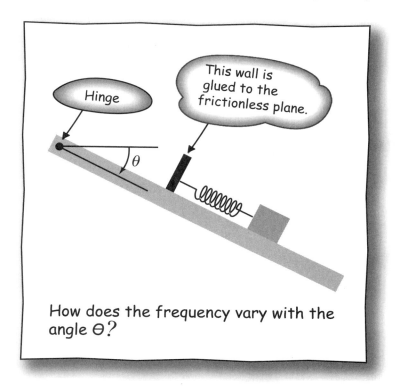

How does the frequency vary with the angle θ?

The inclined plane can be fixed at any angle. Carry out a succession of experiments starting at $\theta = 0$ and increasing θ a little between each one. What happens to the oscillation frequency from one experiment to the next?

(a) Nothing happens.

(b) The frequency increases from one experiment to the next.

(c) The frequency decreases from one experiment to the next.

(d) In going from $\theta = 0°$ to $\theta = 90°$, the frequency increases to a maximum, then decreases.

(e) In going from $\theta = 0°$ to $\theta = 90°$, the frequency decreases to a minimum, then increases.

(p. 310)

Wave Motion?
Version 1

Suppose a point P travels at constant speed v along an elliptical path:

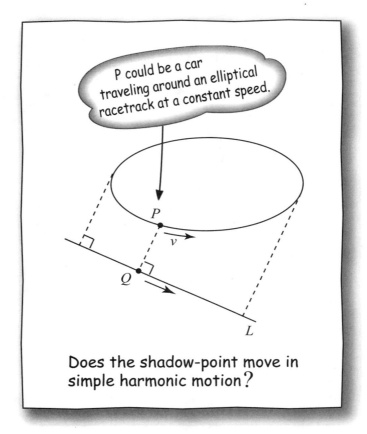

Does the shadow-point move in simple harmonic motion?

If the ellipse is a circle, then Q, the orthogonal projection of P on a fixed line L, describes simple harmonic or wave motion on L. That is, the motion of Q on L is sinusoidal. But what if the ellipse isn't a circle? Then

(a) Q still describes simple harmonic motion on L.

(b) Q describes simple harmonic motion on L if and only if L is parallel to the major or minor axis of the ellipse.

(c) Q does not describe simple harmonic motion on L.

(p. 310)

Wave Motion?
Version 2

Suppose that the angle θ increases at a constant rate, so that P travels counterclockwise around the ellipse:

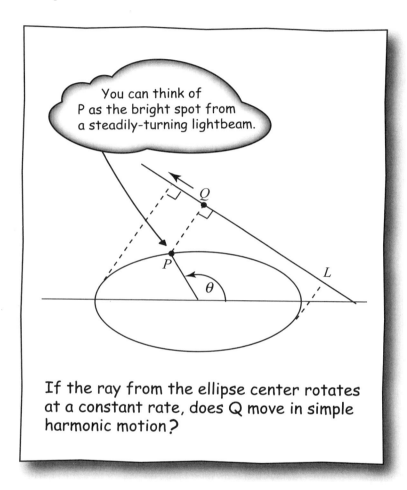

When the ellipse is a circle, Q describes simple harmonic or wave motion on L. But if the ellipse isn't a circle, then what?

(a) Q still describes simple harmonic motion on L.

(b) Q describes simple harmonic motion on L if and only if L is parallel to the major or minor axis of the ellipse.

(c) Q does not describe simple harmonic motion on L.

(p. 311)

Wave Motion?
Version 3

Suppose that P is a planet and that S is a much more massive body like the sun:

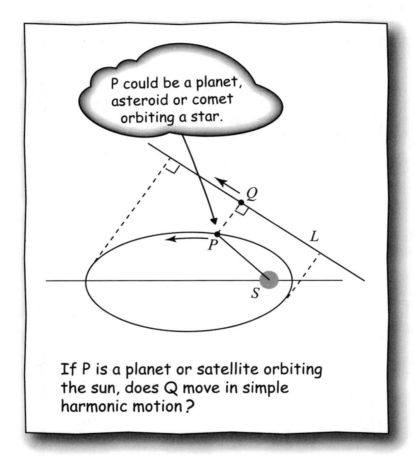

P could be a planet, asteroid or comet orbiting a star.

If P is a planet or satellite orbiting the sun, does Q move in simple harmonic motion?

Kepler's first law of planetary motion tells us that P moves in an ellipse with S at one focus. His second law says that the line segment between P and S sweeps out equal pie-shaped areas in equal times. Does the planet's projection Q on L describe simple harmonic motion?

(a) Yes

(b) It does if and only if L is parallel to the major or minor axis of the ellipse.

(c) Q does not describe simple harmonic motion on L.

(p. 311)

Heat and Wave Phenomena

Is Energy Conserved?

Here's a picture of two sine-shaped blips traveling on a string:

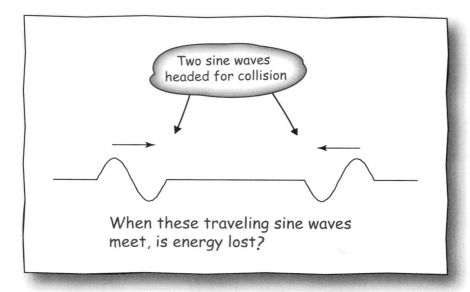

The two mirror-image blips are racing towards each other (at equal speeds). Assume the string is uniform, meaning each traveling shape took the same amount of energy to create initially. Soon the inevitable happens: they meet, and shortly after, we have this situation:

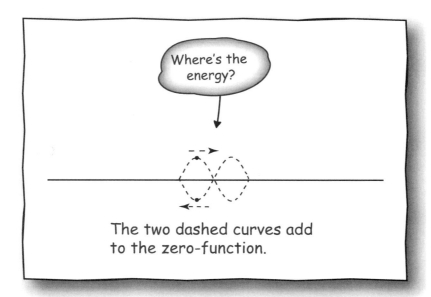

By the superposition principle, the shapes at this instant perfectly cancel each other, making the string straight. Your friend claims that the straight shape implies that energy has disappeared. After all, he says, the string was straight before you put in the energy to create the blips, and it's straight *now*. Your instinct is, "No way—the *velocities* don't cancel!" But he shows you those two heavily-drawn points in the picture, one above the other. In the curve moving to the right, the top point is moving down; in the curve going left, the bottom point is moving up. By symmetry, the velocities are equal and opposite, so they also cancel. Ditto for other points. Now he has you wondering! In these possible ways out, pick the first correct one.

(a) There's something wrong with his velocity-argument.

(b) He hasn't taken into consideration accelerations, and these don't cancel.

(c) Energy is still conserved, but for some other reason. (p. 312)

Which Has More Energy?

Two springs are compressed and a block inserted between them. Assume that the block is at rest and that there's no friction between the floor and block. That means the block is pushed equally hard by both springs.

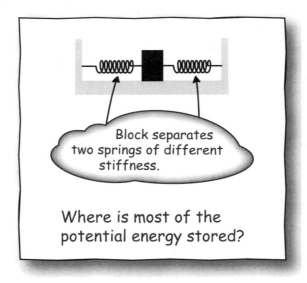

Block separates two springs of different stiffness.

Where is most of the potential energy stored?

We assume that one spring is weak (has a small Hooke constant) and the other is strong (has a large Hooke constant). Which spring stores more potential energy?

(a) The weaker spring   (b) The stronger spring

(c) They both store the same amount of energy. (p. 312)

A Low Orbit

Suppose the earth is a perfect sphere of uniform density and has no atmosphere. Would it then be theoretically possible for a golf ball to travel around the earth forever, skimming only one inch above the surface?

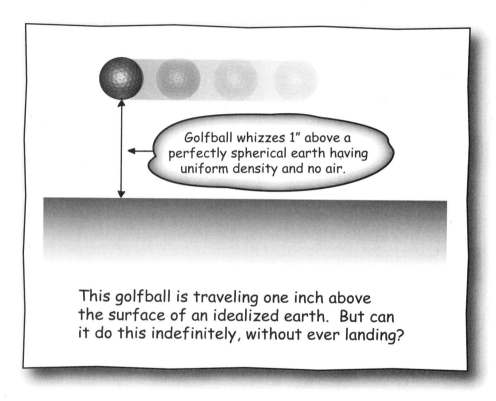

And if this *is* possible, how long would it take to go once around the earth?

(a) No, such a low orbit can't be maintained, even theoretically.

(b) Yes, such a low orbit is possible. It would take substantially less than one day to orbit once.

(c) Such a low orbit is possible. It would take just a fraction of a second less than one day to complete a full revolution.

(d) A low orbit like this is possible, and it would take just a fraction of a second more than one day to make one revolution.

(e) Such a low orbit is possible, but it would take substantially longer than one day to make one revolution.

(p. 313)

Spike

Here's an infinitely-thin, uniform string whose initial shape is a sine curve, but with a downward spike in the middle. The ends are attached to the walls on each side, and the string loses no energy as it vibrates:

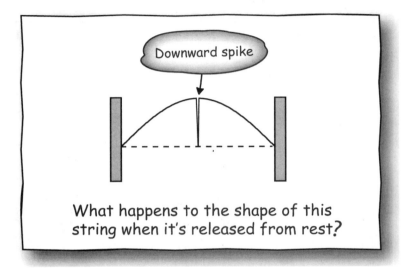

What happens to the shape of this string when it's released from rest?

If the initial shape is released from rest, which one(s) of these six pictures would be possible before the sine-shape first goes through its straight-line shape?

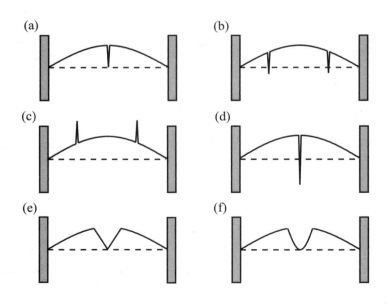

(p. 313)

Spike—Temperature Version

Here's an insulated steel rod with a block of ice at each end, and an initial temperature curve which is sine-shaped except for an ice-cold spot at the center:

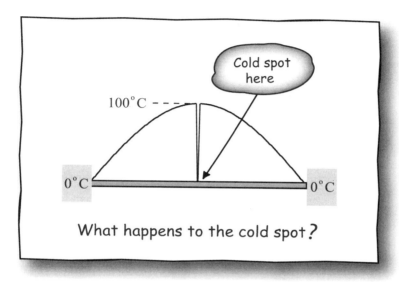

What happens to the cold spot?

Which of these could be a possible temperature curve at a later time?

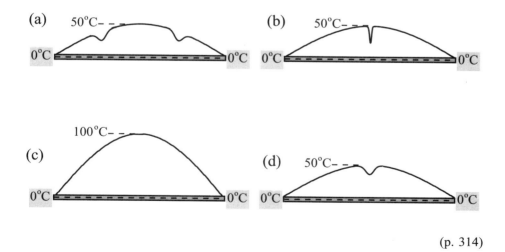

(p. 314)

Can This Happen?

Here's a variant of the conducting rod problem. Instead of holding the end temperatures fixed, suppose they both steadily rise. For example, if all points of the rod start off at 0° and the temperature of each endpoint rises 1° per minute, then after a while the temperature graph could look like this:

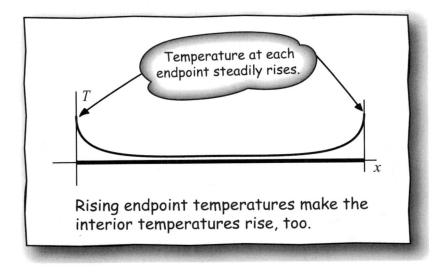

Rising endpoint temperatures make the interior temperatures rise, too.

Here's the question: assuming the endpoints are both initially 0°, is it possible to choose an initial temperature distribution on the rod so that the following can happen? *As we steadily increase both endpoint temperatures by 1° per minute, the temperature at each point of the rod simultaneously rises 1° per minute, too.*

(a) Yes   (b) No   (p. 314)

Beyond Hooke

Hooke's linear law $F = -kx$ assumes an ideal spring, and the law can greatly diverge from reality as the size of x gets large. A nonlinear law involving higher powers of x is often used to extend the range of applicability. Which one(s) of these could be used to refine Hooke's law? Assume A and B are positive.

(a) $F = -(kx + Ax^2)$

(b) $F = -(kx + Ax^2 + Bx^3)$

(c) $F = -(kx + Bx^3)$ (p. 314)

Heat and Wave Phenomena

Exponential?

Remember this picture from "Going, Going, Gone!" on p. 162?

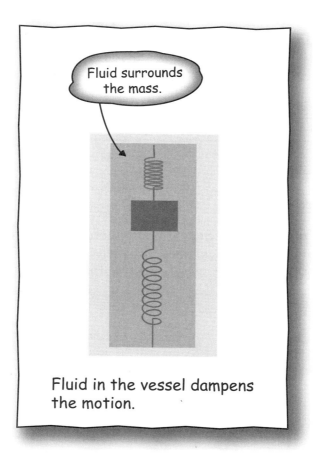

If the fluid is air so that the system's damping is small, the mass will in theory oscillate forever. But increase the fluid's viscosity beyond a certain threshold point, and oscillation ceases: if the fluid is just slightly less viscous than the threshold value, the mass will oscillate forever, while if the fluid is just slightly more viscous than the threshold value, the mass won't oscillate at all.

In the second case, if you release the mass from rest, it heads toward the equilibrium position, never oscillating. Its position approaches the zero equilibrium point, the equation for the position of the mass looking like $\alpha e^{-k_1 t} + \beta e^{-k_2 t}$. This sum of two decaying exponentials decays, too, but is this decay exponential? To be specific, is the function $e^{-t} + e^{-2t}$ exponential? That is, can you write $e^{-t} + e^{-2t}$ as Ae^{-kt} for some A and k?

(a) Yes   (b) No   (p. 315)

Does This Chord Stay Musical?

Here are four strings of different densities and under different tensions:

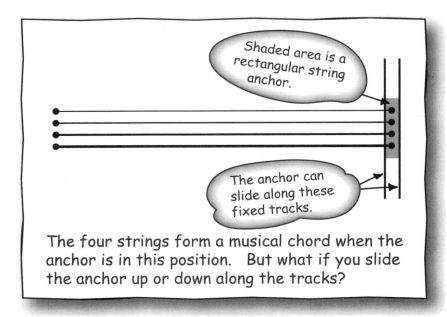

Suppose they sound a musical chord when plucked together. If you slide the string anchor up or down along the vertical tracks shown, will the new chord sound the same, except for a higher overall pitch?

(a) Yes (b) No (p. 315)

> Ideal objects that vibrate come in dimensions 0, 1, 2, and 3. Examples: the point-mass in a mass-spring setup; a vibrating string, a woodwind's column of air, or a triangle; drumhead or cymbal; a wooden block. The last four are found in an orchestra's percussion secion. Most musical instruments are one-dimensional. Their sets of fundamental modes of vibration—overtones—are simpler than those in higher dimension, which are more complicated and noisy. One-dimensional vibrations are predominately transverse or longitudinal. In transverse vibration (as in string instruments, including the piano and harp), the main oscillation is transverse to the string. In longitudinal vibration (as in wind instruments) the oscillation is in the direction of the air column.

The Leaking Tank

Introduction

Take an ordinary empty tin can with no top, punch a small hole in the bottom and stuff a bit of sponge into the hole. Now fill the can with water and let'er drain! In this setup, the rate of draining is nearly proportional to the water height, and proportionality implies that the water height decreases to zero exponentially. The ubiquitous exponential describes a wide range of real world phenomena, and we've already met one example in the last chapter, the cooling cup of coffee. There is an analogy between heat and liquid: the cup of coffee cools like liquid (that is, heat) drains away to the environment. Newton's law of cooling tells us that it's the temperature difference between the object and the environment that drives the flow of heat. In the tin can, that corresponds to the water height: the higher the level, the faster the water leaks out. Physically, the height is proportional to the water pressure at the bottom of the can where the hole is.

The leaking tank can model many other phenomena, such as the flow of electricity, wind (as air moves from a point of high barometric pressure to low), and, as we'll see, even descending musical pitches. And when radioactive carbon-14 decays to ordinary carbon-12, the diminishing amount of C-14 plays the role of the diminishing amount of water in a leaking can. We can also push an empty can bottom-first into a lake and hold it there. Water enters the can and approaches the lake's water level at a decreasing exponential rate, modeling an abandoned cold drink slowly warming to room temperature, or a pizza warming to the high temperature of the oven it's in.

You can also run a leaking tank movie backwards. That could model, say, money (water) in the bank (the tin can) growing as interest continuously compounds. The tank idea is as versatile as it is powerful. For example, we can link together several cans, one leaking into another.

Why do we stuff that little bit of sponge into the hole? Newton's cooling points to the reason: thought of as a liquid, heat has no mass—it moves through a thermal resistor because it's driven by a temperature difference. We mimic this process in our tin can with water. First, minimize the role of water's mass, or inertia, by making the hole small, like a pinhole. That way, very little mass goes through the hole per unit time, which means very little force is needed to accelerate the water. At the same time, we introduce a resistor into the hole, a sponge plug. Then nearly all the force of the water's weight above the plug goes into pushing the water through this resistor. This is analogous to electric current flowing in a flashlight: the current flows in a simple loop, not a coil, so there's not much inductance, the electric analogue of mass. The electrical driver is the battery's voltage V pushing the charge through the resistors (mostly, the bulb's filament). The current is given by Ohm's Law, $i = \frac{V}{R}$, the current i depending only on resistance and electric pressure, not mass. As water flows from our idealized tin can, analogous to heat or electric current flowing through a resistor, the flow rate at any time is the terminal speed determined by the ratio of pressure to total resistance. This is similar to a parachutist's terminal speed, which is essentially the resistance-pressure picture for constant-rate flow of water, heat or charge. But instead of a parachutist falling through an unmoving atmosphere, the atmosphere (water, heat, charge) moves and the parachutist (resistance) stays in place.

The Leaking Tank

How Long?

If it takes 10 sec for a tank to lose 25% of its water, then the half-life of the tank is

(a) 20 sec. (b) more than 20 sec. (c) less than 20 sec. (p. 317)

Again, How Long?

If it takes a minute for a tank to lose 75% of its water, then the half-life of the tank is

(a) 30 sec. (b) more than 30 sec. (c) less than 30 sec. (p. 317)

How Fast?

Suppose that when a tank with a resistive hole is half full, water flows out at the rate of 1 cc/sec. In cc/sec, what was the flow rate when the tank was full?

(a) 1 (b) $\frac{3}{2}$ (c) 2 (d) $e \approx 2.71828$ (e) None of these (p. 317)

Two Philosophies Compared

Two cylindrical tanks are filled with water, and are identical except for their holes. The first has a small hole with a sponge plug. The second has no plug, and the hole is made just the right size so that initially both tanks discharge water at the same rate (that is, the same volume per unit time). Then what happens? As they continue to leak,

(a) the first tank always discharges faster. (b) the second tank always discharges faster.

(c) they always discharge at equal rates. (d) after a while the first catches up, beating the second.

(e) after a while the second catches up, beating the first. (p. 317)

> To model exponential decay using a leaking can, we've idealized the can by assuming it has a small hole with a sponge plug to minimize the effect of water's mass and maximize the role of resistance. In a system with a larger hole and no sponge, resistance becomes less important and mass takes on the major role—water simply pours out of the hole with little friction. Galileo's famous student, Evangelista Torricelli (1608–1647), analyzed this system. Torricelli's law tells us that the discharge rate varies not with height y, but with $\sqrt{y}$. See **Torricelli's Law** in the Glossary.

Two Movies

The picture on the left depicts the first frame of a movie featuring water draining from the top tank to the bottom one, and from the bottom tank to the outside. Same on the right, except that the liquid is oil.

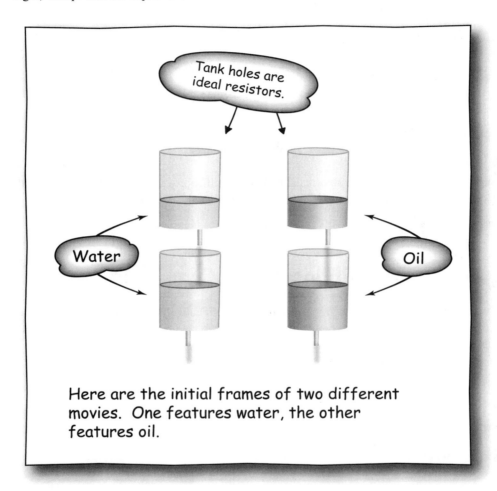

Here are the initial frames of two different movies. One features water, the other features oil.

At the beginning of the movie, the levels in the top and bottom tanks are the same.

Oil drains more slowly than water, but could we speed up the oil movie by some fixed amount so that the liquid heights in the two movies agree at all times?

(a) Yes

(b) Yes, but only if the upper and lower tanks have identical resistive holes and bases

(c) No   (p. 318)

The Leaking Tank

Two Movies, Again

Here are the initial frames of two movies, each movie featuring water draining from two tanks.

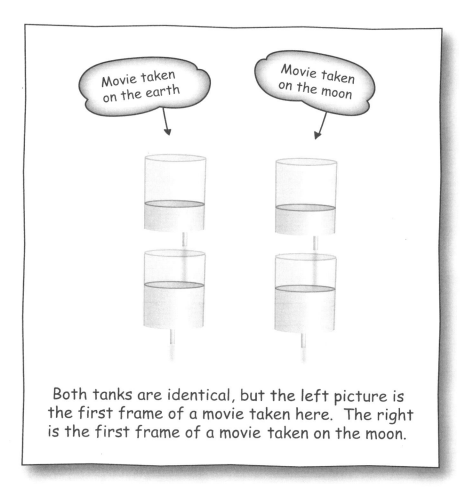

Both tanks are identical, but the left picture is the first frame of a movie taken here. The right is the first frame of a movie taken on the moon.

The first movie is taken here on earth and the second on the moon. Is it possible to run the moon-movie at just the right fixed speed so that the liquid heights in the two movies agree at all times?

(a) Yes

(b) Yes, but only if the upper and lower tanks have identical resistive holes and bases

(c) No

(p. 319)

Who Wins?

Here are three tanks with identical resistive holes and bases. The tanks are held in place so they can't move, and their water levels all start 10 inches from the level line L:

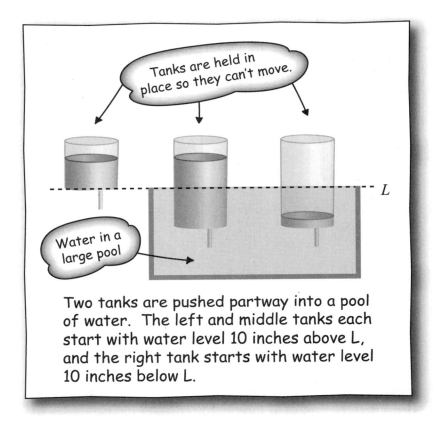

Two tanks are pushed partway into a pool of water. The left and middle tanks each start with water level 10 inches above L, and the right tank starts with water level 10 inches below L.

Which tank's water level first makes it to 1 inch away from L? Ignore the weight of air.

(a) The left tank   (b) The middle tank   (c) The right tank

(d) The left and middle tanks tie.   (e) The left and right tanks tie.

(f) The middle and right tanks tie.   (g) All three tie.   (p. 319)

How Many Half-Lives?

A kettle of boiling water at 100°C is put outside at freezing (0°C). How many half-lives does it take for it to cool down 25°?

(a) A half of a half-life   (b) More than half a half-life

(c) Less than half a half-life   (d) Insufficient information to say   (p. 319)

A Musical Analogue

Half-lives are found even in the world of musical pitches. Take a look at a portion of a keyboard:

A section of a piano keyboard. We assume the piano's tuning is well-tempered.

As you go down an octave, the frequency drops by 50%. The figure depicts this for the pitch C, but it happens for any pitch. If the keyboard extends indefinitely to the right and left, then we can always move to the right one octave to double the frequency, or move to the left one octave to halve it. This suggests that the leaking tank can model a musical keyboard. Frequency plays the role of water height, and going from higher notes to a lower ones corresponds to water draining out as time passes. One octave is covered as one goes down 12 keys (12 semitones), so the keyboard's half life is 12 semitones.

Today's standard turning is well-tempered: the ratio of any note's frequency to the semitone just below it is $2^{1/12}$. In pre-Bach tuning, the frequency ratio was not constant from one semitone down to the next. For example, the frequency ratio of C to B was different from that of B to B♭ or from F to E.

Can we ignore well-tempering in talking about keyboard half-lives? That is, even with non-tempered turning, is it true that no matter where you start on a keyboard, going down 12 semitones still represents a half life?

(a) Yes   (b) No   (p. 320)

Down One Key

In going down a tempered-scale semitone, a note loses about what percent of its frequency?

(a) 4.15%   (b) 5.6%   (c) 5.9%   (d) 8.3%

(e) None of these—it depends the note.   (p. 320)

Raising Frequency

Raising a note's frequency by 33.3% is closest to going up how many tempered semitones?

(a) 4 (a major third)   (b) 5 (a fourth)   (c) 6 (a diminished fifth)

(d) 7 (a fifth)   (e) 8 (an augmented fifth)   (p. 320)

A Full Piano Keyboard

Going from the top key to the bottom on an 88-key piano represents $7\frac{1}{3}$ half lives.

(a) True   (b) False   (p. 320)

> Why did Johann Sebastian Bach (1685–1750) introduce the well-tempered scale? There's a familiar ring to the answer. Often, convention, nomenclature and measurements become established before people see possible problems. Some examples: Benjamin Franklin couldn't have known that his "positive" charge was actually opposite to the natural unit of charge, the electron. Fahrenheit degrees became well established before Celsius introduced his more natural one. Ditto for British units versus metric. And Pythagoras was ecstatic over the discovery that dividing a string's length into simple ratios created harmonious-sounding intervals. He created new tones by taking $\frac{1}{2}, \frac{2}{3}, \frac{3}{4}, \frac{3}{5}, \frac{4}{5}$ of the string's length. The $\frac{1}{2}$ represented the octave, and the others represented new pitches, for a total of five pitches—the number of planets known then. The discovery became imbued with universal significance, and "proved" to him that he was on the right track. By using the simple-ratio trick on Pythagoras' notes, additional notes were created, eventually settling into what is now our 12-note scale.
>
> Pre-Bach scales had a problem: music was not transposable. A piece would sound impossibly sour when played a semitone higher, say. But a singer who couldn't hit a low note would need the piece transposed up. Bach, who tuned his harpsichord, discovered that by slightly retuning scale pitches, he could make the ratio between successive pitches constant (or at least nearly so). No difference was ever unacceptably great, but the $2^{\frac{1}{12}}$-ratio between successive semitones meant that transposition never affected pitch relationships. By these subtle changes, singers suddenly gained great flexibility. So did composers, since a powerful compositional device is repeating a motif, phrase or musical section in different keys. Pre-Bach composers were restricted in this regard. Today, this is not at all a problem. To prove his point, Bach wrote his famous Well-Tempered Clavier.

A Model for Relative Humidity

A tank can model relative air humidity. Air has a capacity to hold water vapor, but the amount is relative. For a given volume of air, the maximum amount of water vapor it can hold before the vapor starts to condense into water droplets depends on the air's temperature and pressure. The amount of water vapor in the air, as a percentage of this maximum, is known as the *relative humidity* of the air.

In our tank model, the amount of water in the tank corresponds to the absolute amount of water vapor in a unit volume of air. If that quantity of water completely fills the tank, the relative humidity is 100%. Dividing the height of this column into 100 parts gives a way of representing the percentage of relative humidity. For example, water at half the tank's height corresponds to 50% relative humidity.

The four examples below give an idea of how the model works.

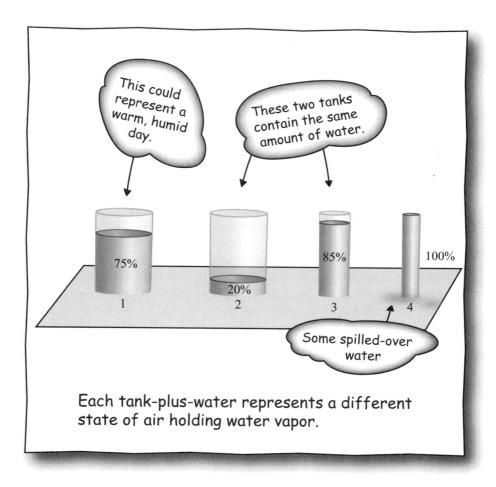

Each tank-plus-water represents a different state of air holding water vapor.

Tank 1 is three-quarters full, and represents air at 75% relative humidity. A tank's cross-sectional area corresponds to the absolute capacity to hold water vapor. Tank 2 models air with the same vapor-holding capacity as the leftmost tank, but is holding only 20% of that capacity. In tank 3 the cross section is smaller. The amount of water is the same, but now fills up 85% of the tank, corresponding to 85% relative humidity. In the atmosphere, a shrinking cross section typically comes from dropping air temperature (say, after the sun goes down). As the base area grows smaller, the height of a fixed amount of water rises. The relative humidity increases as temperature drops.

We now come to our question: The water in tank 4 is not only at its highest possible level, some water has spilled over the top because the base has been reduced so much.

(a) The spilled water means we've overextended our model.

(b) The spilled water can be thought of as part of the model, and corresponds to something physical.   (p. 320)

> The tank model of humidity can also represent other situations. An analogue of relative humidity can be created by going down a step in the solid-liquid-gas totem pole to the solid-liquid level. In analogy to increasing the humidity of a room by squeezing an atomizer a few times to add some water mist, you can stir some grains of table salt into a glass of water. The grains dissolve into the water, the liquid becoming slightly salty. In terms of tank models, in each case we have added some liquid to a nearly empty tank. Just as continually squeezing the atomizer corresponds to slowly filling the tank, so does continually adding more salt to the water. In each case one eventually reaches the saturation point, and in the model the tank is then full. With humidity, warming the room will increase the tank's base, allowing us to add more mist, while cooling the room decreases the base, forcing some moisture to condense back to liquid water. Analogously, for a glass of saturated brine, warming the water means we can dissolve more salt, while cooling it means some salt will crystallize back into solid form. The atomizer could just as well be squirting alcohol, or we could add grains of sugar to the water, or fine particles of another solid soluable in some liquid. The liquid and solid can even be the same material—think of ice dissolving in a glass of warm water, or ice crystallizing out as enough heat is taken away from the glass of water.

There is a dual to the mist picture: instead of a liquid dissolving in a gas, gasses can dissolve in a liquid. For example, fish breathe oxygen dissolved in water. And carbonated water is just water with some carbon dioxide dissolved in it. You can increase the amount of dissolved gas to the point of saturation, which in the model is like filling up the tank. You can likewise increase the model's base so you can add more gas, or decrease the base so some dissolved gas will revert to gas (bubbles). Physically, you increase the base by cooling the liquid, and you decrease the base by warming it. In an abandoned soft drink the model's base decreases to where even when the tank is full, the drink is flat.

Another phenomenon: when the model's tank is full (saturated), you can¹ often trick the physical situation into becoming supersaturated. In the mist example, suddenly cooling the air in its container can do this. In the model, the base of the full column of water decreases, meaning the water actually sticks up higher than the tank. This is unstable, the water requiring only some seed or nucleus around which it can condense. If none is available, an unstable condition can exist for some time. Since temperature is a measure of heat density, if you suddenly increase the size of the chamber, the air temperature inside suddenly falls. Charles T. R. Wilson (1869–1959) used this idea in a clever way. He equipped a chamber with a window on the top and at the bottom, installed a piston that could be moved downward with explosive suddenness, thus instantaneously increasing the chamber's volume. When the chamber was saturated with water vapor, suddenly increasing the chamber's volume decreased the temperature, creating supersaturation. Charged particles passing through the chamber created ions around which water droplets would condense, and the particles' paths could then be photographed. This found an application of historic importance: the theoretical prediction of the positron by Paul Dirac (1902–1984) was confirmed in 1932 by Carl Anderson (1905–1991) using a Wilson cloud chamber. All received Nobel prizes in physics: Wilson in 1927, Dirac in 1933, and Anderson in 1936.

There is a dual to Wilson's cloud chamber: the bubble chamber. Devised by Donald Glaser, superheated liquid in a bubble chamber is on the precipice of boiling or changing into gas form. When a charged particle passes through the liquid, it deposits enough energy along its path so that a string of boiling bubbles forms. Glaser won the Nobel prize in 1960 for his invention.

Two Interconnected Tanks

Consider these two interconnected tanks:

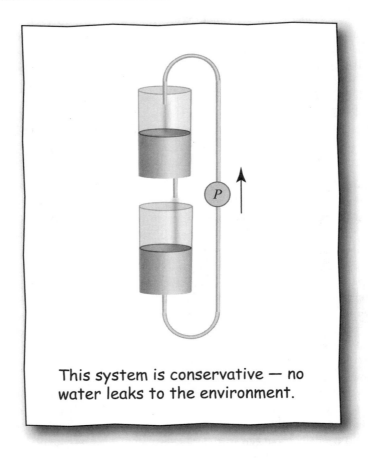

This system is conservative — no water leaks to the environment.

Water draining from the bottom tank is returned to the top tank by the pump. The tanks are identical, except for the size of the purely resistive holes: 1 mm diameter in the top tank, 2 mm diameter in the bottom. Each tank starts with 5 gallons of water. After a long time, the amount of water in the top tank is closest to

(a) 5 gallons.

(b) $6\frac{2}{3}$ gallons.

(c) $7\frac{1}{2}$ gallons.

(d) 8 gallons.

(e) Both levels oscillate forever, never approaching limiting levels.   (p. 320)

The Leaking Tank

Three Interconnected Tanks

Here are three cyclically-connected tanks:

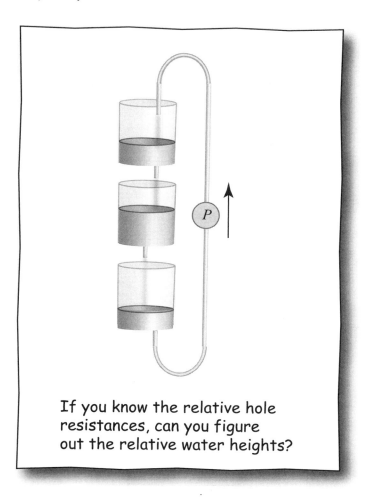

If you know the relative hole resistances, can you figure out the relative water heights?

Water draining from the bottom tank is returned to the top tank by the pump. The only difference in the tanks is hole size: from top to bottom, the tanks' effective hole resistances are 1, 2 and 3. What are the ratios of the respective steady-state water heights?

(a) 3 : 2 : 1

(b) 6 : 3 : 1

(c) 9 : 4 : 1

(d) 6 : 3 : 2

(e) None of these

(p. 321)

Tanks Can Model Current Flow

This picture shows an analogy between water flow and electric current:

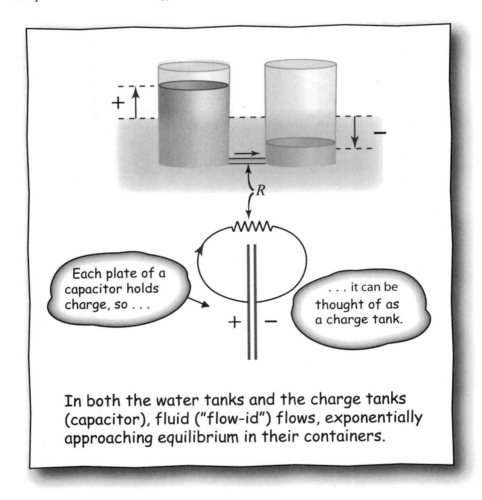

Each plate of a capacitor holds charge, so . . .

. . . it can be thought of as a charge tank.

In both the water tanks and the charge tanks (capacitor), fluid ("flow-id") flows, exponentially approaching equilibrium in their containers.

In the top picture, the surface of the water in a calm lake is taken to be the zero level. The water levels in the tanks start at equal distances above and below this zero level, and hydrostatic pressure makes water flow to bring these levels closer to it. The instantaneous pressure is proportional to the height (positive or negative) of water in the tank, which leads to a differential equation like $\dot{y} = -ky$ whose solution has the form $y = Ae^{-kt}$.

Now for the analogy. In the bottom picture, the two capacitor plates correspond to the water tanks, and start with equal but opposite electric charge. Charge flows to bring the plates' charge closer to zero in response to electrostatic pressure (voltage difference). The voltage difference creates current, and the instantaneous voltage difference is proportional to the amount of charge on the plates. As with water pressure, that means the charge decreases exponentially.

The Leaking Tank

In the water tank picture, for a given water height (depicted by the arrow pointing up), a larger tank base means more water must flow to reach equilibrium. For a common water-flow resistance R, this means that in a tank with a larger base, the height approaches zero more slowly because it is the height, not the tank's base area, that determines the pressure at the hole. Likewise in the capacitor picture, for a given initial voltage difference between the plates, a large plate area means more charge must flow to reach equilibrium. For a common electrical resistance R, the voltage difference between the plates approaches zero more slowly than in a small-plate one. In each case, the water level or amount of charge decreases exponentially.

In the capacitor, suppose that without changing either the resistance or the amount of charge on the plates, the plates are made smaller. Will this affect the rate at which the charge goes to zero?

(a) Yes, the charge difference decreases faster.

(b) Yes, the charge difference decreases more slowly.

(c) No   (p. 321)

Pulling on the Plates

In the last picture, suppose the two charged plates are pulled apart slightly, increasing the air space between them. In the tank picture, what does this correspond to?

(a) Increasing the connecting tube's resistance   (b) Increasing the diameters of the tanks

(c) Decreasing the diameter of the tanks   (d) It would not change the tank picture.

(e) None of the above   (p. 321)

> Since the plate area of a capacitor and the base area of a tank are analogues, the capacitor size "farad" corresponds to, say, square inches for tank base area. The definition of one farad as one coulomb per volt is like defining tank base area as the number of gallons of water you need to pour into the tank to make the water level rise one inch. But a farad is extremely large, so a better analogy might be square miles. A common unit of capacitance is the *puff*—a picofarad, or one trillionth of a farad.
>
> Today, capacitor technology is advancing. Carbon nanotubes have led to 300-farad capacitors the size of a flashlight D cell, with packs of them used as a sort of energy-cache in hybrid vehicles for starting, accelerating and regenerative braking. Unlike a conventional car battery, they can be charged or discharged in less than one second.(!)

Tanks Can Model Heat and Temperature

This picture shows how a tank with water links up with heat-related concepts. The model assumes a large vat with high sides.

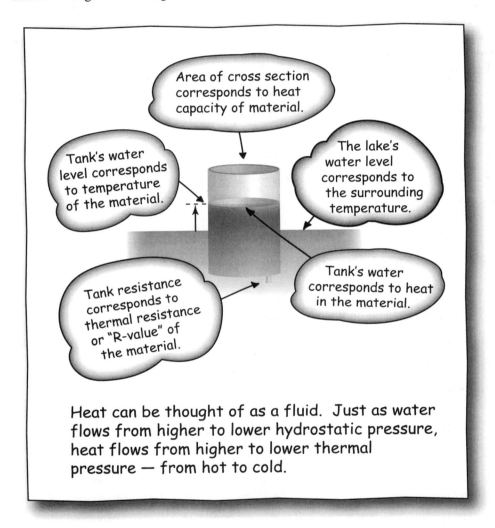

Heat can be thought of as a fluid. Just as water flows from higher to lower hydrostatic pressure, heat flows from higher to lower thermal pressure — from hot to cold.

Relative to the vat's water level, the amount of water in the tank corresponds to the amount of heat in an object relative to the surrounding (ambient) temperature. For a given water height, the tank contains more water if the tank has a large base, and the base size is proportional to the water-holding capacity of the tank. The thermal equivalent of water-holding capacity is heat-holding capacity. Sand, brick and stone have large heat capacities, while iron and aluminum have much smaller heat capacities.

One type of solar house can be modeled using this picture. The day getting hot corresponds to the vat's water level rising. When the level exceeds the water level in the tank, water

The Leaking Tank

flows into the tank through the resistive hole at its bottom, and the tank takes on water. In the solar house, when the day's temperature gets higher than that of the stone or sand, that material absorbs heat. Since these materials have high heat capacity, corresponding to a large water tank cross section, quite a bit of heat (water) will be absorbed. When it's cold outside that night, one can adjust the thermal resistance R (akin to a water spigot) to keep the house cozy inside.

The question: In a diesel engine, each cylinder rapidly compresses air in its chamber, raising its temperature so much (usually to at least 1000° F) that when a bit of diesel-fuel mist is injected into the hot air, the air-mist mixture spontaneously combusts. The explosion pushes the cylinder down for its power stroke. Does this fast compression stage of the diesel cycle correspond in our model to quickly changing the size of the tank's base?

(a) Yes   (b) No   (p. 321)

The diesel effect was used long before Rudolph Diesel (1858–1913) applied it to engines. In 1865, European explorers chanced upon a fire-making device in Indonesia, the fire piston. Handcrafted for generations, it's a cylinder with a tightly-fitting plunger, and it works like this: place a tiny bit of dry tinder into a small concavity at the end of the plunger and place the loaded piston into the cylinder. Then plunge the piston into the cylinder as quickly as you can. If the tinder is dry enough and the compression fast enough, you get a glowing ember. Fire pistons are available today as survival tools.

As a young man already intensely interested in thermal engines, Diesel once saw a cigarette-lighter version of a fire piston. This triggered his imagination and changed the course of his life. If extreme compression could boost temperature enough to ignite tinder, couldn't it just as well ignite engine fuel? This question led him to turn on its head the day's conventional thinking on internal combustion engines. Instead of moderately compressing a fuel-air mixture and then igniting it for the power stroke, why not *greatly* compress pure air, and only then inject the fuel? Even thicker, cheaper fuel would self-ignite. In addition, theory told him that high compression meant better efficiency.

Diesel saw in his engine a way to make society fairer: at the time, big industry's virtual monopoly on energy production made it hard for small companies and individuals to compete. His engine could use readily-available vegetable oil as fuel. This early visionary already thought green—in the 1900 World's Exhibition in Paris, his engine ran on peanut oil.

Tank With Constant Water Supply

In this tank with a resistive hole, the pipe supplies water at a constant rate:

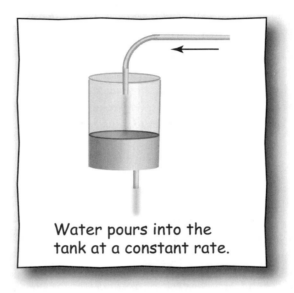

Water pours into the tank at a constant rate.

Is there an initial water level so that the water level never changes?

(a) Yes, always (b) Never

(c) Only sometimes. It depends on the rate of water supply and hole size.   (p. 322)

Increasing the Water Supply

In the last example, suppose that the spigot is steadily opened so that its flow rate increases in proportion to elapsed time t. Choose units so that at time t, the flow rate from the spigot is t, and so that the flow rate from the tank's bottom is equal to the water's height y. This means $\dot{y} = t - y$ is the rate of increase in the amount of water in the tank. There is exactly one solution curve C of this equation passing through the point $(t, y) = (1, 1)$.

The slope of C at $(1, 1)$ is

(a) positive. (b) negative. (c) zero.

At $(1, 1)$, C is

(a) concave up. (b) concave down.

(c) neither concave up nor concave down.   (p. 322)

Linear Algebra

Introduction

The equation $y = ax$ of a line through the origin is simple indeed, but generalizing this to many linear equations in many variables? To those unfamiliar with linear algebra, that might sound like a recipe for a mess! However, linear algebra's powerful ideas of vectors and matrices allow such a generalization to be written in essentially the same way as $y = ax$. By virtue of compact notation, linear algebra lets us back away and look at a large forest, and a possibly bewildering number of leaves can be much more easily understood and managed.

We begin by briefly shaking hands with a few basic ideas. Linear algebra is important because, for one thing, many natural laws are linear: for fixed resistance R, Ohm's law $V = Ri$ is really just a rewriting of $y = ax$, and so is Hooke's law, $F = -kx$. Newton's law of cooling $\frac{dT}{dt} = -KT$ is linear in the temperature difference T and the rate of temperature increase $\frac{dT}{dt}$. Some basic physical laws are actually clusters of $y = ax$, amounting to vectorial laws. For example, the vector law $\mathbf{F} = m\mathbf{a}$ is really an instance of $y = ax$ in each of three space Cartesian coordinates.

Although many natural laws are linear, they often involve several variables. For example, instead of a single loop as in a flashlight, there might be several light bulbs and batteries, all part of one fancy circuit interconnected by, say, 20 wires. Nature does little to keep such a system trivially solvable. It's easy to find the current in a simple circuit consisting of one resistor and one battery: in Ohm's law $V = Ri$, R is the bulb's resistance, V is the battery's voltage and we divide by R to find the current i. It would also be easy to find the currents in ten separate circuits by doing the division for each one. But in a complicated, interconnected problem, everything is mixed up, with different batteries and bulbs interacting, and the currents could have a very complex traffic pattern. When you apply Ohm's law to a set of simple subcircuits, the complication is reflected in the equations—the variables (currents) can look as if they've been mixed up like the ingredients of a tossed salad.

The notation and methods of linear algebra often allow you to massage the algebra describing such a messy system into the simpler algebra describing a number of separate one-resistor, one-battery circuits. Two tools are especially important. One is Gaussian elimination, and the other is the eigen-approach of finding eigenvalues and eigenvectors.

(See the Glossary for more about these.) These tools are very general, and in addition to electrical circuits, they also apply to a variety of Hooke's law and Newton's cooling law analogues. In many cases Gaussian elimination is ideal for solving static or steady-state problems, while eigen-methods are useful in solving dynamical systems—ones that change with time.

Just as we find the current in a single circuit using Ohm's law, we can find the displacement x when a weight F is hung from a single spring of stiffness k: divide $F = -kx$ by $-k$. As with ten circuits, so also for ten springs-with-weights. Ten single mass-springs, each individually hanging from a ceiling, act independently and their displacements can be found individually. But springs and weights can be mixed up, too. We could hang all ten mass-springs, one below another, and allow everything to come to rest. In this vertical arrangement, the amount that any one spring stretches is affected by how much others stretch. Since nothing is moving, the signed sum of all forces on each mass is zero, which leads to a set of linear equations. Once again, Gaussian elimination solves the system.

One can play this same game with Newton's law of cooling, and we can use our leaking tin can model to visualize what's happening. At the "decoupled" end of things, the arrangement just a line of individual leaking tanks, each doing its thing. The amount of water in each tank can be found one tank at a time—there's no interaction between the tanks and the solution is easy. But we could, in analogy with the mass-springs, arrange a bunch of leaking tanks in a vertical line, each tank leaking into the one below it. The bottom one could leak out to the environment, or the water might be pumped back up to the top. Now the tanks *don't* act separately. By looking at which tank is feeding which and using Newton's cooling law, we can write down a linear system of differential equations. The tank levels will generally be rising or falling, making the system dynamic. The eigen-approach can be used to solve this system.

Let's look at the example of the simplest coupled tank setup—two coupled tanks, each feeding into the other, like this:

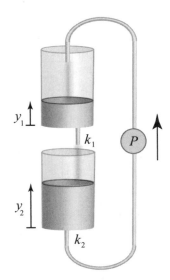

Linear Algebra

The tank bases are equal, the water heights are $y_1(t)$ and $y_2(t)$, and the purely-resistive tank holes have constants k_1 and k_2. The time derivatives $\dot{y}_1(t)$ and $\dot{y}_2(t)$ measure the rate of water-height *increase*. The net flow rate into either tank is rate in minus rate out. This translates to

$$\dot{y}_1 = -k_1 y_1 + k_2 y_2,$$
$$\dot{y}_2 = k_1 y_1 - k_2 y_2.$$

This can be written more compactly as $\dot{Y} = KY$, where

$$\dot{Y} = \begin{pmatrix} \dot{y}_1 \\ \dot{y}_2 \end{pmatrix}, \quad K = \begin{pmatrix} -k_1 & k_2 \\ k_1 & -k_2 \end{pmatrix}; \quad Y = \begin{pmatrix} y_1 \\ y_2 \end{pmatrix}.$$

With any single first-order differential equation $\dot{y} = ky$, we substitute a trial solution $y = ce^{\lambda t}$ and see what information falls out. For a system of two linear differential equations, we do the analogous thing: substitute a trial solution $Ce^{\lambda t}$ and see what happens. C is a column 2-vector of constants, and substituting $Ce^{\lambda t}$ into $\dot{Y} = KY$ gives $\lambda C e^{\lambda t} = K C e^{\lambda t}$. Because $e^{\lambda t}$ is never zero, we can cancel it. Writing λC as $\lambda I C$ leads to $(K - \lambda I)C = 0$. To insure a non-zero solution, we must assume that $\det(K - I\lambda) = 0$. When this determinant is evaluated and simplified, we get $\lambda(\lambda + (k_1 + k_2)) = 0$, which has solutions (eigenvalues) $\lambda = 0$ and $\lambda = -(k_1 + k_2)$. When $\lambda = 0$, $(K - \lambda I)C = 0$ has a solution vector $Y = \begin{pmatrix} k_2 \\ k_1 \end{pmatrix}$, and when $\lambda = -(k_1 + k_2)$, $(K - \lambda I)C = 0$ has a solution vector $Y = \begin{pmatrix} 1 \\ -1 \end{pmatrix}$. Therefore the general solution to our tank problem is

$$Y(t) = \alpha \begin{pmatrix} k_2 \\ k_1 \end{pmatrix} e^{0t} + \beta \begin{pmatrix} 1 \\ -1 \end{pmatrix} e^{-(k_1+k_2)t}.$$

If tank i has water height h_i when $t = 0$, then we can substitute this information into the last equation and solve to get α and β. Doing this and neatening up a bit gives this final, explicit solution for the heights y_i at any time t:

$$y_1(t) = k_2 \cdot \left(\frac{h_1 + h_2}{k_1 + k_2} \right) + \left(\frac{k_1 h_1 - k_2 h_2}{k_1 + k_2} \right) \cdot e^{-(k_1+k_2)t},$$

$$y_2(t) = k_1 \cdot \left(\frac{h_1 + h_2}{k_1 + k_2} \right) - \left(\frac{k_1 h_1 - k_2 h_2}{k_1 + k_2} \right) \cdot e^{-(k_1+k_2)t}.$$

There you have it—a full, exact solution. Often, with no knowledge of this solution, common sense can lead to an answer. This is not only nice in its own right, but represents another route to a solution, and therefore as a check. As just one example, the system is conservative—that is, there are no leaks—so the water heights must always sum to what they were at the beginning, just $h_1 + h_2$. Though this follows by adding our two solution equations, we didn't need to know these equations, or about differential equations or linear algebra.

There's an ongoing competition between $y_1(t)$ and $y_2(t)$, because as one increases, the other decreases. How much does each tank contain at any time? It's revealing to plot $y_2(t)$ against $y_1(t)$, obtaining a state-diagram:

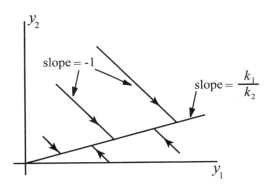

The lines with arrows show the paths traced out by $(y_1(t), y_2(t))$. As time goes on, any path approaches a steady-state solution, and all these steady states lie on the line $y_2 = \frac{k_1}{k_2} y_1$. The diagram sheds light on the physical meaning of both eigenvalues and eigenvectors. In the above full solution, time t is always multiplied by an eigenvalue, so the larger the eigenvalue, the faster any path approaches its steady state. Even an eigenvalue of 0 tells us something. It's associated with the steady state itself, and right at the steady state, the experiment doesn't progress at all—the levels just sit there, steady and unchanging. An eigenvector associated with eigenvalue 0 generates a line consisting of all the steady states. Any eigenvector generates a line that we'll call an *eigenline*. For $\lambda = 0$, this line extends the eigenvector $\begin{pmatrix} k_2 \\ k_1 \end{pmatrix}$ we found earlier. The eigenline is determined by its slope $\frac{k_1}{k_2}$. This rise over run corresponds physically to the ratio of steady-state water height in tank 1 to that in tank 2. The other eigenvalue $\lambda = -(k_1 + k_2)$ also gives meaning to the lines with arrows on them, depicting how fast the point $(y_1(t), y_2(t))$ moves along a line. And the slope of the paths with arrows? Their slope is that of the eigenline generated by an eigenvector $\begin{pmatrix} 1 \\ -1 \end{pmatrix}$ corresponding to that eigenvalue. Those are all lines of slope -1 and physically, it means that as one tank gains water, the other loses the same amount.

There is a physical linkup between these two eigenvectors. If we introduce a very tiny leak into one of the tanks, then the progress of $(y_1(t), y_2(t))$ will for a long time look very much like one of the above paths with arrows, but when the system gets close to the steady state and nothing much seems to be happening, it is then that the main attraction is the slow, drop by drop water leak from the system. The system still maintains a nearly constant ratio of water heights, but the heights themselves gradually drift toward zero. The curve makes a quick turn, following what appears to be a straight line to the origin. The graph of $(y_1(t), y_2(t))$ looks like two line segments, but they are actually connected by a tight, smooth turn.

What if the leak is larger? Then the turn is less tight, and we get something like the picture on the next page. We see the arrow indicating increasing time on the curve, and now the curve bends more gently. We can still read off eigenvectors from this curve: its limiting slope near the origin (that is, as $t \to +\infty$) determines one eigenvector, and tells

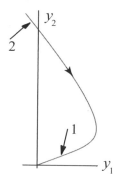

us what the limiting ratio of water heights is. The other eigenvector is determined by the limiting slope "at the other end" of the curve—that is, the slope's limit as $t \to -\infty$. It says that this eigenvector gives the limiting ratio of water-heights as we mathematically continue the solution back in time.

A Matter of Distance

Which of these lines is closest to the origin?

(a) $3x + 4y = 5$ (b) $-3x + 4y = 5$

(c) $3x - 4y = 5$ (d) $-3x - 4y = 5$

(e) All four lines are the same distance from the origin. (p. 323)

A Matter of Distance (3D Version)

Which of these planes is closest to the origin?

(a) $1x + 2y + 3z = 4$ (b) $11x + 12y + 13z = 14$

(c) $101x + 102y + 103z = 104$ (d) $1001x + 1002y + 1003z = 1004$

(p. 324)

> For many years, "good" functions were considered to be those that were differentiable. They have graphs that look smooth. Increasingly magnify such a graph about any of its points, and you get "approximate self-similarity" in that what you see looks more and more like part of a line, plane or $\mathbb{R}^n$, the geometric objects of linear algebra. We now know that fractals play a major role in the real world, and there is a basic connection between them and graphs of differentiable functions in that fractals also display self-similarity. It is usually approximate, becoming ever more exact at increasingly high magnifications. What you see is not straight, but rather some particular bent, curved or fancy shape. Some fractals are iteratively constructed by directly using a given bent-line shape as a seed. The popular Koch snowflake is an example. Other times, fractals arise as an indirect consequence of an iteratively-defined geometric object such as the Mandelbrot set.

What's its Area?

The area of the needle-shaped triangle having vertices $(0, 0)$, $(2, 1)$ and $(100, 2)$ is

(a) exactly an integer. (b) slightly more than an integer (within .01).

(c) slightly less than an integer (within .01). (d) None of these (p. 325)

Linear Algebra

> The familiar dot product formula $\mathbf{v} \cdot \mathbf{w} = \|\mathbf{v}\| \cdot \|\mathbf{w}\| \cos \theta$ connects its algebraic side $\mathbf{v} \cdot \mathbf{w}$ with the geometric counterpart $\|\mathbf{v}\| \cdot \|\mathbf{w}\| \cos \theta$, and is one of the most useful tools in linear algebra. For example, the equation $ax + by + cz = 0$ of a plane factors into the dot product $(a, b, c) \cdot (x, y, z) = 0$. Because $\cos \theta = 0$ implies that $\theta = 90° + n \cdot 180°$, this reveals that the coefficient vector (a, b, c) is orthogonal to the plane. This holds in any dimension. You can even specify an angle other than $90°$. If, for example, you fix a vector (a, b, c) and angle θ, then the equation
>
> $$(a, b, c) \cdot (x, y, z) = \|(a, b, c)\| \cdot \|(x, y, z)\| \cdot \cos \theta$$
>
> describes all vectors (x, y, z) making an angle of θ with (a, b, c). The points (x, y, z) lie on a cone whose generators make the angle θ with the line through the origin and (a, b, c). Looked at this way, a plane is a cone whose generators are perpendicular its normal. Squaring both sides of the displayed equation leads to a quadratic equation for the double cone.

What's the Volume?

The pyramid with vertices $(1, 1, 1)$, $(1, -1, -1)$, $(-1, 1, -1)$, $(-1, -1, 1)$ has volume

(a) $\frac{3}{2}$ (b) $\frac{5}{3}$ (c) $\frac{11}{6}$ (d) $\frac{8}{3}$ (e) 3 (p. 326)

> It can be useful to develop a certain "thirst": when you see something algebraic, you wish to see its geometric counterpart; and when you see something geometric, you want to translate it into algebra. Both counterparts are there for almost everything—dot product, cross product, determinants, $\cdots$. One example: in a nonsingular $n \times n$ matrix, the row (or the column) vectors form one corner of an n-dimensional parallelepiped, and the absolute value of the determinant is the n-volume of this parallelepiped. The determinant's sign also has meaning—the order of vectors can provide orientation ($+$ or $-$) to the n-space.
>
> Another thing: the parallelepiped's volume is base times height—the base being the $(n-1)$-parallelepiped defined by some $n-1$ of the vectors. If you rotate the remaining vector into the space of this $(n-1)$-parallelepiped, the altitude decreases and the volume—the determinant—goes to zero. This is precisely the condition that its row (and also column) vectors are linearly dependent, and those vectors become dependent after rotating one of them into the space defined by the others.

Stretched From Both Ends

On a foot-long ruler, a rubber band stretches from the 1-inch mark to the 10-inch mark. Suppose the ends of the rubber band are further pulled so that the rubber band lies exactly over the entire 12-inch ruler. During the pulling, points near the left end of the rubber band move left, and those near the right end move right:

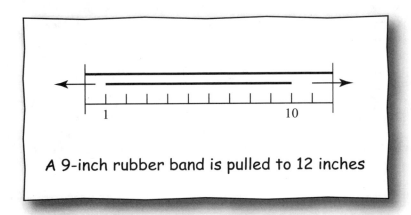

A 9-inch rubber band is pulled to 12 inches

Somewhere in between, there has to be some point that ends up where it started. Where on the ruler is that point?

(a) At the 3-inch mark

(b) At the 4-inch mark

(c) At the 5-inch mark

(d) At the 8-inch mark

(e) None of these (p. 327)

Which Way Do They Point?

A rubber band stretches between $P = (2, 0)$ and $Q = (0, 1)$ in the plane. P and Q simultaneously begin moving at 1 unit per minute, P heading due south and Q going due west. During the first minute, does any point of the rubber band always head directly toward the origin?

(a) They all do   (b) Two do

(c) One does   (d) None of them do   (p. 327)

Linear Algebra

Pyramid Peak

A right pyramid 100 ft high has a square base 200 ft on a side. How many steradians are in the solid angle at the top?

(a) $\frac{3\pi}{4}$ (b) $\frac{2\pi}{3}$ (c) $\frac{\pi}{2}$ (d) $\frac{\pi}{3}$ (e) $\frac{\pi}{4}$ (p. 328)

As with the dot product, the cross product can be looked at both algebraically and geometrically. With $\mathbf{i} = (1, 0, 0)$, $\mathbf{j} = (0, 1, 0)$ and $\mathbf{k} = (0, 0, 1)$, the cross product of the two vectors $\mathbf{v} = (a_1, a_2, a_3)$ and $\mathbf{w} = (b_1, b_2, b_3)$ algebraically is the determinant

$$\mathbf{v} \times \mathbf{w} = \begin{vmatrix} \mathbf{i} & \mathbf{j} & \mathbf{k} \\ a_1 & a_2 & a_3 \\ b_1 & b_2 & b_3 \end{vmatrix}.$$

Geometrically it is a vector perpendicular to the plane through $\mathbf{v}$, $\mathbf{w}$ and the origin. Its length is numerically equal to the area of the parallelogram that has $\mathbf{v}$ and $\mathbf{w}$ as edges, and it is directed the same way that an ordinary wood screw penetrates wood when $\mathbf{v}$ turns to $\mathbf{w}$ in less than 180°. If it must turn exactly 180°, then $\mathbf{v}$ and $\mathbf{w}$ are scalar multiples of each other, and $\mathbf{v} \times \mathbf{w}$ is zero. Notice also that $\mathbf{v} \times \mathbf{w} = -\mathbf{w} \times \mathbf{v}$.

Coupled with the dot product, the cross product can be impressive. As just one example, it can find the area of spherical triangles . . .

Cross Product

Is there some natural generalization of cross product to all higher-dimensional spaces $\mathbb{R}^n$?

(a) Yes (b) No (p. 328)

A Special Cone

Which of these is the equation of a right circular cone tangent to each coordinate plane?

(a) $z^2 = x^2 + y^2 + x + y - z$

(b) $x^2 + y^2 + z^2 = xy + xz + yz$

(c) $x^2 + y^2 + z^2 = 2xy + 2xz + 2yz$

(d) $z^2 = x^2 + y^2 + x + y + z$ (p. 329)

Discriminant versus Area

Any conic centered at the origin has an equation of the form $Ax^2 + Bxy + Cy^2 = 1$. A famous quantity associated with the quadratic form $Ax^2 + Bxy + Cy^2$ is its discriminant $B^2 - 4AC$. It does indeed discriminate, because it is positive if the conic is a hyperbola and negative if it's an ellipse. For an ellipse, therefore, $4AC - B^2 > 0$.

It turns out there's a close relationship between $4AC - B^2$ and the area of an ellipse. What is it?

The expression $4AC - B^2$ varies

(a) directly with the ellipse area.

(b) as the reciprocal of the ellipse area.

(c) as the square of the ellipse area.

(d) as the reciprocal-squared of the ellipse area.

(p. 329)

On a sphere, go down from the north pole along a line of longitude to the equator, then travel an angle of ϕ radians along the equator, and finally go directly to the north pole along another line of longitude. This round trip defines a spherical triangle. Both angles at the equator are $\frac{\pi}{2}$, so the angles add up to $\pi + \phi > \pi$, and the excess ϕ is proportional to the area of the triangle. What is remarkable is that this is true of *any* spherical triangle. On a unit sphere, the area of any spherical triangle is exactly its excess in radians over π.

Suppose that you're given three vectors $\mathbf{v}_1$, $\mathbf{v}_2$ and $\mathbf{v}_3$ based at the origin. The rays extending these vectors intersect the unit sphere in three points P_1, P_2, P_3, and the question is, what's the area of the spherical triangle having these vertices? The answer can be found if we know the three angles of the triangle. A little thought shows that the angle at P_1—that is, the angle between the two sides $P_1 P_2$ and $P_1 P_3$—is the angle between two planes: the plane through $P_1 P_2$ and the origin, and the plane through $P_1 P_3$ and the origin. This angle is in turn the angle between normals to these planes. Now one normal to the first plane is $\mathbf{v}_1 \times \mathbf{v}_2$, and one to the second plane is $\mathbf{v}_1 \times \mathbf{v}_3$. The dot product can be used to find the angle between these normals, which means we can find the angles at P_1, P_2 and P_3. Now add them up and subtract π to get the area! This can be automated as a little program in Maple, for example, with three vectors as input and the triangle area as output.

Linear Algebra

How Much of the Heavens?

In outer space, the three vectors $(2, 1, 1)$, $(1, 2, 1)$, $(1, 1, 2)$ define this megaphone shape with three faces:

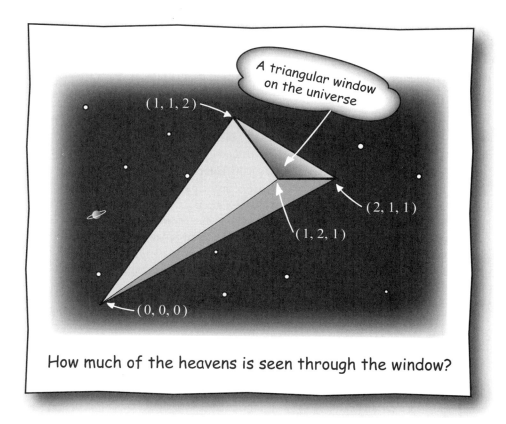

How much of the heavens is seen through the window?

Suppose your eye is at the origin and you peer upon the cosmos through the mouthpiece of the megaphone. Which of these best describes how much of the heavens you see?

(a) 1% (b) 2% (c) 3% (d) 4% (e) more than 5% (p. 329)

Degrees in a Spherical Triangle

Of the following, which of these best describes the greatest number of degrees a spherical triangle can have?

(a) 270 (b) 360 (c) 540 (d) 720 (e) 900 (p. 330)

Shearing an Ellipse

Suppose the ellipse $\frac{x^2}{a^2} + \frac{y^2}{b^2} = 1$ is horizontally sheared 45° by the linear transformation sending $(1, 0)$ to itself and sending $(0, 1)$ to $(1, 1)$. The x-axis therefore stays fixed, and the y-axis rotates 45° into the line $y = x$. The image of the ellipse under this transformation

(a) is never an ellipse.   (b) may or may not be an ellipse.

(c) must be an ellipse, but major and minor axes are not at right angles.

(d) must be an ellipse, with major and minor axes at right angles.   (p. 330)

> Any ellipse or hyperbola has a center of symmetry, and if that center is $(0, 0)$ then its equation can be written as $Ax^2 + Bxy + Cy^2 = 1$. If $B = 0$, then the principal axes of the conic are the coordinate axes. As B varies from 0, the conic's pair of principal axes rotates, and one can directly find the amount of this rotation. To carry this out, find how much you need to rotate the conic $Ax^2 + Bxy + Cy^2 = 1$ to remove the mixed term Bxy, then rotate the other way to put it back in. Replacing x by $x \cos\theta + y \sin\theta$ and y by $-x \sin\theta + y \cos\theta$ rotates the conic counterclockwise by $\theta > 0$. Make this substitution and collect terms. The xy-term then has coefficient
>
> $$2(A - C) \sin\theta \cos\theta + B(\cos^2\theta - \sin^2\theta),$$
>
> which can be written as $(A - C) \sin 2\theta + B \cos 2\theta$. Equating this to 0 leads to $\tan 2\theta = \frac{B}{C-A}$, which says how much rotation is needed to remove the mixed term. Therefore B itself creates the negative of this rotation. Since the tangent function is odd, the amount that B rotates the pair of principal axes is given by
>
> $$\tan 2\theta = \frac{B}{A - C}.$$

> Suppose that instead of varying B, we fix B and vary A and C. The last formula in the above box shows that this also affects the orientation of the conic. For example if A is fixed and C approaches A (or if C is fixed and A approaches C), then the conic's inclination approaches 45°. In the limit when $A = C$, varying B simply stretches or shrinks the conic in the 45° direction.

Linear Algebra

Geometry in a Mixed Term

Suppose that to the equation $\frac{x^2}{2^2} + \frac{y^2}{1^2} = 1$ we add a mixed term Bxy, getting

$$\frac{x^2}{2^2} + Bxy + \frac{y^2}{1^2} = 1.$$

The first box on p. 208 says that varying B from 0 rotates the conic's pair of principal axes. Is the rotation clockwise or counterclockwise? And as B increases, what happens to the rate of rotation?

As we steadily increase B from 0 (choose all that apply):

(a) the rotation is clockwise about the origin.

(b) the rotation is counterclockwise about the origin.

(c) the rate of rotation increases.

(d) the rate of rotation remains constant.

(e) the rate of rotation decreases.

(p. 331)

How Long are the Axes?

We learned from the last question that varying only B rotates the conic's pair of principal axes. Since A, B and C determine the conic $Ax^2 + Bxy + Cy^2 = 1$, the coefficients also contain all the information needed to decide whether the conic is an ellipse or a hyperbola, as well as the dimensions of the fundamental rectangle (the rectangle whose sides are parallel to the principal axes and tangent to the conic). Remarkably, for the right choice of $\bigcirc$, the solutions s_1 and s_2 of

$$(B^2 - 4AC)s^4 + \bigcirc s^2 - 4 = 0$$

determine those dimensions: the dimensions are just the real and/or imaginary parts of s_1 and s_2. The conic is an ellipse if and only if both s_1 and s_2 are real.

Which one of these is $\bigcirc$?

(a) $A + C$ (b) $2(A + C)$

(c) $4(A + C)$ (d) $A - C$

(e) $2(A - C)$ (f) $4(A - C)$

(p. 331)

Comparing Times

Here is a state diagram for the two-tank system shown on p. 198. The lines with arrows depict how the system's state $(y_1(t), y_2(t))$ changes as time increases:

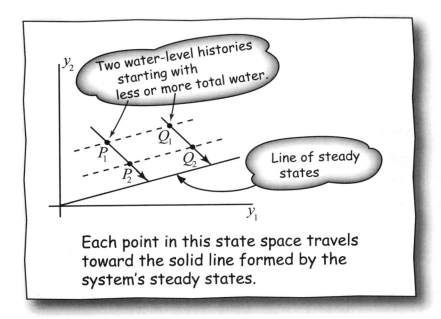

Each point in this state space travels toward the solid line formed by the system's steady states.

The two dashed lines are parallel to the line of steady states. Does it take longer for a point $(y_1(t), y_2(t))$ to go from P_1 to P_2, or from Q_1 to Q_2?

(a) It takes longer to go from P_1 to P_2.   (b) It takes longer to go from Q_1 to Q_2.

(c) It takes the same time either way.   (p. 332)

Smaller Drain Holes

In the two-tank setup shown on p. 198, what does cutting each k_i in half do? Choose all that apply.

(a) It decreases the magnitude of an eigenvalue.

(b) It increases the magnitude of an eigenvalue.

(c) It has no effect on the eigenvalues.

(d) It rotates the eigenlines clockwise.

(e) It rotates the eigenlines counterclockwise.

(f) It has no effect on the eigenlines.   (p. 332)

A Leak to the Environment

Suppose we drill a little hole through the bottom of one of the two tanks in the picture on p. 198. The water leaks to the outside environment, making the system nonconservative. Choose all that apply:

(a) Both eigenvalues must be negative.

(b) Only one eigenvalue must be negative.

(c) The sum of the eigenvalues must be zero.

(d) The sum of the eigenvalues must be negative. (p. 332)

Eigenvectors have connections with steady states going beyond that illustrated in this chapter's Introduction. For example, consider the linear transformation

$$T: \begin{pmatrix} x \\ y \end{pmatrix} \longrightarrow \begin{pmatrix} \cos 30° & \sin 30° \\ \sin 30° & \cos 30° \end{pmatrix} \begin{pmatrix} x \\ y \end{pmatrix}.$$

T rotates the x-axis $+30°$ and the y-axis $-30°$, squeezing the diagonal line $y = x$ equally on both sides. Although points in this line may move, the line as a whole doesn't. Both sides of the other diagonal line $y = -x$ are stretched equally, so it doesn't move, either. These two stationary 1-subspaces are the eigenlines of T.

What does this have to do with steady-states? T maps lines to lines, and a little thought shows that except for the two lines $y = \pm x$, any other line L through the origin rotates toward $y = x$ and away from $y = -x$. That is, $y = x$ is an attactor and $y = -x$ is a repeller. If you repeatedly apply T to L, the successive image lines move closer and closer to the attracting line $y = x$. In this sense, $y = x$ represents a steady state and it is stable in that it's the line the images approach as this process continues indefinitely. The line $y = -x$ is also a steady-state because it doesn't move when T is applied to it, but it's unstable: if L is near $y = -x$, then successive images of it move further away from $y = -x$.

Although we rotated by $\pm 30°$, the linear transformation T^* defined by rotating by $+\theta$ and $-\psi$ (where $\theta + \psi < 90°$) likewise produces one fixed attracting line and one fixed repelling line. All the other 1-spaces will move toward the steady-state attractor under repeated applications of T^*. . . .

Behavior of Water Heights
Version 1

If the two-tank setup is conservative, then the water levels in the top and bottom tanks approach steady-state levels that are proportional to k_2 and k_1, respectively. But just how do they approach these levels? Is the approach monotone, one level always increasing, and the other always decreasing? Or can one level rise, then fall toward the steady-state? Or perhaps might the levels oscillate forever with smaller and smaller amplitude as they go to their steady states? Choose all that apply:

(a) The approach must always be monotone.

(b) The approach in one tank can initially rise, then fall towards steady state.

(c) With appropriate choices of water levels and hole constants k_i, it is possible for the levels to oscillate indefinitely with ever-diminishing amplitude. (p. 332)

> ... but what if the transformation T rotates both the x- and y-axes by, say, $+30°$? The geometry says the whole plane is rotated $+30°$, so there can't be any fixed lines. But the algebra says "Not so! You can actually *solve* for a fixed line: just find $\lambda \neq 0$ so that $Tv = \lambda v$, which comes down to solving $\det(Tv - I\lambda) = 0$." Who wins the argument? The answer is, both do. There *are* two fixed lines, but they live in complex 2-space and are themselves complex, each a copy not of the real line $\mathbb{R}$ but of the complex line $\mathbb{C}$. If we could see easily in four dimensions, those two fixed complex lines would reveal themselves for all to observe!

Behavior of Water Heights
Version 2

Suppose that one of the tanks has a little hole drilled through its base, with water leaking out to the environment. Then the two-tank setup isn't conservative, so the steady-state levels are both zero. How does the water empty out? Choose all that apply:

(a) Both water levels must always decrease monotonically.

(b) The level in one tank can initially rise, then fall towards zero.

(c) With appropriate choices of water levels and hole constants k_1, k_2, k_3 (k_3 describing the leak to the environment), it is possible for the levels to oscillate indefinitely, with ever-diminishing amplitude. (p. 333)

Behavior of Water Heights
Version 3

What are the possible ways that the water levels can approach their steady states in a conservative arrangement of three tanks, like this?

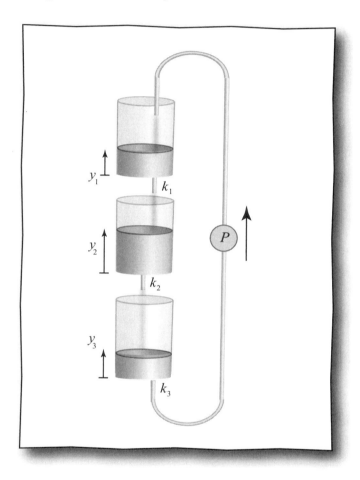

Choose all that apply:

(a) The approach must always be monotone in all three tanks.

(b) The level in at least one tank can initially rise, then monotonically fall toward steady state.

(c) With appropriate choices of water levels and hole constants k_i, it is possible for all three levels to oscillate indefinitely with ever-diminishing amplitude. (p. 333)

Ellipses and Hyperbolas

Any ellipse or hyperbola centered at the origin can be written as

$$Ax^2 + Bxy + Cy^2 = 1.$$

If $B = 0$, it has horizontal and vertical principal axes. As B changes from 0, the principal axes rotate about the origin. Now A, B and C completely determine the conic, and in the first box on p. 208 we see a formula relating the rotation angle θ with A, B and C: $\tan 2\theta = \frac{B}{A-C}$.

It turns out that A, B and C determine something else, too: a quadratic equation whose two solutions are the slopes of the principal axes. The equation's coefficients are in terms of A, B and C.

What's the right choice of $\square$ so that the two solutions of

$$Bm^2 + \square\, m - B = 0$$

are the slopes of the two perpendicular principal axes?

(a) $A + C$ (b) $2(A + C)$
(c) $4(A + C)$ (d) $A - C$
(e) $2(A - C)$ (f) $4(A - C)$ (p. 335)

Interpolation

Suppose $(a_1, b_1), \ldots, (a_n, b_n)$ are n points in $\mathbb{R}^2$. If the a_i are distinct, then there is exactly one interpolation polynomial $p(x)$ of degree at most $n - 1$ whose graph goes through all n points. This leads to a nice application of linear algebra, because we can find $p(x) = c_1 x^{n-1} + \cdots + c_n$ by solving the n linear equations $\{b_i = p(a_i)\}$ for the n unknowns $c_1, \ldots, c_n$.

For example, let $(a_i, b_i) = \left(\frac{\pi i}{5}, \sin(\frac{\pi i}{5})\right)$, $(i = 0, \ldots, 5)$. Then there is exactly one polynomial $p(x)$ of degree five going through the six points. The first three terms of the Maclaurin series for $\sin(x)$ produces the polynomial $q(x) = x - \frac{x^3}{3!} + \frac{x^5}{5!}$. Is $q(x)$ the interpolation polynomial $p(x)$ of degree five?

(a) Yes (b) No (p. 335)

Adding Equal Weights

Three empty boxes hang from four identical springs:

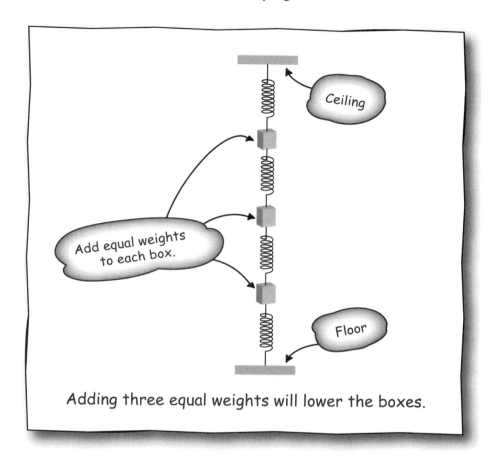

Adding three equal weights will lower the boxes.

The bottom spring just touches the floor and is glued to it. Assume the boxes and springs are weightless. After a one-pound weight is placed in each box, all the boxes have moved down. When each box contains a one-pound weight, which box has dropped the farthest?

(a) The top box moves down the most.

(b) The center box moves down the most.

(c) The bottom box moves down the most.

(d) They all move down the same amount.

(e) None of these

(p. 335)

What Do You Think? Answers

Cube-Shaped Fishing Weight

As the cube is lowered into the water, we see a point becoming a tiny equilateral triangle, and that continues to grow until the water wets the next set of cube vertices—three of them. As the cube descends a little more, small triangular corners get trimmed off, leaving a hexagon. The trimmed-off triangles grow, making the long sides of the hexagon shorter and the short sides longer. When the cube is half-submerged, the hexagon has become regular. After that, the sequence reverses. Here's the whole sequence:

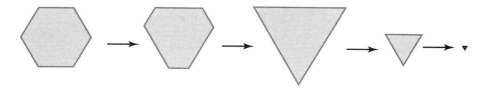

Therefore the answer is (d).

From the illustration, it appears that the largest area isn't the regular hexagon. Is that perception right? If not, then as the cube is lowered, when *do* we get the greatest area?

Subtracting by Adding?

The method always works, so the answer is (d). First, observe that the nines' complement of any n-digit integer N is $(10^n - 1) - N$. In the second step, 1 is subtracted from the leading integer of N and added to its final digit. If M denotes the minuend, then this way of finding $M - N$ is encapsulated in the identity

$$M - N = M + (10^n - 1 - N) - (10^n) + (1).$$

The method's steps correspond to carrying out what is in parentheses, going from left to right. Looked at this way, we see that the terms can be rearranged, giving other versions of the algorithm. To illustrate, the 2876 in the third example could be immediately changed to 1877. That added to the nines' complement 321 at once gives 2555, a method some people prefer.

This generalizes to any base, using B in place of 10 and taking the $(B-1)$'s-complement in place of the nines' complement. In base two, this would be the ones' complement, which is the same as simply interchanging zeros and ones. In designing a calculator, for example, its base-two adder circuit could easily be modified to perform subtraction.

Would This Tempt <u>You</u>?

If you chose (a), think some more—this is serious money!

You can do better than 50%, and here's how: take 49 yellow balls and transfer them to the other basket. When you choose a basket, you already have a 50% chance of getting the basket with a single yellow ball in it, and in that case, you're home free. Should you choose the other one, you still have 49 chances in 99 of getting a yellow ball—and that's about 49.5%. This way you have a nearly 75% chance of walking away with $200,000, making the answer (b).

Friday the 13th

Every month has a 13th, and it falls on one of the seven days of the week. So over a long period of time, the 13th falls on any particular day of the week with a probability very close to, but not exactly, $\frac{1}{7}$. That makes the answer (c). The problem "Friday the 13th One More Time" on p. 44 supplies some hints on getting the exact probabilities for each of the seven days.

How Turns the Earth?

The answer is (c). If you're at a point P on the earth, high noon occurs when P directly faces the sun. That is, the plane containing P and the north and south poles passes through the sun's center. Orient a statue directly facing the sun this way and let a month and a half go by—about 45 days. If each day the earth turned exactly 360° around its north-south axis, then after the 45 days the statue wouldn't be looking right at the sun, but 45° away from it. If P is on the Greenwich meridian, then here is the earth, magnified to show the Greenwich meridian at these two times:

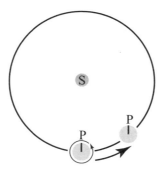

You need to add an extra degree each day to account for the earth's annual (≈ 360 days) rotation of 360° around the sun. Actually, since there are about 365 days in a year, the exact answer is slightly less than 361°.

What Will Happen?

The answer is (c). If δ is the chain's mass per unit length and g is gravitational strength, then a length s of the chain weighs $\delta g s$, which we write as ks. So the left side weighs ka and the right side kb. These two downward forces have components along the left and right triangle sides, and the picture shows these components to be $ka \cdot \sin\theta$ and $kb \cdot \sin\phi$:

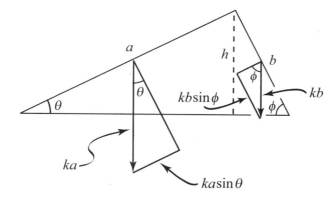

Because $\sin\theta = \frac{h}{a}$ and $\sin\phi = \frac{h}{b}$, the component forces are $ka \cdot \frac{h}{a} = kh$ and $kb \cdot \frac{h}{b} = kh$, which are the same! So, like two equally-strong people in a tug of war, gravity pulls each section equally hard, resulting in a stalemate. Therefore the chain continues to just sit there.

How Close?

The answer is (c). That's because the slope of $y = e^x$ at $x = 0$ is 1 and the slope of $y = \ln x$ at $x = 1$ is 1, so their tangent lines at these points are parallel. This picture completes the story:

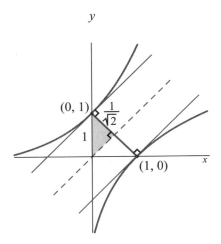

The line segment between $(1, 0)$ and $(0, 1)$ is perpendicular to both curves. Its length of $\sqrt{2}$ is the minimum distance between them.

Travel Time

Each of the wires has the same resistance and is subjected to the same voltage, so by Ohm's law $i = \frac{V}{R}$. That is, the currents through the thin wires are equal. If it takes 10 minutes for electrons to travel the length of one of them, then it takes that long for them to travel the length of any of the others. That makes (c) the answer.

Sinking 2-by-4

The answer is (c). Water is heavy: a cubic foot weighs about 62.4 lbs. This means that for every additional 33 ft of depth, the water pressure increases by one atmosphere, or 14.7 lbs per square in. By the time the 2-by-4 has descended a mile, it's being crushed by a pressure of more than a *ton* per square inch! Wood floats courtesy of millions of tiny air pockets trapped in its cellular structure. Subjected to this tremendous pressure, the air gets squeezed to a small fraction of its normal volume, increasing the wood's density. The 2-by-4 will have no trouble in continuing its journey to the ocean bottom. And there's no particular reason for the string to break, because as pressure increases, the 2-by-4's volume decreases, and therefore so does the buoying force on the 2-by-4. The tension actually decreases as it makes its downward journey.

Some woods have less air in their cells and correspondingly higher density. A number are more dense than water and sink. Ebony is an example.

By the Way . . .

When the assemblage is held in the air, the string holds up the entire 50 lbs. As the iron weight is lowered into the water, the increasing volume of displaced water increases the buoying force on the weight, effectively making it weigh less and less. The string tension needed to hold it in place decreases, and reaches a minimum when the entire weight is submerged. It stays at this minimum as the string gets lowered into the water. But what about when the 2-by-4 stud enters the water? The water displaced by the stud means increasing buoying force on it. The string doesn't know what's supplying the constant minimum force needed to hold up the submerged weight, but *we* do: it's the buoying force on the stud plus the extra force you supply to keep it from slipping from your fingers. As the stud descends, the buoying force grows on it and your contribution diminishes until you hold the assemblage fully submerged. No matter where you hold the assemblage in this process, the string continues to feel that same minimum force, making the answer (a).

Up or Down?

The answer is (c). Current flowing through a wire is like water flowing down a trough, and voltage, which drives the current, is like potential energy coming from gravity that drives the water. Just as current flows from places of high voltage to low, water flows from high elevations to low. The current expends some voltage every time it works its way through a resistor, and at the end of the trip from the positive terminal of the battery to the negative, the voltage has decreased from an initial V_0 to zero, ready for the battery to pump it up so it can remake the journey. Similarly, the flowing water spends some potential energy each time it works its way through a water-resistance, its potential energy decreasing from an initial E_0 to zero—ready for a water pump to elevate the water so it can remake the journey.

At the top of the next page is a hydrodynamic analogue of the current problem, a picture of water flowing down an arrangement of sluices or troughs. Like our electrical current, water enters at the left, flowing rightward and branching. In analogy to the two current branches each encountering a resistance, the two streams of water flow down the sloping ramps representing, say, beds of fine gravel or sand—something the water must expend energy to get through. The horizontal troughs are frictionless and sheets of water just glide along them. From extreme left to extreme right, each stream of water falls the same amount, but one wall has three ramps while the other has two (corresponding to three and two resistors in the top and bottom branches in the electrical circuit). Since all resistances are equal, the water must fall *further* when it goes down a ramp on the two-ramp side.

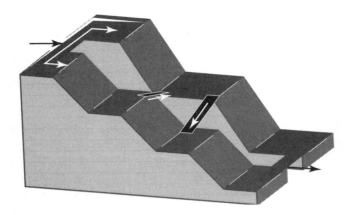

We can now read off the answer to the problem. The two cross ramps correspond to the vertical resistors in the circuit. Since water flows from high to low, the water travels in the directions of the arrows on them. So on the left cross ramp the water flows from the three-ramp side to the two-ramp side. Electrical translation: in the left vertical resistor, the current flows up (from the three-resistor side to the two-resistor side). In the right vertical resistor the current flows down.

The analogy could qualify as a "proof without words." Here is another pictorial argument, instructive because it shows a mode of thinking that mathematicians frequently use: continuously morphing from one form to another to show their essential sameness, the new form making some property more apparent.

- Here's the initial circuit:

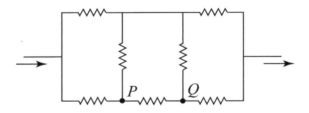

- To go to the next picture, deform the circuit. Doing this doesn't alter what's going on in each wire. Here, both middle wires, one with resistor and one without, have drifted downward:

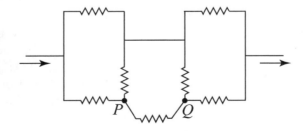

- Now the junction points P and Q are drifting further apart. The vertical resistor above each of P and Q follows along, turning its respective corner:

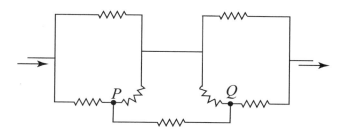

- Points P and Q have completed their drifting, and the resistors have finished turning their corners:

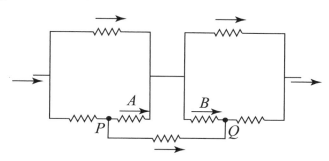

With this circuit, it is easy to see that the currents through resistors A and B are flowing in the directions of the arrows there. Now work backwards through these pictures to the original one and see what happens to the arrows! The vertical arrow on the left goes up, and the one on the right goes down. This demonstrates the power of topological thinking.

We were lucky to be able to answer this question, because suppose the resistances were not all the same? Or what if the circuit had many more branches and interconnections? It's easy to imagine how difficult it would be to find the current through each resistor. A tool like calculus is useful because it can solve a wide range of problems. Is there a tool that can do the same here? That is, is there a method requiring only that you be a good secretary—that you write down equations carefully and solve them correctly? Yes, it's linear algebra. Here's the idea: Add a battery of known voltage to complete the circuit, and divide the circuit into "irreducible" (smallest-possible) loops. Assign a signed "virtual current" to each such loop. This picture illustrates:

Each little loop is closed, and as we go once around that oriented loop the various drops in voltage all sum to zero. In the ramp picture on p. 222, that's like following a molecule of water through an entire cycle: after the pump has returned it to the top, the height lost (voltage drop) in going down the ramps is made up by the pump which supplies a boost (negative drop) to take the molecule back to the starting position. The voltage drop as current flows through a resistor is in the direction of current flow, and is given by Ohm's law: $V = Ri$. V is signed because i is.

Our problem contains four irreducible loops. Each is assigned an orientation, each has

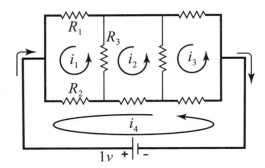

its own virtual current, and each produces one linear equation expressing that the voltage drops around the loop add to zero. To illustrate this, call the top left resistor R_1. Go around the i_1-loop counterclockwise and let R_2 and R_3 be the other resistors in that loop. The voltage drop at the top of that loop is $R_1 i_1$. Notice that there are *two* directed currents through R_2. Therefore, relative to the loop's orientation, the total current through the i_1-loop is $i_1 - i_4$. The equation for loop 1 is therefore

$$R_1 i_1 + R_2(i_1 - i_2) + R_3(i_1 - i_3) = 0.$$

Take each resistance R_j to be 1. Here are the four equations, in raw form, that come directly from the four loops:

$$i_1 + (i_2 - i_4) + (i_1 - i_2) = 0,$$

$$(i_2 - i_1) + (i_2 - i_4) + (i_2 - i_3) = 0,$$

$$i_3 + (i_3 - i_2) + (i_3 - i_4) = 0,$$

$$(i_4 - i_3) + (i_4 - i_2) + (i_4 - i_1) + 1 = 0.$$

The constant $+1$ in the last equation comes from the battery. Relative to the orientation of virtual current loop i_4, the voltage drops $+1$ as the current passes through it. Now collect terms in each of the four equations to get

$$3i_1 - 1i_2 + 0i_3 - 1i_4 = 0,$$

$$-1i_1 + 3i_2 - 1i_3 - 1i_4 = 0,$$

$$0i_1 - 1i_2 + 3i_3 - 1i_4 = 0,$$

$$-1i_1 - 1i_2 - 1i_3 + 3i_4 = -1.$$

These lead to the augmented matrix

$$\begin{pmatrix} 3 & -1 & 0 & -1 & 0 \\ -1 & 3 & -1 & -1 & 0 \\ 0 & -1 & 3 & -1 & 0 \\ -1 & -1 & -1 & 3 & -1 \end{pmatrix}$$

Solving this system (say, using Gauss-Jordan or Maple's **rref** command) gives

$$i_1 = -\tfrac{4}{8}, \qquad i_2 = -\tfrac{5}{8}, \qquad i_3 = -\tfrac{4}{8}, \qquad i_4 = -\tfrac{7}{8}.$$

The choice of V for the battery was arbitrary—any nonzero voltage will do. If we'd used $8V$, then the denominators would disappear, giving currents

$$i_1 = -4, \qquad i_2 = -5, \qquad i_3 = -4, \qquad i_4 = -7.$$

Using $8V$ and algebraically adding the various is in the circuit leads to a pretty picture, labeled with directed currents; a stands for amps.

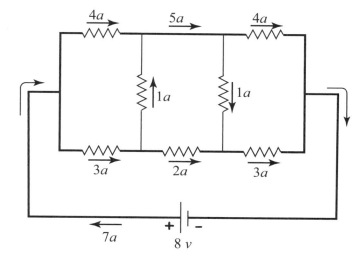

At each junction point of the circuit, the total current entering equals the total current exiting. So the analogy between electric current and water flow goes a bit further: both can be regarded as incompressible fluids. This solution reconfirms our answer since here, too, the current in the left resistor goes up, and in the right resistor it goes down.

Here are a few other points of interest:

- Our technique works for any finite circuit in the plane. The circuit could have hundreds of resistors, batteries and irreducible circuits. In adding up voltage drops, as you pass through a battery having V volts, the voltage drop is $+V$ if you go from the positive to the negative terminal. The voltage drop is $-V$ if you go from negative to positive.

- The basic (non-augmented) 4×4 matrix of i-coefficients is symmetric. Even for very complicated circuits in the plane with lots of resistors and batteries, this will be true provided all the irreducible circuits are assigned the same orientation. This serves as a useful check on the equations.

- Although we've used irreducible oriented circuits (which we called loops), this was just a way of insuring that we get enough circuits, and that they are all "independent." The idea is that independent circuits yield independent equations, and their definition is similar. In our example, if we refer to the circuit having current i_j as C_j, then we can add oriented circuits side-wise. For example in $C_1 + C_2$, the two common vertical sides cancel out because their orientations are opposite, and $C_1 + C_2$ itself is a larger circuit having four resistors and a counterclockwise orientation. We can now see that any two of C_1, C_2 and $C_1 + C_2$ are independent, and that all three are dependent. (The "zero circuit" is the empty circuit—no sides!) It turns out that there are as many maximal circuits as our minimal ones, and they would do just as well. If there are n minimal or maximal circuits, then any n linearly independent circuits will do.

Yellow Points, Blue Points

In the lattice, moving one unit from any point to another means we must go horizontally or vertically. But that's not true in $\mathbb{R}^2$—one can move one unit in *any* direction. This leads to the coup de grâce: draw in the plane any equilateral triangle having side length one. No matter how you color the plane's points, at least two of its vertices have the same color. The property fails at both these vertices, making the answer (b).

Geometry Answers

Archimedean Spirals

In polar coordinates (r, θ), for each θ there's a one-to-one correspondence sending the point (θ, θ) on the spiral $r = \theta$ to the point $(k\theta, \theta)$ on the spiral $r = k\theta$. Applied to the whole plane, this correspondence scalar multiplies each vector (r, θ) by k, giving a uniform expansion by k about the origin. Since k is positive, every object in the plane gets magnified or contracted by the transformation. So any two Archimedean spirals with $k > 0$ are similar to the spiral $r = \theta$, and therefore to each other.

Spiral Length

Take a look at this picture:

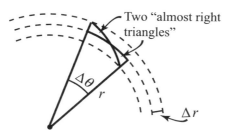

The three dashed circles have radii $r - \Delta r$, r and $r + \Delta r$. Within the angle $\Delta \theta$, which we measure in radians, the arcs cut off are $\Delta \theta$ times their respective radii (think "$s = r\theta$"). So the length of the solid circular arc is the average of the other two circular arcs. But what about that little segment of an Archimedean spiral? It crosses the solid arc in the middle, forming two small "almost right triangles." The lengths of their two *circular* legs add up to our average. In each triangle, the hypotenuse is part of the spiral, and therefore has length greater than the average, making (b) the answer. Calculus lets us put some numbers on this: the total length of the spiral $r = \theta$ from 2π to 4π is $\int_{\theta=2\pi}^{4\pi} \sqrt{1 + \theta^2} d\theta \approx 59.563$, slightly more than the distance around a circle of radius 3π, which is $3\pi \cdot 2\pi \approx 59.218$.

Church Window

The left picture shows a familiar arrangement of seven pennies:

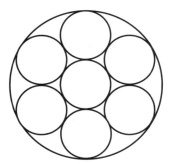 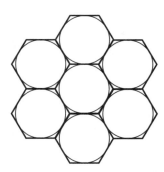

The right picture shows part of a plane tiling by regular hexagons, common in older floors. Inscribing circles in them shows that a seventh penny really does fit perfectly in the center of the ring of six pennies.

In the left picture, each penny has one third the diameter of the surrounding large circle, and therefore also a third of its circumference. That means the six small circles have, in all, $6 \times \frac{1}{3} = 2$ times the circumference of the large circle, or 20 ft. Therefore the total perimeter of the church window is $10 + 20 = 30$ ft, making (c) the answer.

Church Window Again

The answer to the first question is (b). The above seven-penny picture makes this easy. Since all three questions ask about relative sizes, we can simplify computation by re-defining the penny's radius to be 1. The penny's area is then π, and the big circle has area $\pi 3^2 = 9\pi$. Therefore the shaded area is $9\pi - 6\pi = 3\pi$, one-third of 9π. The answer to the second question is also (b). To see why, look at the regular hexagon having penny centers as vertices:

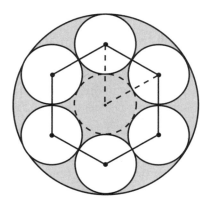

Since the sides of the hexagon connect the centers of unit circles, each side has length 2. Therefore the hexagon's area is six times the area of the triangle having two dashed sides. This triangle is equilateral with side 2, making its area $\sqrt{3}$, so the hexagon's area is $6\sqrt{3}$. This hexagon covers one-third of each of the six unshaded pennies, and that's equivalent to covering two full pennies. So the central shaded area is $(6\sqrt{3} - 2\pi) \approx 4.109$. This is $\frac{6\sqrt{3}-2\pi}{9\pi} \approx 14.53\%$ of the big circle's area of 9π.

The answer to the third question is (b). We've just seen that the hexagon covers the equivalent of two of the six unshaded pennies. Therefore, outside the hexagon the unshaded area totals up to four pennies' worth, or 4π. So from the area of the large circle we subtract the area of four pennies, as well as the hexagon's area. That gives a total of $9\pi - 4\pi - 10.392 \approx 5.316$ for the outer shaded area. The inner shaded area is only 4.109.

Mastic Spreader

This is an application of the distributive law. The perimeter of a semicircle is $\frac{\pi}{2}$ times its diameter, and in our case, the top semicircle's diameter is the sum of the many small diameters. So we can write, in obvious notation,

$$\frac{\pi}{2} \cdot D = \frac{\pi}{2} \cdot (d_1 + \cdots + d_n) = \frac{\pi}{2} \cdot d_1 + \cdots + \frac{\pi}{2} \cdot d_n.$$

The leftmost expression is the perimeter of the top semicircle and the rightmost expression is the perimeter of the bottom, serrated edge. The middle expression links these using the distributive law, making (c) the answer.

This shows that even if the little semicircles along the bottom had different sizes, the spreader's bottom perimeter would still equal the perimeter going along the top.

Does This Curve Exist?

Look at the graph of $y = e^x$. After stretching it vertically by a, the equation becomes $y = ae^x$. Since a is positive, $a = e^\alpha$ for some real number α, so $y = ae^x = e^x e^\alpha = e^{(x+\alpha)}$—the original curve shifted to the left by α. The same argument works for the graph of $y = Ae^{Bx} + C$, where A, B and C are any real numbers, so the problem's answer is (a). When $A = 0$ we get horizontal lines.

Elliptical Driveway

The answer is (b), and we can construct a counterexample this way: first, the inner oval is an ellipse with equation $\frac{x^2}{25^2} + \frac{y^2}{15^2} = 1$. If the outer oval were also an ellipse, its equation would be $\frac{x^2}{35^2} + \frac{y^2}{25^2} = 1$. Our mission is simply to find a point on the inner ellipse so that the distance along the normal there to the outer ellipse isn't 10 feet! For elegance, it would be nice to choose a point having integer coordinates. We can do this by making the inner ellipse equation $(\frac{x}{25})^2 + (\frac{y}{15})^2 = 1$ look like $3^2 + 4^2 = 5^2$ rewritten as $(\frac{3}{5})^2 + (\frac{4}{5})^2 = 1$. That is, set $\frac{x}{25} = \frac{3}{5}$, $\frac{y}{15} = \frac{4}{5}$ and solve for (x, y). This gives $(15, 12)$ for an integral point on the inner ellipse. Now find the equation of the normal line through this point and determine where the line meets the larger ellipse.

To get the equation of the normal line, implicitly differentiate $\frac{x^2}{25^2} + \frac{y^2}{15^2} = 1$, rearrange and simplify to get $\frac{dy}{dx} = -\frac{x}{y}\frac{9}{25}$. From this we find that the slope of the ellipse at P is $-\frac{9}{20}$, so the normal at P has equation $y = \frac{20}{9}(x - 15) + 12$. Substituting into the equation of the large ellipse leads to $Q \approx (19.04, 20.98)$. The distance between P and Q turns out to be approximately 9.84. Conclusion: the outer oval isn't an ellipse.

Perpendicular at Both Ends

The answer is (b). Being perpendicular at both ends has a lot to do with extremal length. To see the connection, imagine that both ellipses are made of rigid steel wire and are fixed in place. Suppose that on each ellipse there's a small bead that can slide smoothly. To each bead, attach one end of a short length of rubber band. When you let go, the rubber band shrinks to become as short as possible, with its ends at resting points P and Q. *In this position, each end of the rubber band is perpendicular to its ellipse.* Why should this be? Suppose they weren't. Then the tangent lines at P and Q would not be parallel, meaning they converge. Since each ellipse closely follows its tangent line for a little distance, the rubber band could shorten by moving appropriately. By moving the opposite way from P and Q it would get longer, meaning the length is intermediate. If the distance between P and Q is a local maximum, the beads won't move and the tangent lines are parallel. This resting position is an unstable equilibrium. When the distance is a local minimum, the equilibrium is stable. The distance going perpendicularly from points of the inner ellipse to the outer varies continuously. From the previous problem this distance becomes less than 10, while being 10 on the axes. So at some points P and Q in each quadrant, not on either axis, the distance attains a local minimum and the two beads will rest comfortably there. Thus on our two ellipses, there are at least four additional point-pairs P and Q where the connecting line is perpendicular at both ends. They are stable. (Are there any other stable or unstable point pairs besides those four?)

Star Area

The star is covered by ten copies of the triangle shown at the top of the next page. Ten triangles fill out 360° at the circle's center, so this triangle has angle 36° at the star's center. And why is the triangle's *small* angle 18°? Any two star sides forming a star vertex, when extended, intercept an arc of $\frac{360°}{5} = 72°$ on the bounding circle. The star

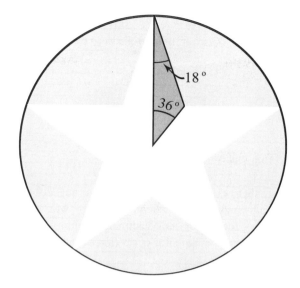

vertex angle is half that, or 36°. Half again gives the 18° marked in our shaded triangle. If our circle has radius 1, then the longest side of the triangle has unit length. Now apply the law of sines to get the length of the shortest side: $\frac{\sin 18°}{\sin 126°}$. Therefore

$$\text{Shaded triangle area} = \frac{1}{2} \cdot \frac{\sin 18°}{\sin 126°} \cdot \sin 36° \approx 0.112.$$

The area of the entire star is ten times this, which is about 35.73% of the bounding circle's area π, making (c) the answer.

Stereo Camera

Kodak places the right and left images three frames apart and has the camera advance two frames between shots. This way, just one frame at each end of the film is unused and all others are single-exposed. The answer is therefore (a). For a 24-exposure roll, this is about 92% efficient, and for a 36-exposure roll, over 94% efficient.

Diagonals of a Parallelogram

This is an exercise in using the law of cosines. Let a be the horizontal side at θ, and b the other side. Applying the law of cosines at θ gives D_2^2:

$$D_2^2 = a^2 + b^2 - 2ab \cos \theta.$$

Applying it at $\pi - \theta$ gives $D_1^2 = a^2 + b^2 - 2ab \cos(\pi - \theta)$, which can be written as

$$D_1^2 = a^2 + b^2 + 2ab\cos\theta.$$

Adding the equations gives

$$D_1^2 + D_2^2 = 2(a^2 + b^2),$$

which is a constant. This makes the answer (c).

Our argument suggests that because there are both sum and difference diagonals at θ, there are analogously plus and minus versions of the law of cosines. The plus version gives the length-squared of the vector-sum diagonal, the minus version gives the length-squared of the vector-difference diagonal. So we can write

$$\|\mathbf{a} \pm \mathbf{b}\|^2 = a^2 + b^2 \pm 2ab\cos`.$$

Half and Half?

The answer is (a). Choose a line L which is to the left of our finite set S and not parallel to any of the finitely many lines joining pairs of points in S. Now, always keeping its slope constant, let it sweep rightward over S. This line can never contain two points of S, so it crosses these points one at a time. If n is even, stop just after the line has crossed over exactly $\frac{n}{2}$ points. If n is odd, stop exactly when the line has met the $\left(\frac{n+1}{2}\right)^{\text{th}}$ point.

This idea generalizes to any finite set S in $\mathbb{R}^n$. Pass a line through a point-pair of S, and let H be the hyperplane through the origin and perpendicular to that line. There are finitely many such hyperplanes. Because these don't fill $\mathbb{R}^n$, we can select a line L that is not contained in any of them. Now sweep the hyperplane normal to L across S, as we did in the plane.

Half and Half Again

The answer is (a). All points of the finite set are at least some distance $\epsilon > 0$ from the dividing line. Form a circle by bending the line so slightly that all points of the finite set remain at least $\frac{\epsilon}{2}$ from the circle. In short, replace the dividing line by a circle so large that the points hardly know the difference!

A Platonic Solid

The number of vertices of the solid is 20. Every pentagon has five vertices, and there are 12 pentagons. That sounds like 60 vertices, but from the picture we see that each vertex is really three pentagon vertices all at the same place. So the actual number of distinct vertices is $60 \div 3 = 20$.

Why should all 20 touch the circumscribed sphere? We're given that the *inscribed* sphere touches the 12 regular pentagon faces at their centers. The inscribed sphere is tangent to each dodecahedron face, so the face is perpendicular to the sphere's radius to that point of tangency. The distances from the center of a regular pentagon to its vertices are all the same. Therefore all right triangles having one vertex at the sphere center, a second at a pentagon center and a third at a pentagon vertex are congruent. The hypotenuse of each triangle extends from the center of the sphere to a dodecahedron vertex, and all the hypotenuse lengths are the same. Therefore if one vertex touches the sphere, then all 20 do, making (d) the answer.

The *only* answer is (d), although it's easy to think there's a mistake in the picture. Actually, what you see is a consequence of projecting a three-dimensional object into a two-dimensional screen. The vertices appearing at the 3 o'clock and 9 o'clock positions, for example, are actually closer to you than the ones touching near the top or bottom of the sphere.

A Chip Off the Old Block

The volume of a pyramid is $\frac{1}{3}$ base $\times$ height. From the picture we see that our pyramid, cut from a cube of edge length a, has base $\frac{a^2}{2}$ and height a. Therefore the chip has volume $\frac{1}{3} \times \frac{a^2}{2} \times a = \frac{1}{6} \times a^3$, making (e) the answer.

Pyramid Volumes

Assume the cube is one unit on each edge. In the problem's picture, notice that one altitude of the chip pyramid is the vertical edge closest to us, which has length 1. One altitude of the square-based pyramid extends from the center of a cube face to the center of the cube, so it has length $\frac{1}{2}$. On the other hand, the chip-pyramid base is half that of the square-base one. Since the volume of a pyramid is $\frac{1}{3}$ base $\times$ height, both pyramids have the same volume—that is, one-sixth the volume of the cube. This makes the answer (c).

Ring-Shaped Fishing Weight

The answer is (a). To see why, lower the torus horizontally half-way into the water. Then begin tipping the torus by increasing θ:

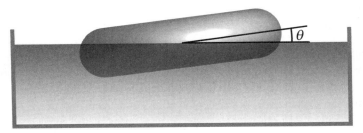

When θ is 0, the waterline consists of two concentric circles, the first picture in the following sequence. The second in this sequence corresponds approximately to the θ in the picture above.

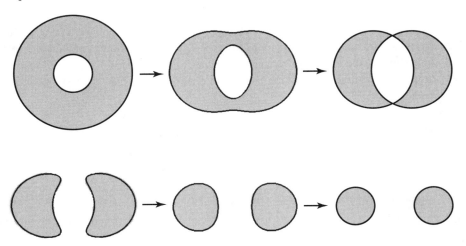

The third picture in the sequence is the magic moment. It corresponds to the amount of tipping shown here:

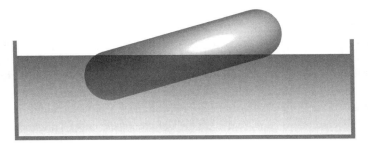

At this special angle, the intersections of the circles are the promised waterline crosspoints.

The two circles arising from slicing a torus at this angle were discovered by the French astronomer and clock-maker Antoine Joseph François Yvon Villarceau (1813 - 1889). They are sometimes used by architects in designing spiral staircases.

Largest Cylinder in a Sphere

One can approach this in a couple of ways. On the one hand, there's a common sense argument that requires no calculus but provides an intuitive justification of the answer, (c). One can also use calculus in the traditional way.

The intuitive approach shows that slightly increasing the radius increases the volume. To see this, assume the circle and square are each centered at the origin, and that the square has $(1, 1)$ as a corner. Inspect the square-in-circle picture near this corner:

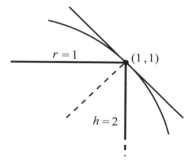

We take the square's corners to be $(\pm 1, \pm 1)$. Very near $(1, 1)$, we can barely see any difference between the circle and its tangent line $x + y = 2$ at $(1, 1)$. Therefore if the radius $r = 1$ increases to $1 + \epsilon$, then the full height decreases by nearly $2(1 - \epsilon)$. This means the cylinder's volume of base × height is $\pi(1 + \epsilon)^2(2 - 2\epsilon)$, or

$$\text{Volume} = 2\pi(1 + \epsilon - \epsilon^2 - \epsilon^3) \approx 2\pi(1 + \epsilon).$$

This shows that a small increase in the radius produces a small increase in the cylinder's volume.

Calculus delivers more in the way of specifics. The method is standard, so we leave carrying out the computation to the reader. Some of these specifics are:

- Compared to the square's height, the *largest* cylinder's height is $\sqrt{\frac{2}{3}}$ as great, a decrease of about 18% in height.

- Compared to the square's base, the largest cylinder's base diameter is $\frac{2}{\sqrt{3}}$ times as great, an increase of about 15.5%. The largest cylinder's radius is equal to its height.

- Compared with the rotated square's volume, the largest cylinder's volume is nearly 9% larger.

Largest Cone in a Sphere

The answer is (c). As in the previous problem, one can give an intuitive argument not involving calculus. For convenience, we draw the problem's triangle upside down in a unit circle:

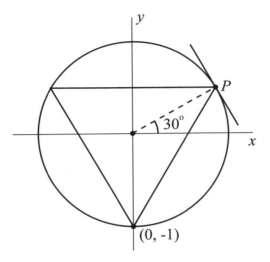

It is easily checked that the coordinates of P are $(\frac{\sqrt{3}}{2}, \frac{1}{2})$, making the slope of the dashed radius $\frac{1}{\sqrt{3}}$. The slope of the tangent line is the negative inverse of this, $-\sqrt{3}$, so the equation of the line tangent to the circle at P is $y - \frac{1}{2} = -\sqrt{3}\left(x - \frac{\sqrt{3}}{2}\right)$.

Now argue as in the solution to the last question: if you increase the radius $\frac{\sqrt{3}}{2}$ of the cone to $\frac{\sqrt{3}}{2} + \epsilon$, then its height $\frac{3}{2}$ is decreased to approximately $\frac{3}{2} - \sqrt{3}\epsilon$ and the corresponding volume is

$$\text{Volume} = \frac{1}{3}\text{Base} \times \text{Height} \approx \frac{\pi}{3}\left(\frac{\sqrt{3}}{2} + \epsilon\right)^2 \left(\frac{3}{2} - \sqrt{3}\epsilon\right).$$

When we discard higher powers of ϵ, this simplifies to

$$\text{Volume} \approx \frac{\pi}{8}\left(3 + 2\sqrt{3}\epsilon\right).$$

Therefore increasing the cone's radius by ϵ leads to a volume increase and a chubbier shape for the cone.

As in the above problem, calculus gives specifics:

- The largest cone's height is $\frac{8}{9}$ as long as the equilateral triangle's altitude, a decrease of about 11%.

- Compared to the triangle's base, the largest cone's base diameter is $\frac{4\sqrt{2}}{3}$ times longer, an increase of about 9%. The largest cone's diameter is $\sqrt{2}$ times its height.

- The largest cone's volume is slightly more than 5% greater than the rotated triangle's volume.

Spin It Again!

The answer is (d). In this picture, the circle and sphere have unit radius:

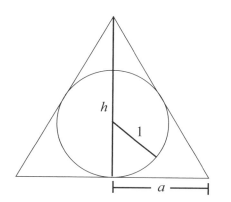
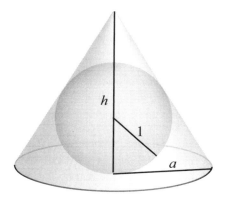

The circle has area π and the isosceles triangle has area ah, so the ratio of circle area to triangle area is $r = \frac{\pi}{ah}$. At one dimension higher, the sphere has volume $\frac{4}{3}\pi$ and the cone has volume $\frac{1}{3}\pi a^2 h$, so the ratio of sphere volume to cone volume simplifies to $R = \frac{4}{a^2 h}$. Therefore $\frac{r}{R}$ is $\frac{\pi}{ah} \times \frac{a^2 h}{4} = \frac{\pi a}{4}$, which means

$$r = \frac{\pi a}{4} R.$$

From the picture we see that a can be chosen to be any real number larger than 1. Because $\frac{\pi}{4} < 1$, when a is a little bit larger than 1, $\frac{\pi a}{4}$ is still less than 1. So from the displayed equation, we have $r < R$. On the other hand, a large a makes $r > R$. Some intermediate a makes $r = R$.

Solid Angles of a Tetrahedron

The answer is (b). Instead of pushing P down, pull it up—*way* up! In the limit, you get a half-infinite triangular prism. In approaching this limit, the solid angle at P becomes smaller than any preassigned positive number. Each of the remaining three solid angles is at a vertex of the triangular base. Each is enclosed by two vertical walls and a horizontal base, and one can see geometrically that if the angle at a vertex Q of the triangular base is ϕ radians, then the solid angle is ϕ steradians. This means that the three solid angles sum to π steradians, which is less than the 2π obtained by pushing P down. Backing off a little from these two degenerate tetrahedra produces nondegenerate tetrahedra with close to 2π steradians, and nondegenerate ones with close to π radians. This dovetails with a general fact: the sum of the solid angles of any nondegenerate tetrahedron lies strictly between π and 2π.

Disk in the Sky

We begin this answer with a question: should we bother to solve the problem exactly, or is an approximation good enough? Since successive answer choices decrease by a factor of ten, an approximation seems justified. Both ways are instructive, however, so we include both.

- **Approximate:** Although what we see of the moon is actually the cap of a sphere, it's relatively small in the sky, so we approximate it by a flat disk of area $\pi r^2 = \pi(1,080)^2$, and divide that by the area of the hemisphere, $\frac{1}{2} \cdot 4\pi R^2 = 2\pi(240,000)^2$ to get 0.000010125, making (d) the answer.

- **Exact:** Take a look at this picture of a unit hemisphere:

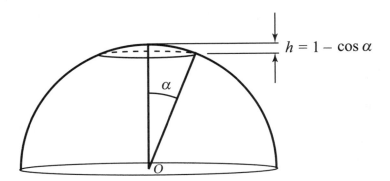

We've drawn a cap at its top depicting an exaggerated view from O of the full moon. From what was mentioned in this chapter's Introduction on p. 11, the area of the cap varies linearly with h, and in the picture $h = 1 - \cos \alpha$. Half the moon's diameter is 1,080 miles, and the moon is 240,000 miles from us. So $h = 1 - \cos\left(\arctan\left(\frac{1,080}{240,000}\right)\right) \approx 0.000010124846$—close indeed to 0.000010125.

A Right Turn

Begin by snapping together the long edges of a flat sheet to make a tube, whose radius we take to be 1. In this picture, imagine an ant starting a journey from the bottom of the tube and walking along the circular edge at an end, indicated by the arrow:

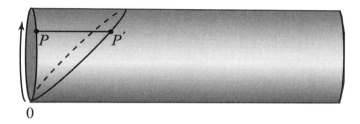

At each point of its round trip (which has length 2π) the ant measures off a horizontal distance on the cylinder and plots a point at the ellipse edge created when the tube is cut to make the 90° bend. We can look at this ellipse in two ways: as the edge when a round dowel running along the x-axis is cut at 45°, or as the edge when a round dowel running along the z-axis is cut at 45°. The first way makes it clear that the projection along the x-axis to the (y, z)-plane is a circle. The second shows that the projection along the z-axis to the (x, y)-plane is also a circle. From this, one can write parametric equations for the space curve ellipse:

$$x = \cos t, \quad y = \sin t, \quad z = \cos t.$$

After making the cut, unwrap the tube to a flat sheet. The ant's circular path of length 2π is now a straight line of length 2π, and since the ant plotted in the x-direction, those distances are $x = \cos t$. The answer is therefore (d).

The amplitude of the cosine curve is important. For example, decreasing its amplitude by the same amount on each of the two sheets results in a bend that is less sharp. Decreasing the amplitude to zero means there's no bend at all—the tubes are joined in a straight line.

A Limit – or Chaos?

Look at the top illustration on the next page. The picture on the left depicts a regular n-polygon inscribed in a unit circle. Its perimeter is an estimate from below of the circle's circumference 2π. The picture shows that this estimate of 2π is $2n \sin\left(\frac{\pi}{n}\right)$. The picture on the right similarly depicts the corresponding polygonal estimate from above of 2π, and that's $2n \tan\left(\frac{\pi}{n}\right)$.

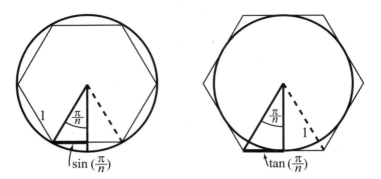

For any n, 2π is sandwiched between these lower and upper estimates, and one can sketch a picture like this:

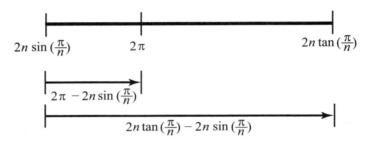

Both endpoints of the interval at the top approach 2π as n increases, but we can compensate by magnifying the interval at each stage to keep the endpoints fixed as we gaze at the spectacle. And what about 2π? Its location wiggles around as n increases. We can quantify its location in the interval using the ratio of the two lengths shown with arrows. Answering the problem's question comes down to deciding whether the limit of this ratio exists, and if so, what it is. In short, analyze

$$\lim_{n \to \infty} \left(\frac{2\pi - 2n \sin\left(\frac{\pi}{n}\right)}{2n \tan\left(\frac{\pi}{n}\right) - 2n \sin\left(\frac{\pi}{n}\right)} \right).$$

Multiply numerator and denominator by $\cos\left(\frac{\pi}{n}\right)$ to get sines and cosines but no tangent, then write out the Maclaurin expansions and multiply out the first few terms. This leads to an expression of the form

$$\lim_{n \to \infty} \left(\frac{\frac{\pi^3}{3!n^2} + O(\frac{1}{n^4})}{\frac{\pi^3}{2!n^2} + O(\frac{1}{n^4})} \right)$$

where, as $n \to \infty$, $O(\frac{1}{n^4})$ goes to zero like n^{-4}. We can therefore ignore these terms, leaving us with

$$\lim_{n \to \infty} \left(\frac{\frac{\pi^3}{3!n^2}}{\frac{\pi^3}{2!n^2}} \right) = \frac{1}{3}.$$

Therefore 2π approaches a limiting position one-third the way from the left to right endpoints, making the answer (b).

What About Areas?

The area of a unit circle is π, and the areas of regular inscribed and circumscribed polygons give our lower and upper estimates of π. We use the same strategy as in the perimeter version. From this picture

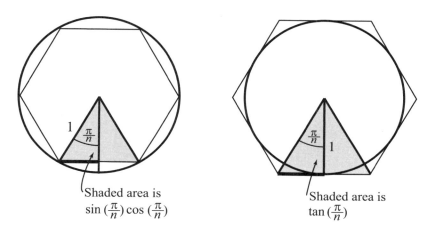

Shaded area is $\sin\left(\frac{\pi}{n}\right)\cos\left(\frac{\pi}{n}\right)$

Shaded area is $\tan\left(\frac{\pi}{n}\right)$

we see that the area of a regular inscribed n-gon is $n\sin\left(\frac{\pi}{n}\right)\cos\left(\frac{\pi}{n}\right)$, and the area of a regular circumscribed n-gon is $n\tan\left(\frac{\pi}{n}\right)$. The sandwich picture now looks like this:

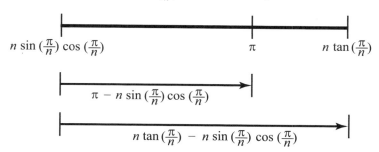

The analogous limit to analyze is

$$\lim_{n\to\infty}\left(\frac{\pi - n\sin\left(\frac{\pi}{n}\right)\cos\left(\frac{\pi}{n}\right)}{n\tan\left(\frac{\pi}{n}\right) - n\sin\left(\frac{\pi}{n}\right)\cos\left(\frac{\pi}{n}\right)}\right),$$

which we evaluate in the same way as above. We arrive at

$$\lim_{n\to\infty}\left(\frac{\frac{4\pi^3}{3n^2} + O\left(\frac{1}{n^4}\right)}{\frac{2\pi^3}{n^2} + O\left(\frac{1}{n^3}\right)}\right) = \frac{2}{3},$$

making (c) the answer.

A Glass Skeleton

Replace the artisan's glass rods by bread sticks—*before* they're baked! Gently pull the bottom square outward, letting the vertical sides and top square slowly fall into the baking pan. This is a topological transformation, and after being coated with beaten egg and baked, the sticks expand into a glossy pretzel having five holes:

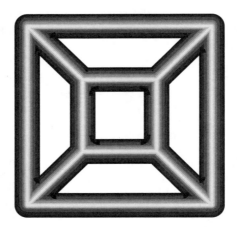

Therefore (c) is the answer.

Adding a Room

Use the same idea. This time you get a fancier pretzel with nine holes:

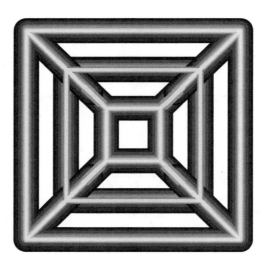

The answer is therefore (d).

Genus of a 4-Cube

If you take two lumps of clay of genus m and n and connect them with a slender clay rope, the new object has genus $m + n$. If instead you connect them with *two* ropes, you introduce one additional hole, so the new object has genus $m + n + 1$. Connecting with r ropes adds $r - 1$ holes, making the genus $m + n + r - 1$. Looking at a 4-cube as two 3-cubes connected by eight edges or ropes adds seven to the total. We saw from "A Glass Skeleton" that the genus of a 3-skeleton is five, so the genus of a 4-cube is $5 + 5 + 7 = 17$, making (e) the answer.

Based on this idea, the genus of a 5-cube, being two 4-cubes connected by 16 new edges, is $17 + 17 + 15 = 49$. Can you find a formula for the genus of an n-cube made from glass-rod edges?

Numbers Answers

A Curious Property

The answer to the first question is (a), and to the second question, (b). An example gives the idea: scramble the digits in 83601 to get, say, 68103. Now

$$83601 = 8 \cdot 10^4 + 3 \cdot 10^3 + 6 \cdot 10^2 + 0 \cdot 10^1 + 1 \cdot 10^0,$$
$$68103 = 6 \cdot 10^4 + 8 \cdot 10^3 + 1 \cdot 10^2 + 0 \cdot 10^1 + 3 \cdot 10^0.$$

Collect terms of $83601 - 68103$ having the same coefficients:

$$8 \cdot (10^4 - 10^3) + 3 \cdot (10^3 - 10^0) + 6 \cdot (10^2 - 10^4) + 0 \cdot (10^1 - 10^1) + 1 \cdot (10^0 - 10^2).$$

Each term has a factor that is either $10^m(10^n - 1)$ or $-10^m(10^n - 1)$ for some integers m, n. Therefore each term, and thus the sum, is divisible by 9. This observation is the heart of a general proof—the notation becomes a little fancier, but the idea is the same. This has an analogue in any base B, because $B^n - 1$ always has $B - 1$ as a factor: $B^n - 1 = (B - 1)(B^{n-1} + \cdots + B + 1)$. The argument makes no assumption that digits can't be 0, so the answer to the second question is also (a).

Multiplication from Another Era

In any base B, the largest two digits we ever need to multiply are just $B - 1$ with itself. The square $(B - 1)^2$ of the largest digit $B - 1$ is always less than the value of the next-larger place value, B^2. That is, $(B-1)^2 < B^2$. For example, the square of $(10-1)$ is 81, which is less than 10^2. Therefore (a) is the answer to the question.

How Often Does Lightning Strike?

The answer is (b). The earth's surface area is approximately 200 million square miles, which means roughly 65 million strikes a year. With 31,536,000 seconds in a non-leap year, the rate works out to slightly over 2 strikes each second.

A Never-Ending Decimal

Two real numbers $a_0.a_1a_2a_3\cdots$ and $b_0.b_1b_2b_3\cdots$ are equal if

$$\lim_{n\to\infty}[(a_0.a_1a_2a_3\cdots a_n) - (b_0.b_1b_2b_3\cdots b_n)] = 0.$$

So for $a_0.a_1a_2a_3\cdots = 1.000\cdots$ and $b_0.b_1b_2b_3\cdots = 0.999\cdots$, we see that

$$(a_0.a_1a_2a_3\cdots a_n - b_0.b_1b_2b_3\cdots b_n) = 0.000\cdots 1,$$

and this approaches 0 as $n \to \infty$. Therefore the answer is (a).

What is This?

This is an application of the geometric series

$$\frac{1}{1-r} = 1 + r + r^2 + \cdots,$$

which is obtained by the long division of 1 by $r - 1$. The series converges if $|r| < 1$. The problem's expansion is of $0.1 \cdot \frac{1}{1-r}$, with $r = -0.1$. That is, what appears in the question is the geometric series expansion of

$$\frac{1}{10} \cdot \frac{1}{1-(-0.1)} = \frac{1}{10 \cdot 1.1} = \frac{1}{11},$$

making (c) the answer.

A Wild Sequence

The answer is (a). The pattern is

$$\sin x = x - \frac{x^3}{3!} + \frac{x^5}{5!} - \frac{x^7}{7!} + \cdots,$$

where x is 10π. Since $\sin 10\pi = 0$, the numbers in the sequence converge to 0.

Those numbers don't *look* as if they'll converge, much less to 0. As we extend the list, things only get wilder—the size increase becomes astonishing. For example, the 10th term has size greater than 100 quadrillion, and the 15th term has grown to over a quintillion.

As we watch these terms grow, we are witness to a fight-to-the-death war between two determined forces—a battle fought between the numerators (powers of 10π) and denominators (factorials). The numerators start off with a decisive advantage, and although the underdog denominators constantly grow in relative power, the numerators hold that advantage for some 30 terms. During this, the terms $|S_n|$ skyrocket. By the 30th term, this battle reaches a fever pitch, the sum being over 50 *trillion trillion*—that's 50 octillion—and up to this point, the numerators win every battle. But then the balance of power tips. The denominators begin to show their strength in the gigantic arm-wrestle, and they soon grow dramatically faster than the numerators. By the 50th term, the sum is "only" a few billion trillion, by the 70th term, it's down to 15 billion, by the 80th term, about 368, and—here goes—by the 90th term, it is 0.00000072. Here is a plot of the struggle, showing the magnitudes of the S_n as n increases:

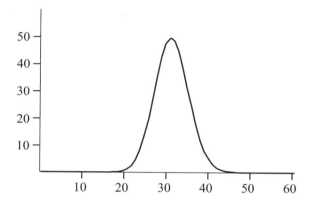

To put things into perspective, the familiar power series expansions of the elementary functions surely rank as one of the great natural mathematical wonders, but battles of unimaginable intensity can take place to bring about convergence. In the end, the series not only converges (remarkable in itself), but converges to just the value it's supposed to! If the battle above is impressive, consider the internal struggles as $\sin(10^{100}\pi)$ converges to 0, or for $\cos(10^{100}\pi)$ converging to 1.

Body-Builder

The answer is (a). That's because the range between 320 lb and 10 lb is the same as between 0 lb and 310 lb, and the problem asks for this in 10-lb increments. The question then becomes this: using 10 lb as a unit, can the body-builder range between the integers 0 and 31 without any gaps? Now think base two: in this base, counting from 00000 to 11111 uses the place values 2^0 through 2^4, so counting in base two from 00000 to 11111 translates to counting from 0 to 31 in base ten. The barbell itself weighs 10 pounds, so all 10-pound increments from 10 to 320 lbs are in fact covered. As an example, suppose he wants to lift 200 lb. That's 190 lb worth of weights, or 19 10-lb units. In base two, 19 is 10011, shorthand for $1 \cdot 2^4 + 0 \cdot 2^3 + 0 \cdot 2^2 + 1 \cdot 2^2 + 1 \cdot 2^0 = 16 + 2 + 1$. In pounds, that means $190 = 160 + 20 + 10$, or, on each side of the barbell, $80 + 10 + 5 = 95$.

Decimal versus Binary

All the examples in the problem avoid using the digits 8 and 9, which convert digitwise to blocks of *four*: 1000 and 1001. These larger blocks make for a larger conversion. For example, 88 converts digitwise to $10001000 = 136 > 88$, or 98 converts to $10011000 = 152 > 98$. If the starting integer has three or more decimal digits, the base two block sizes can add up to the point where not all digits need be 8 or 9. Example: 884 converts digitwise to $10001000100 = 1092 > 884$. The conjecture is clearly false, making (b) the answer.

Hexadecimal versus Binary

Any integer N written in hexadecimal as $N = a_n a_{n-1} \cdots a_0$ has the decimal value $N = a_n \cdot 16^n + \cdots + a_0 \cdot 16^0$, which can be written as $N = a_n \cdot 2^{4n} + \cdots + a_0 \cdot 2^{4 \cdot 0}$. The powers of 2 jump by 4 (that is, $0, 4, 8, \cdots$), and these jumps provide exactly the room to replace any a_i (a hexadecimal digit from 0 to F) by its binary equivalent (ranging from 0000 to 1111). So the answer to the problem is (a). It's easy to see that this works just as well in reverse: we can partition any binary integer into blocks of four and convert each block into a hexadecimal digit from 0 to F. In such a partition, the integer's unit position is rightmost in its block.

This natural "commensurability" between binary and hexadecimal is useful, because although long strings of binary digits are fine for machines, but they can require unnecessary effort for us to grasp them. Because translating back and forth between binary and hexadecimal is so effortless, the hexadecimal representation is a great tool. In hexadecimal it's easy to conceptualize the number, and if needed, we can at once write down its binary equivalent. As a real world example, the additive RGB (Red, Green, Blue) color scheme used on a computer monitor specifies how intensely red, green and blue phosphors on the screen are energized. This intensity is divided into 256 levels, from 00 (no electron beam hitting the phosphor) to $FF = 255$ (maximum beam strength). So $C800C8$ represents a fairly intense purple since red and blue phosphors are quite strongly (and equally) energized, while green is left out entirely. This is much easier to grasp than its binary form of 110010000000000011001000.

> There are 10 kinds of people in the world—those who understand base two, and those who don't.

Gelosia Again

To multiply two integers in base B, we must be able to multiply base digits up through $(B-1)^2$. An integer N greater than 1 increases when you square it, so for any base $B \geq 3$, the square of the largest base digit $B - 1 \geq 2$ is greater than that largest digit. This means we must open up an additional place value to represent it. So for any base larger than two, we must divide the gelosia squares to accommodate extra digits, making (a) the answer to the question.

Do Their Behaviors Correspond?

Assume that $\frac{N}{M}$ is written in lowest terms, and that $N < M$. For any base B, let $p_1^{n_1} \cdots p_r^{n_r}$ be the prime-power factorization of B. It is a fact that if $\frac{N}{M}$ has a finite expansion in base B, then the only prime divisors of M are $p_1, \cdots, p_r$, and conversely. For example, in base ten, $B = 2 \cdot 5$, so the fractions $\frac{N}{M}$ having finite decimal expansions are precisely those with only 2's or 5's in the denominator. Thus we'd predict that $\frac{147}{1250} = \frac{145}{2^1 \cdot 5^4}$ has a finite expansion, which it does: .1176. We'd equally well predict that $\frac{3}{13}$ *isn't* finite. It is, in fact, unending: $.\overline{23076}$. And base two? Our basic fact tells us that $\frac{N}{M}$ has a finite binary expansion if and only if $M = 2^m$ for some m. So $\frac{13}{16} = .1101$, while $\frac{1}{5}$ is unending: $.\overline{0011}$.

We can now answer our question. Base $B = 2$ has only 2 as a factor, while $B = 10$ has both 2 and 5. So any fraction finite in base two (for example, $\frac{1}{5} = .2$) is finite in base ten, but a fraction finite in base ten ($\frac{1}{5} = \frac{1}{101} = .\overline{0011}$) may not be finite in base two. That makes (b) the answer.

Ingenious Student

The answer is (c). Suppose the left arm length is $1 + \epsilon$ and the right arm length is $1 - \epsilon$. To be specific, let's agree that the unknown has unit weight and that $0 < \epsilon < \frac{1}{2}$. With the unit weight in the left pan, what weight W must the student place in the right pan to balance it? W needs to satisfy the fulcrum law $1 \cdot (1 + \epsilon) = W \cdot (1 - \epsilon)$, so

$$W = \frac{1 + \epsilon}{1 - \epsilon}.$$

When he next puts the unit weight into the right pan, he balances it with a weight W^* in the left pan. Since now $W^* \cdot (1 + \epsilon) = 1 \cdot (1 - \epsilon)$, we have

$$W^* = \frac{1 - \epsilon}{1 + \epsilon}.$$

The average weight is

$$\frac{W + W^*}{2} = \frac{1 + \epsilon^2}{1 - \epsilon^2},$$

which for any of our ϵ's is *strictly greater than 1*. So the average of even two perfectly accurate measurements using this scale always yields an answer that's too big, and even a thousand careful weighings can be expected to be very, very close to precisely this too-large estimate $\frac{1+\epsilon^2}{1-\epsilon^2}$. The bias is inherent in the method!

Our ingenious student's instincts did point him in the right direction, however. By the way we've set up the problem, $W > W^*$. Their average is sandwiched in between: $W > \frac{W+W^*}{2} > W^*$. We can write this as

$$\frac{1+\epsilon}{1-\epsilon} > \frac{1+\epsilon^2}{1-\epsilon^2} > \frac{1-\epsilon}{1+\epsilon}.$$

Using a common denominator, this becomes

$$\frac{(1+\epsilon)^2}{1-\epsilon^2} > \frac{1+\epsilon^2}{1-\epsilon^2} > \frac{(1-\epsilon)^2}{1-\epsilon^2}.$$

This in turn means

$$1 + 2\epsilon + \epsilon^2 > 1 + \epsilon^2 > 1 - 2\epsilon + \epsilon^2.$$

If ϵ is small, then ϵ^2 is very tiny, and $1 + \epsilon^2$ is closer to the actual weight of 1 than adding or subtracting the relatively large 2ϵ in $1 + 2\epsilon + \epsilon^2$ or $1 + 2\epsilon - \epsilon^2$.

An example: with $\epsilon = 0.1$, $W = \frac{1.1}{.9} \approx 1.22$ and $W^* = \frac{.9}{1.1} \approx 0.819$, while the average is $\frac{1.01}{.99} \approx 1.02$.

Gain or Lose?

If you think he gains, think again! The scales work against him. Suppose that he weighs out a pound of peanuts a thousand times. Using a scale as in the last problem with left arm length $1 + \epsilon$ and right arm length $1 - \epsilon$, suppose he first puts the standard 1-pound weight in the left pan and adds peanuts to balance. The fulcrum law tells us that

$$W = \frac{1+\epsilon}{1-\epsilon} > 1$$

is the weight of peanuts actually placed in the empty pan, so he loses in this transaction. When he puts the 1-pound weight in the right pan, he weighs out

$$W^* = \frac{1-\epsilon}{1+\epsilon} < 1,$$

so he gains this time. But the average is

$$\frac{W + W^*}{2} = \frac{1+\epsilon^2}{1-\epsilon^2} > 1,$$

so he parts with more peanuts than he should. He loses, making (b) the answer.

What Sort of Graph?

The initial decrease multiplies the pre-sale price P by $1-x$. The subsequent increase multiplies that by $1+x$, for a final price of $(1-x)(1+x)P = (1-x^2)P$. In the example, $x = 10\% = 0.1$, so the final price is $(1-x^2)P = \$(1-.1^2) \cdot 100 = \$99$. Even if the item was first marked up and then down by the same percent, the final price is still $(1-x^2)P$. For any positive P, the graph of $y = (1-x^2)P$ is a concave-down parabola, making (b) the answer.

Round Trip

This is a denominator analogue of the previous problem. Suppose that the distance each way is D, and that units have been chosen so that the boat's speed relative to the water is 1. If the water's speed is x, then the time traveling against the river's flow is $\frac{D}{1-x}$, while the time traveling with the river's flow is $\frac{D}{1+x}$. The total round trip time is therefore $D\left(\frac{1}{1-x} + \frac{1}{1+x}\right)$. That's $\frac{2D}{1-x^2}$, and we see $1-x^2$ appearing in the denominator. Since $\frac{2D}{1-x^2}$ has a minimum at $x=0$, (a) is the answer. If the river flows faster than the boat's speed, then the boat can't even complete a round trip.

A Reality Test

The answer is (c), because $\sin(.01°)$ is about 0.0001745, while $.01$, the purported approximation, is over 57 times larger! The approximation $\sin(\theta) \approx \theta$ assumes that θ is measured in radians, and the sine of $.01$ radian is $.00999983\cdots$, off by less than two thousandths of a percent.

Why must we use radians in $\sin\theta \approx \theta$, and why is the approximation so good?

One radian is that central angle of a circle subtending an arc equal in length to its radius. Equivalently, it's that unit of angle measurement which makes $s = r\theta$ true for a circle of radius r. If $r = 1$, this becomes $s = \theta$, illustrated here:

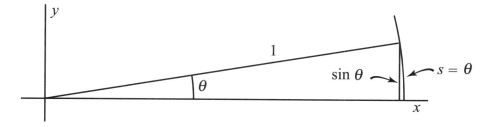

Because θ is small, the length $\sin\theta$ of the vertical line segment (a half chord) is a tiny bit less than $s = \theta$. For θ small enough, the arc in the above picture could be magnified to room size, yet the human eye could not detect any difference between the arc and the half chord.

The above picture allows you to see geometrically that as θ grows, the approximation $\sin \theta \approx \theta$ gets worse.

Friday the 13th Again

The answer is (a): whenever Friday the 13th occurs in a 28-day February, it also occurs in March, since 28 is evenly divisible by 7. This can happen in no other month because none of 29, 30 and 31 is evenly divisible by 7.

Friday the 13th One More Time

Since there's a 13th in each of the $12 \cdot 400 = 4800$ months in this cycle, one can now simply count! Mathematica or Maple can make this into a doable programming exercise. The winner is Friday, making the answer (c). Friday the 13th always occurs exactly 688 times in any 146,097-day period. Sunday the 13th and Wednesday the 13th tie in second place, each occurring 687 times. Then come Monday and Tuesday at 685, and finally Thursday and Saturday at 684.

What Comes Next?

There's an interesting story to how the author stumbled upon this quirky problem. It took place at the home of mathematician Hassler Whitney in Princeton. After an evening of playing string quartets, we were all seated at a table enjoying cheese, crackers and juice. At one point the discussion turned to puzzles and problems. The second violinist, Kerson Huang, had just learned he'd been promoted to Professor of Physics at MIT and was in high spirits. On a napkin, he wrote this problem as a puzzle. I stared at the sequence, and had absolutely no idea what came next. It was then handed to Whitney, sitting opposite me. He stared too, and saw nothing obvious. But he kept staring and thinking, and I was also looking at the numbers, but from my vantage point they were upside down and, more importantly, reversed. Because of that, something eventually clicked. The answer is (e).

Astronomy Answers

Which Way?

The answer is (b). Think of the figure as showing someone gazing at a spinnable tabletop globe. If the sun is rising in New York, then, with the sun fixed, you need to turn the globe counterclockwise to have the sun rising on the west coast three hours later.

Our Spinning Moon

The answer is (b). This picture depicts the situation viewed from over the earth's north pole:

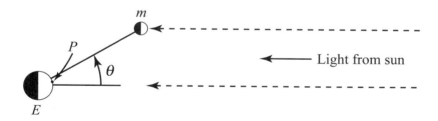

The angle θ has increased counterclockwise, and an observer at P on earth sees the illuminated part of the moon as a closing parenthesis. If θ had increased clockwise, the observer would see instead an opening parenthesis.

How High is High Noon?

The answer is (d). The latitude of New York City is about $N\,40.7°$, which is $49.3°$ down from the north pole. The earth has a $23.5°$ tilt, and in the summer the north pole is inclined toward the sun by that amount. This produces the shortest possible shadows in New York. At the summer solstice, the sun rises to a maximum inclination of $49.3° + 23.5° = 72.8°$ above the horizon. If the sun were ever actually directly overhead, the inclination would be $90°$, and we see that can't happen. Here's a picture showing these angles:

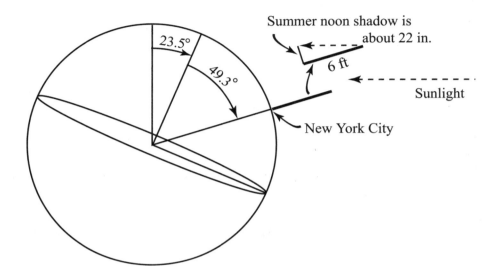

From the diagram, we can calculate the length of a six-footer's noontime shadow. It is $6 \cot(72.8°)$ ft, which is about 22.3 inches. In the winter, the tilt in the northern hemisphere is directly away from the sun, and the $72.8°$ gets replaced by $49.3° - 23.5° = 25.8°$. This means, in turn, that on December 21 in New York, the 6-footer standing upright has a noontime shadow over 12 feet long.

Flagpole Sundial

The answer is (b). The picture at the top of the next page shows that if we neglect the earth's $23.5°$ tilt, then the shadow of a flagpole in the northern hemisphere has a north component, and at high noon the shadow extends due north. The shadow shown in the south has a south component, and at high noon it extends due south. Since the sun-chariot travels westward no matter where we are, the shadow always moves eastward, so in the north, the clock numerals would be placed in the usual clockwise order. In the south, one reads the numerals facing south, so the shadow moving east means the numerals have to go counterclockwise.

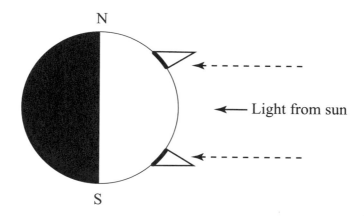

The Greatest Number of Stars

The plane containing the earth's equator (called the equatorial plane) divides the heavens into northern and southern halves. At the north pole, you can see stars lying in the northern half, and at the south pole you can see those that are in the southern half. This never changes throughout the year, except that half the year it's night, and the other half it's day. It's a different story if you are standing at one spot on the equator. You can always see half the sky on a clear night, but that half doesn't stay the same. Fast-forward six months after a great night's look at the heavens, and you see a nighttime view of the opposite half of the sky because the earth has rotated halfway around its orbit:

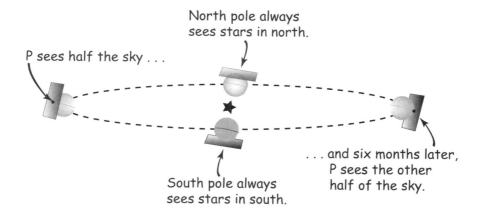

So throughout a full year's observation, you can see everything from a point on the equator. You can't do that at either pole, so (c) the answer.

Is This Possible?

The answer is (a). The reasoning: our moon spins about its own north-south axis, taking slightly over 27.3 days to make one full rotation. If you start the experiment with the sundial bathed in the sun's light and casting no shadow, it will be directly facing the sun again one rotation later. The sundial's shadow length (and which way it points) could be used during the moon's approximately two-week "day" to record what part of the month you're in.

Sunspots

If you travel perpendicularly outward from the equator a few thousand miles, you'd see points on the earth move from left to right—both you and the sun see the western U.S. later than the eastern. From the sketch of how the solar system formed (in this chapter's Introduction), you can see that the sun and earth spin the same way about their polar axes. This means the sunspots travel from left to right, making (b) the answer.

Not a Plumb Plumb

The answer is (d). The earth's rotation about its polar axis creates a centrifugal force on the bob, aimed perpendicularly away from the north-south axis. This force destroys the much better perpendicularity that gravity alone would produce. In New York, this centrifugal force kicks the string line southward about a tenth of a degree. This outward force is greatly exaggerated to show what happens:

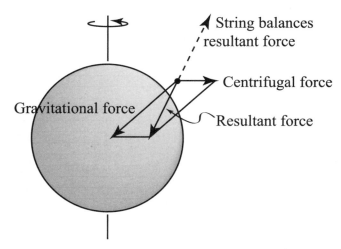

Not including the antenna on top, the Empire State Building is around 1250 feet tall, and that tenth-degree angular difference means that the bob is pulled southward by over 26 inches!

Foucault's Pendulum

How fast a Foucault pendulum knocks over the pins, and whether that happens in a clockwise or counterclockwise direction, depends on what circle of latitude the pendulum is on. Imagine a Foucault pendulum at the north pole. As you observe it, you are rotating counterclockwise around a minuscule circle of latitude. This picture depicts the situation, looking down upon the north pole:

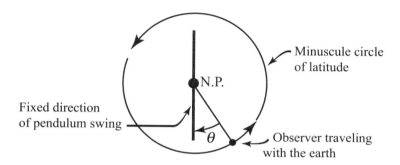

As an observer moves counterclockwise, the angle θ increases since the direction of swinging stays fixed. The observer believes he or she is not moving, and interprets the increasing θ as the pendulum swing direction increasing clockwise. This is another application of the symmetry principle given in this chapter's Introduction: the pendulum sees the observer rotating counterclockwise, so the observer sees the pendulum's plane of swinging rotating clockwise.

Let's move down to some latitude in the northern temperate zone, like this:

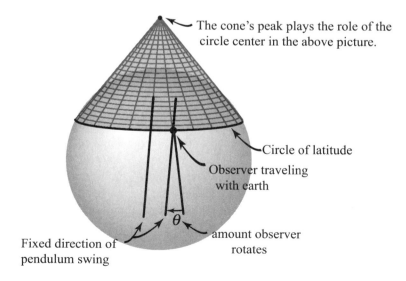

The observer's relation to the circle is now different. Instead of gazing at its center, the observer is looking at the peak of a cone tangent to the sphere along that circle of latitude. As before, the line of sight is perpendicular to the circle, but the peak has now moved up, and that makes the angle θ smaller. (Think of an isosceles triangle whose two equal sides have become longer.) So as the earth rotates, the observer still sees the swing direction change, but not as fast. At latitudes really close to the north pole, it takes just a bit more than 24 hours to knock down the circle of pins. As you move south, it takes longer.

What about when you get all the way to the equator? Here's the corresponding picture:

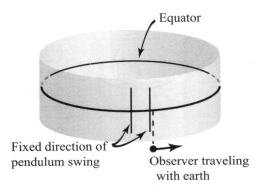

The observer is now looking in a direction perpendicular to the plane of the circle. It's as if the equal sides of the isosceles triangle have become infinitely long and the amount θ of the observer's turning has diminished to zero—neither the pendulum nor the observer are changing direction. The cone's peak, having moved upward to infinity, has transformed the cone into a cylinder tangent to the earth at the equator. On the equator, the Foucault pendulum will knock down one pin if it happens to be in the way, but no others. It certainly will never knock down a circle of pins.

What if we start at the south pole and move north? For an observer looking down on the north pole, the earth rotates counterclockwise. At the south pole, you're upside down relative to the north pole, so when you look down upon the earth from the south pole, you see the earth rotating clockwise. The pins therefore go down counterclockwise in 24 hours, and as you begin to go north, the length of time to knock down an entire circle increases. (The tangent cone is now below the equator.) On the equator, clockwise and counterclockwise are the same since the rotation there is zero. By drawing a picture with a triangle, it's easy to see that the time in hours needed to knock down the circle of pins is $\left|\frac{24}{\sin\alpha}\right|$, where α is the latitude of the Foucault pendulum.

The answers are therefore (a), (b) and (d).

Attaching the Wire

The answer is (b), unless the pendulum is on the equator, in which case it wouldn't work at all. The direction of the pendulum's back and forth motion never varies, and if the pendulum is located at any point not on the equator, the earth's constant turning means that as the days go by the chain would become impossibly twisted, eventually affecting the motion of the ball.

A <u>Lunar</u> Foucault Pendulum?

The moon coalesced from a ring of matter thrown into orbit by a huge impact early in the earth's history, and it rotates around the earth in the same direction that the earth spins. However, the moon takes nearly a month to complete its orbit, and since it always faces us, it rotates about its own north-south axis at that rate. A Foucault pendulum at the earth's north pole knocks over the pins clockwise, so it does the same at the moon's north pole. On the moon, the pendulum shows that the moon is spinning about its own axis. Since that spinning rate is about one rotation per month, it will take a month to knock over the moon pins. So the answers are (a) and (c).

Retrograde Motion?

The answer is (a), an application of the symmetry principle on p. 49. Whenever an earthling sees Mars go retrograde, an observer on Mars sees earth do the same, but with all directions reversed.

Moonday

The answer is (b). As the moon orbits the earth, the moon spins about its axis. As for moon-days, the earth-orbiting doesn't matter, but the moon's spin about its own axis does. Think of the earth or our moon as a turkey slowly rotating on a spit, with the sun as the fire. A day on any of them is the period of rotation *about its own axis*. A moonday is approximately a month because, over many millions of years, our moon has become gravitationally locked to us, causing it always to face us. (See **gravitational locking** in the Glossary.) This ties the length of our moon's day to its rotation rate around us.

Full Moon

The answer is (a). Look again at this view from over the earth's north pole:

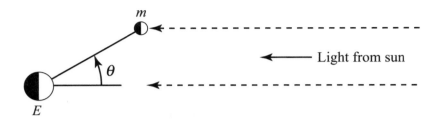

From one new moon to the next, which is approximately 28 days, θ increases $360°$. That's less than $13°$ a day, or $0.54°$ an hour. So whatever θ is for Londoners, it wasn't much different in China a few hours earlier.

Full Moon Again

The answer is (b). To us, the moon seems to move, but that's mostly because the earth spins so fast. The earth rotates about our north-south axis, and the moon revolves around the same axis. However, the rotations have very different rates: our own rotation is about 28 times faster. The result? From one day to the next, the moon's position doesn't change a lot. Even in the daytime, unaided eyes have seen the moon in all its phases (except possibly the new moon, when there is almost no illumination).

Moondial

The answer is (b). To see why, suppose for the sake of argument that the moon is highest in the sky at midnight. The moon orbits the earth in the same direction that the earth spins about its north-south axis, but at about $\frac{1}{28}$ as fast. By the next midnight, the earth has completed one rotation, and the moon has likewise progressed in that direction about $\frac{360°}{28} \approx 13°$. At that midnight, the earth must spin more than another $13°$ to catch up with the moving moon so that the moon appears at its highest point. That takes over 51 minutes. So the moondial loses that much in just that one day—a pretty lousy timepiece!

Moonrise

The answer is (c). The earth's fast, 24-hour spin means that the sun and moon travel across the sky in the same direction. So the moon rises in the east and the sun sets in the west. Therefore if the moon is rising just as the sun is setting, then the sun and moon are on opposite sides of the earth. Here's a figure depicting this. The camera looks down on our north pole, revealing that Q sees a full moon.

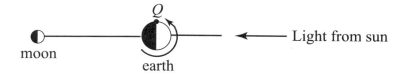

The earth spins counterclockwise, so an earthling at point Q is enjoying the end of a day and the beginning of a moon-filled night.

Moonstruck?

The answer is (b). On the moon, you can see the stars because the sky is always black there, and that's because the moon has no atmosphere to scatter sunlight. Although on a clear dry day our atmosphere is quite transparent, it's nonetheless good at scattering blue wavelengths of sunlight, which effectively turns the sky into trillions of little blue light bulbs, making our planet the "blue dot" in space. From our position on earth, all this scattered light decreases the contrast between stars and sky to the point that it drowns out the stars' weak light.

The moon has had plenty of time—over four billion years—to develop at least *some* atmosphere. For example, radon gas seeping up through its interior is constantly contributing. But Nature conscripts would-be atmospheric molecules into a war—a war endlessly fought throughout the universe. It pits the outward, expansive forces of heat against the contracting, consolidating force of gravity. Gas molecules on the moon get heated up by the sun, speeding them up, "pumping them up with break-away juice." The opponent, moon's gravity, is not strong enough to win this one, and most molecules eventually escape into space. Earth's gravity is six times stronger, giving it a decisive advantage. A minute proportion of the lightest of our atmospheric molecules do manage to escape each year, but the much stronger gravity keeps most of them under leash.

To put some numbers on this, the escape speed on the moon is about 1.47 mi/sec while on earth, it's 7 mi/sec, nearly five times faster. Few molecules in the earth's atmosphere are both high enough and fast enough to make the jail-break.

Earth's Phases

The symmetry principle on p. 49 tells us that when you're on the moon, the earth seems to be going around the moon in the opposite direction and at the same orbiting rate.

Both the earth and moon are bathed with light from one source—the sun. Conclusion: the cycle takes about a month, making (d) the answer.

Opening or Closing Parenthesis?

The symmetry principle on p. 49 tells us that from the moon's perspective the earth is orbiting the moon in the opposite direction, so the phases appear in reverse order. On earth, a new moon looks like a closing parenthesis while on the moon, a new earth looks like an opening parenthesis, making (a) the answer.

Light Lumps, Massive Spheres

The answer is (b), and that's mostly because gravity is the great leveler. On earth, a hypothetical mountain having a base of one square mile and a height of 100 miles would exert such phenomenal pressure at the base, it would simply sink downward, if it didn't crumble first. Our highest mountain, Mt. Everest, is a bit under 5.5 miles high. That's less than 1.4% of the earth's 4000-mile radius.

On bodies with really strong gravity, deviations from perfect sphericity can be breathtakingly small. For example, a neutron star is a once-was sun whose nuclear furnace has burned out. Gravity on a neutron star, unopposed by the expansive pressure of a powerful nuclear furnace, squeezes mercilessly, and the matter of the dead sun compresses to a sphere perhaps only 15 miles in diameter. Gravity, coming from so much mass and such a small radius-squared, is powerful to the point of squeezing together the electrons and protons in atoms, forcing them to combine into neutrons—hence the name. At the surface of a typical neutron star, a chunk of material the size of a sugar cube weighs some 100 million tons, more than the combined weight of two thousand Titanic ships.

And neutron-star sphericity? We're talking about a 15-mile diameter sphere looking something like a giant steel ball bearing *whose overall deviation from a perfect sphere is less than one-tenth of a millimeter.* The highest mountain on a neutron star could be no more than a thousandth of a millimeter high.

Dynamite Asteroids

The answer is (d). An asteroid entering our atmosphere at speed v has kinetic energy $\frac{1}{2}mv^2$, so an asteroid with 10 times the speed has $10^2 = 100$ times the kinetic energy. After an asteroid slams into the earth, nearly all its kinetic energy has converted to heat, so the amount of heat released by the faster-moving asteroid is close to 100 times that of the slower one.

Impact Angle

The answer is (b), since no matter what the impact angle is, virtually all the asteroid's energy gets changed into heat.

A Manned "Deep Impact" Mission?

The answer is (a). To see why, translate the given information into a figure:

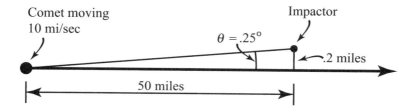

The right triangle has angle θ of only $.25°$, so if the side adjacent to it has length 50 miles, then the opposite side is about 0.218 miles long. Since the comet is approaching at 10 mi/sec, the pilot has only 5 seconds to correct. A maximum acceleration of 32 ft/sec^2 means that the greatest distance he can move the Impactor in t seconds is $\frac{1}{2} \cdot 32 \cdot t^2$ ft. If $t = 5$, that's $16 \cdot 5^2 = 400$ ft, less than .076 mi. So even going the wrong way, the Impactor will still be less than 0.3 miles from the comet's center. The comet has radius 1 mi, so the Impactor will impact.

It's the size of the comet that saves the mission. If the comet had a diameter of only .2 miles, the answer would be (d) and the situation would have been hopeless. No matter what the pilot did, the closest he could get to the comet's center is .142 mi, so the Impactor would have missed it by nearly half a mile.

Vaporizing Asteroid

The answer is (c). In the vaporizing stage, energy is released more gradually, while in an impact, energy is released suddenly, and that can be a lot! Auto refurbishers, when sand-blasting off old paint, must be careful not to aim the nozzle too long at one point: as grains of sand hit the auto body, they lose speed, and their lost kinetic energy converts to heat. The heat buildup can be so great that the auto's metal warps out of shape. At a more extreme level, one of the tell-tale signs of a meteoric impact is little spherules of glass formed from sand grains that melted from the intense heat caused by the impact. At a yet more extreme level, if an asteroid one tenth the mass of the moon were to smash into the earth, the heat generated would melt the entire surface of the earth.

Vernal Equinox

The answer is (b). The two cities are in opposite hemispheres, so their winters are a half-year apart, meaning that their vernal equinoxes are too.

Martian Equinox?

The answer is (a)—Mars' polar axis tilts 24°, about the same as Earth's 23.5°, so circles of Martian latitude receive the sun's light in nearly the same way they do here. Even the length of a Martian day is close to ours, being just 37 minutes longer. A Martian year is about 687 earth days, so Martian seasons vary more slowly, but the cycle is the same: the planet experiences four seasons, and they're reversed in the northern and southern hemispheres, just as on earth. The same continuity argument shows there are two equinoxes in each Martian year, vernal and autumnal. The rover experiences seasons just as we do.

An Ancient "Aha!"

Eratosthenes' solution is both simple and brilliant, using only $s = r\theta$ relating a circle's arc length to its radius and subtended angle. For our question, the formula gives $500 = 3820r$, or $r \approx .131$ radians. That's close to 7.5°, making (e) the answer.

Smiley Face?

The answer is (a). The photo on the next page was taken near the north pole during winter. Relative to the direction of the low-lying sun, the earth's 23.5° tilt gives the plane through our equator an upward tilt. Since the moon travels close to that plane, it appears above the sun, and we see the backlit moon as a smiley-face crescent.

Archimedes' Principle Answers

Visual Demonstration?

The answer is (b). Instead of suspending the connecting bar, balance it on a pencil point. The bar won't feel any difference, but doing this emphasizes that the equilibrium is unstable, since if you unbalance the bar ever so slightly, it will tumble off the pencil point. But unbalancing is exactly what happens when you lower the arrangement into water: the lump of gold is more dense than the wreath, so the lump experiences less buoyancy, meaning a greater net downward force on the right side. This initiates a clockwise turning of the connecting bar. In the bottom picture, resolve the end force vectors into components along and perpendicular to the bar. The greater force perpendicular to the bar is still on the right, so the turning continues, and only when the chunk of gold is hanging straight down does the system attain stable equilibrium.

Another Way?

Archimedes' problem was to determine the wreath's density so it could be compared with the density of gold. Density being mass per volume, it would have been enough to weigh the wreath and to determine its volume. To get its volume, place it in a vessel and fill it to overflowing with water so that the wreath is completely submerged with no air bubbles clinging to it. Then remove the wreath, letting any excess water drip back in, and find how many units of volume it takes to return the bucket to overflowing. Nowhere is buoyancy used, so the answer is (b). Although this method works in theory, from a practical standpoint it is prone to error.

Rain Gauge

The answer is (a). To see why, suppose that in place of the shot you use a big block of metal having dimensions just an eighth of an inch less than the baking pan's inside dimensions. The rainwater would run off the block and into the narrow moat. It wouldn't take much rain to get a reading of one or two inches!

Another way of thinking about this is to dispense with the shot and just glue the pan down so it won't blow away. You get a fair reading. Placing the shot in the pan after the rain would simply artificially raise the water level.

Beachball

The beach ball's volume is $\frac{4}{3}\pi 1^2 \approx 4.2$ cubic feet. When submerged, it would displace that many cubic feet of water, and water is heavy, weighing about 62.4 lb/ft^3. Archimedes' principle tells us that keeping it submerged would require a downward force of over 260 lbs. This makes (b) the most reasonable answer.

Top versus Bottom

Water pressure increases with depth. At only six feet down, the water undergoes almost no compression, but that's not the case for air. A beach ball at the bottom of the pool feels the pressure of the water above it, and that pressure slightly compresses the air in it. This decreases the beachball's volume, meaning less water is displaced. Therefore its buoyancy is decreased, making the ball easier to hold down. The answer is (a).

Pouring Buckshot

The answer is (a). Iron is nearly eight times heavier than water. If you drop a 1-inch cube of iron into the water, it sinks to the bottom, displacing one cubic inch of water. But place it on the block of ice and, assuming the block of ice stays afloat, *all the iron's weight goes into pushing down the ice.* The mass consisting of the ice and iron remains floating, so by Archimedes' principle, the water supplies a buoying force equal to its entire weight. Therefore adding the iron cube results in displacing a volume of water equal to its weight, a volume of about 8 cubic inches. That much displaced water of course raises the water level more than just one cubic inch of displaced water.

It's easy to see that it doesn't matter where on the block you put the iron, just as long as it stays on. (Would it matter if it were suspended from *below* the block of ice, like a fishing weight?)

The only property of ice we use is that it remains floating. If the iron cube were set on a big block of floating wood or Styrofoam instead of ice (or even placed in a toy boat that doesn't sink) that iron cube would still displace the same amount, about 8 cubic inches of water.

Wooden Spheres

The answer is (c). Since the wood floats, the additional water displaced by adding the balsa is just the weight of the wooden spheres.

Hot Buckshot

The weight of water stays the same as it toggles between liquid and solid state, so on the one hand the berg itself won't change the water level, no matter how it melts. And on the other hand the berg will continue to support the buckshot, with or without melting. As long as this freely floating berg entirely supports the buckshot, the volume of displaced water stays the same. Therefore (c) is the answer.

Hot Buckshot Melts Through

As the buckshot falls, the ice, relieved of its load, bounces back up. Alone, this action would make the water level fall. But the buckshot enters the water, and that's a vote for making the water rise—but only a little bit. The dense buckshot occupies a much smaller volume than the water it originally displaced did. The net effect of the two actions? The level suddenly falls, making (b) the answer.

Air Pocket

As long as any 50-lb object is freely floating, it displaces 50 pounds of water. The object could be a 50-pound floating boat made of lead, or just as well, a chunk of Styrofoam. And that needn't even be in one piece. 50 pounds of Styrofoam "peanuts" would make the water level rise the same amount as one 50-pound piece. The answer is (d).

Melting Ice Cube

As water cools from room temperature, its density increases, reaching a maximum at around 4°C. With only one ice cube melting, the water's temperature never falls to anywhere near 4°C, so the density has increased, meaning that the slightly cooler water is more effective in buoying up the ice. Therefore less water need be displaced to keep it afloat, making the answer (a). In an extreme case, an ice cube floating in a pool of liquid mercury would be almost totally exposed.

Float Capacity

The answer is (b). Let the berg be a slowly growing cylinder. If, while growing, 10% of the cylinder's volume always sticks above the water level, the cylinder freely floats since there's just the right amount of displaced water to provide balancing buoyancy. This is true when there's only $\frac{1}{8}$ in of liquid water surrounding the submerged part, or even just 0.01 in. Therefore the percentage of floating ice approaches 100%.

By the way, if the glass started completely full of water, you could carefully let a frozen cylindrical core grow, always keeping 10% of it exposed so it stays afloat. (If you ever see more than 10% exposed, it isn't floating.)

Carbon Monoxide Detector

The fairest answer is (d). There are a number of competing issues, and no one of them dominates decisively. On one hand, carbon monoxide is slightly lighter than air, provided both are at the same temperature: from the periodic table, the atomic weight of CO is about $12 + 16 = 28$, and that is proportional to its density. This is a bit less than the density of dry air, which is about 78% N_2 and 21% O_2, giving an atomic weight of $.78 \cdot 28 + .21 \cdot 32 \approx 28.56$. Also, CO leaking from a furnace will be warmer than the surrounding air, so it will tend to rise, a vote for (a). On the other hand, air is moving and the gasses will mix, which tends to decrease how much weight this vote should be given. Also, where is the furnace? If it's in the basement but the detector is on the first floor so you can hear it should it go off, then it would be better to put the detector near the floor so it will pick up CO first as it makes its way upward from the basement, a vote for (b). Probably more important than any of these considerations is to put it where you'll hear it!

Arctic Blast

The answer is (c). This is just a fancier version of what we've seen before. As long as the berg floats, the water level stays the same.

Liquid Nitrogen

The answer is (c). Ice forms around the capsule, and water expands as it freezes. The heavy capsule keeps the ice at the bottom of the glass, so water there, as it expands into

Archimedes' Principle Answers

ice, makes the water level rise. Since the ball eventually floats to the top, it must have become more buoyant than needed to just sit there. We conclude that the ice-encrusted capsule became slightly less dense than liquid water, and that happened because more ice formed than was necessary just to balance. As the ball rises *above* the water surface, the buoying force on the ball decreases to bring buoyancy and weight into balance. The decrease means that less water is displaced, so the level then falls a tiny bit. The story continues even after the ball rises to the top: the nitrogen is so cold that ice continues to form around the ball, so the proportion of ice in the berglet continues to rise. Since the capsule is denser than the overall density of the ball, the ball's density decreases as the ball grows, meaning an increasingly greater proportion of exposed ice. This means that the water level will continue to fall slightly after the ball floats to the top. Of course both the capsule and any ice eventually warm to room temperature. With the capsule at the bottom, the water returns to its initial level.

Archimedes' Principle?

If you answered (c), read the box below the problem! The air between the ball bearings decreases the effective density of steel, multiplying 7.85 by at most $\frac{\pi}{\sqrt{18}} \approx .74$. Since the ball bearings are randomly distributed, multiplying by $.74 - .1 = .64$ gives a more realistic answer. Because $7.85 \times .64 \approx 5.02$, the answer is (b). The argument justifying Archimedes' principle given in this chapter's Introduction applies equally well here. The "molecules" aren't as small, but the buoying force is created in essentially the same way. Solid tin is too heavy for the ball bearings to support it since the density of the steel soup is diluted by about 25% air.

A Tale of Two Boats

Both boats have a total surface area of 100 in^2 and have known heights, so we can determine the length of the edge of each square base and from that, the boat's volume. We find that the six-inch high boat has base edge about 3.62 in and volume about 78.6 in^3. The two-inch high boat has base edge close to 6.77 in and volume of approximately 91.67 in^3. The boat with the larger volume wins—it can displace over 91 in^3 of water, and therefore can carry that weight of cargo before sinking, making (a) the answer. Actually, using 100 in^2 of material, we can construct a boat displacing more than the two-inch boat. In terms of the base edge x, the volume V is $25x - \frac{x^3}{4}$, and at the top of the next page we see a plot of $V = 25x - \frac{x^3}{4}$.

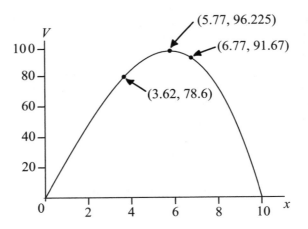

Differentiate this cubic and set it equal to zero to find the boat dimensions giving the largest volume. This volume turns out to be $\frac{500}{3\sqrt{3}} \approx 96.225$ in^3 and is attained at $x = \frac{10}{\sqrt{3}} \approx 5.77$ in. With this x, the height of the box is $\frac{5}{\sqrt{3}} \approx 2.887$ in.

Probability Answers

Red and White

Since the bin contains so many balls and only six are removed, the probability is very close to the analogous "with replacement" experiment, in which any ball is tossed back into the bin after its color is noted. This is equivalent to repeatedly dropping six fair pennies on the floor and asking for the probability of getting three heads and three tails. The experiment is binomial, with $n = 6$, $r = 3$ and $p = q = \frac{1}{2}$. The binomial probability $\frac{n!}{r!(n-r)!} p^r q^{n-r}$ becomes $\frac{6!}{3!3!} \frac{1}{2^6} = \frac{20}{64} = 31.25\%$. Notice that in the Pascal Triangle, $\frac{6!}{3!3!}$ is the middle number in the row 1 6 15 20 15 6 1. Since the answers are so different (increasing by 10% increments) the probability of the original problem is closest to 30%, making (c) the answer.

Red, White and Blue

Just as the above two-color problem can be approximated by a binomial experiment, this three-color one can be approximated by a *tri*nomial experiment. As suggested in this chapter's Introduction, if we write $n = r + s$ then the binomial probability $\frac{n!}{r!(n-r)!} p^r q^{n-r}$ can be written symmetrically as $\frac{(r+s)!}{r!s!} p^r q^s$. This suggests a trinomial analogue: Let the three single-draw outcomes have probabilities p_1, p_2, p_3 and in n draws, let there be respectively r, s, t occurences. Then $n = r + s + t$, and the generalization is $\frac{(r+s+t)!}{r!s!t!} p_1^r p_2^s p_3^t$. With three colors, an equal number of each color in 6 draws means $r = s = t = 2$, and since there are 1000 of each color, $p_1 = p_2 = p_3 = \frac{1}{3}$. Our trinomial formula then gives $\frac{6!}{2!2!2!} \cdot \frac{1}{3^6} = \frac{720}{8 \cdot 729} \approx 12.35\%$, making (d) the answer.

Dropped Penny

The answers are (a) and (b). To see how 3 is a possibility, start from the center of a tile and slide the penny along either diagonal. At one stage in this process, the penny becomes tangent to two sides of that tile. Then push just a bit further and the penny will touch exactly three tiles. Push some more to make it touch four tiles.

What are the Chances?

Suppose the penny lands on at least part of tile T. The penny's radius is $\frac{3}{8}$ in, so it will land entirely within T exactly when the penny's center is not less than $\frac{3}{8}$ in from any edge of T. That condition defines a square with $\frac{3}{8}$ in trimmed off from each side, which is a square 11.25 in on a side. So the probability is $\frac{11.25^2}{12^2} \approx 87.89\%$, making (b) the answer.

Teachers versus Students

You should distrust average class size calculated from the students' responses because in large classes, there are more students reporting large numbers, and that will result in an average that is too large. When you poll only the teachers, you're likely to get a fairer estimate of average class size because there is usually one teacher per class. Doing things this way better approximates the ideal of adding all class sizes and dividing by the number of classes. The answer is therefore (a).

Exit Polls

The poll doesn't report the undecideds, so we can't draw any conclusion about how many of them there were. You might poll 100 voters and find that 50 were for the war and 50 were against it. Or, you might poll 100 voters and find that 10 were for the war, 10 were against the war, and 80 were undecided. In addition to this, we don't know the ratio of Republican to Democratic voters. The answer is (e).

Ideal Tetrahedron

When tallying tetrahedron tosses, ignoring the tetrahedron's blank face means there are only the three possible outcomes: 1, 2 and 3. The tetrahedron is fair, so from the tally-sheet's viewpoint the probability of getting any one of these is $\frac{1}{3}$. When you tally the outcomes for the cube, only 1, 2 and 3 are possible. The cube is fair, so the probability of landing on any particular face is $\frac{1}{6}$. Therefore the probability of landing on either of the two 1's is $\frac{1}{6} + \frac{1}{6} = \frac{1}{3}$. The same is true for the two faces marked 2, and also for the two faces with 3. So in both the tetrahedron game and the cube game the probability of getting any one of 1,2,3 is $\frac{1}{3}$. The experiments are equivalent, making (a) the answer.

Sex and Money

The answer is (b). One reason figuring probabilities can be tricky is that it's often hard to determine exactly how much is known at various times, and which of those times actually matter. Take the couple with two children. The children are a *fait accompli*—the parents had the children, resulting in one of the four before-any-children outcomes of $\{GG, GB, BG, BB\}$. In the problem all you know after meeting the first child and before you meet the second is that the outcome wasn't BB, and that leaves three other possibilities. Assuming they're equally likely, the probability of GG is $\frac{1}{3}$.

You may wonder, "How's that different from flipping a coin?" The sample space for flipping two coins is similar: $\{HH, HT, TH, TT\}$, four possibilities, just as with the children. In the coin flip, as with the children, the question is, when did you know what you knew, and what was it you knew? The "when" is after the first coin flip and before the second flip. You knew then that the outcome was heads, so that eliminates two possibilities as outcomes: TH and TT. There are two possibilities remaining, HT and HH, for a probability of $\frac{1}{2}$.

You may scratch your head and say "I still don't get it! Why couldn't I just interchange girl-boy and heads-tails and get the same result?" The reason is that when you met the first child, you knew her sex but not whether she was the older or younger child. You didn't know her position in the sample space $\{GG, GB, BG, BB\}$. You were present at the coin-flipping at a more informative time: you knew then not only that the coin was a heads, but that it came first, that it was "first-born." You have more information, so you an improved chance of correctly guessing the second outcome. You could make the situations analogous: Only after the two coins are flipped do you get to see how one of them landed, but you're not told *which* flip, the first or the second, you're looking at. If it's heads, the probability that the other is heads is $\frac{1}{3}$. One could also transform the girl-boy question to make the answer $\frac{1}{2}$: if you'd heard ten years ago that the family had a girl, and five years later, that they had another child, then when you meet the ten-year-old girl, you easily see whether it's the younger or older one. You can therefore guess with improved odds on the sex of the other child.

Three Children

After the last problem, you're wiser and write the entire sample space:

$$\{GGG, GGB, GBG, GBB, BGG, BGB, BBG, BBB\}.$$

After you're told that at least two children are girls, but before you see the third child, you can eliminate four possibilities, leaving $\{GGG, GGB, GBG, BGG\}$. So the probability that the third child is a girl is $\frac{3}{4}$, making (e) the answer.

Average Distance

The sample space for this problem is $[0, 1] \times [0, 1]$. A point (x, y) in it corresponds to the two points x and y sitting in $[0, 1]$. The distance between these points is $|x - y|$. The area of the sample space is 1, so the average value of $|x - y|$ over this unit square is $\int_{y=0}^{1} \int_{x=0}^{1} |x - y| dx dy$. Using symmetry, we can evaluate this by integrating over the area above the 45° diagonal, thereby eliminating the absolute-value sign. That is, $\int_{y=0}^{1} \int_{x=0}^{1} |x - y| dx dy$ is $2 \int_{y=0}^{1} \int_{x=0}^{y} (y - x) dx dy = \frac{1}{3}$, making (c) the answer.

Short Sticks

For random points x and y, getting any particular length L means $|x - y| = L$, and this condition can be rewritten as $y = x \pm L$. These are both lines of slope 1, so the set of all outcomes with $L < \frac{1}{2}$ corresponds to the shaded region in $[0, 1] \times [0, 1]$:

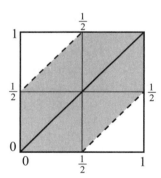

We've divided the sample space into eight congruent triangles, and six of them form the shaded region. That makes the answer $\frac{3}{4}$, so (e) is correct.

Average Distance-Squared

In this case the sample space is $[0, 1] \times [0, 1] \times [0, 1] \times [0, 1]$, and we say a point (x, y, u, v) in it corresponds to two points (x, y) and (u, v) in the unit square $[0, 1] \times [0, 1]$. The

Probability Answers

4-volume of the unit 4-cube sample space is 1 and the distance-squared between these points is $(u-x)^2 + (v-y)^2$, so its average value over the 4-cube is

$$\int_{v=0}^{1}\int_{u=0}^{1}\int_{y=0}^{1}\int_{x=0}^{1} \left((u-x)^2 + (v-y)^2\right) dx\,dy\,du\,dv.$$

This is easily found to be $\frac{1}{3}$, so (c) is the answer.

High-Noon Shadows

If you chose (e), think again! If one had to pick an average angle, one might guess $45°$ from the vertical, which gives a shadow length of about .707 ft. But the randomness is in not in two dimensions, but three: angles are measured not in a plane, but in three-space. There are more points close to the equator than close to the north pole, so this suggests an answer greater than .707. We can construct a sample space for the experiment this way: once the ruler is frozen in mid-air, parallel-translate its lower end to the origin. The other end then lies on a hemisphere with radius one foot. So it's as if points are randomly chosen on the hemisphere. To each point x is associated a number, its shadow-length, and the problem asks for their average. That's no different from asking for the average selling price of 100 cars in a used-car lot: you add up their prices and divide by 100. To get the exact answer, however, even a trillion randomly chosen points are not enough. We need a strategy.

In the used-car example, for car x you multiply its price $f(x)$ by $\frac{1}{100}$ and sum from $x=1$ to $x=100$. Writing that small number $\frac{1}{100}$ as Δx puts things in the suggestive form $\sum_{x=1}^{100} f(x)\Delta x$. That's clearly on the way to an integral.

Now for our hemisphere. Imagine a huge number of randomly-chosen points on it, and visualize a large number of horizontal planes, closely and equally spaced, each intersecting the hemisphere in a circle of latitude. From a remark on p. 11, it follows that each section of the hemisphere trapped between two successive planes has the same area. That means each has about the same number of random points on it, and we even know how long their associated shadows are: all the points having a particular shadow length form a latitude circle, and if the circle has altitude z, then by the Pythagorean theorem, all the shadows have length $\sqrt{1-z^2}$. In the car example, $\Delta x = \frac{1}{100}$ is the probability of choosing a point in the sample space of all 100 cars, and all 100 of these $\frac{1}{100}$'s sum to $1 = 100\%$; analogously on the hemisphere, the dz's from $z=0$ to $z=1$ sum to $1 = 100\%$. In the car example the car price was $f(x)$ and the average car price was $\sum_{x=1}^{100} f(x)\Delta x$. Correspondingly, the shadow length is $\sqrt{1-z^2}$ and the average shadow length is $\int_{z=0}^{1} \sqrt{1-z^2}\,dz$. This is $\frac{\pi}{4} \approx .7854$, making (h) the answer. It is indeed larger than .707, agreeing with our initial observation.

Seeing Two Sides

Look at the left picture:

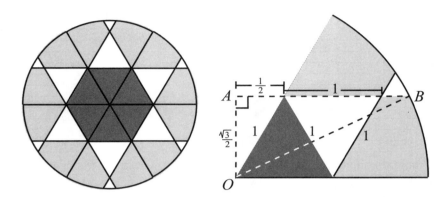

Within the circle, the locations where you see exactly two sides of the hexagon are shaded light grey. The probability we want is then the area of the light grey part divided by the area outside the hexagon and within the circle. The picture on the right shows one of six congruent 60° sectors of the left picture. That sector's area is $\frac{1}{6} \cdot \pi 2^2 = \frac{2\pi}{3}$. Our plan is to find the unshaded area in this sector, for then we'll essentially have solved the problem. This unshaded area is composed of a 1-1-1 triangle, a smaller equilateral triangle, and a tiny sector between the circle and the small triangle. We can begin by finding the area of the small triangle. For that, look at the dashed right triangle OAB in the picture to the right. Its hypotenuse is 2 since it is a radius of the circle, and its vertical leg is $\frac{\sqrt{3}}{2}$, so its horizontal leg is $\frac{\sqrt{13}}{2}$. The picture then tells us that the leg of the smaller equilateral triangle is $\frac{\sqrt{13}}{2} - \frac{3}{2} \approx .303$. The area of an equilateral triangle of side s is $\frac{\sqrt{3}}{4}s^2$, so the areas of the large and small equilateral triangles are about .4330 and .0397, which add to .4727. The tiny segment between the circle and the small triangle makes a minuscule contribution of about .00116. All this shows that the light grey area divided by the area outside the hexagon and inside the circle is $\approx .7148$, or about 71.5%, making (c) the answer to the problem.

How Grim the Reaper?

Compared to just one true positive in 1000, the number of *reported* positives is 50, and that's a lot! It means that out of 51 positive letters mailed, only one recipient actually has the disease. So our worried friend should relax: at 1 out of 51, there's less than a 2% chance that he actually has the disease. The answers are therefore (c) and (e).

Can You Trust This Headline?

There's nothing like getting the full story. Here are the specifics. Big Al ended up with a .300 average in the first half because he hardly ever went to bat. He went up 10 times in all, with 3 hits. His rival was up ten times as often, with 28 hits out of 100 at bats. In the second half, the workload was the reverse: Big Al was up 100 times with 25 hits, while his rival was up only 10 times with 2 hits. But the final average is over the whole season, and during the big-load times, Al's rival performed better—just enough better to put him in first place: Al made a total of 28 hits out of 110 times at bat, for a season average of .255. His rival hit a total of 30 out of 110 for a season average of $30/110 = .273$, making (b) the answer and making, we hope, the fan a bit wiser.

Common Birthdays

Thinking on a grand scale, from millions and millions of people, select many groups of 730. Or from a huge bin containing millions of ping-pong balls with equal numbers each of 365 different colors, scoop up many buckets of 730 balls, see what you've got, and then dump the ping-pong balls back into the bin. In any scoop, what is the probability of getting exactly two of each color? Intuitively, it's extremely small, but taking our cue from the trinomial problem "Red, White and Blue" we can, by analogy, write the exact answer:

$$\frac{730!}{2!\,2!\,\cdots\,2!} \cdot \left(\frac{1}{365}\right)^{730}$$

To get an idea of its size, we can arrange these factors like this:

$$\left(\frac{730}{365}\right) \cdot \left(\frac{729 \cdot 1}{365^2} \cdot \frac{728 \cdot 2}{365^2} \cdot \cdots \cdot \frac{366 \cdot 364}{365^2}\right) \cdot \left(\frac{365}{365}\right) \cdot \left(\frac{1}{2^{365}}\right)$$

Each factor enclosed by the second set of parentheses is less than 1 because its numerator is $(365 + i)(365 - i)$, with i going from 364 down to 1, and $(365 + i)(365 - i) = 365^2 - i^2 < 365^2$. Also, the first factor $\frac{730}{365}$ is 2, so the entire expression is less than $\frac{1}{2^{364}}$, which is the same order of magnitude as 10^{-108}. Therefore (d) is the answer.

A Serious Waiting Game

The answer is (c). The problem tells us that the odds increase to 50-50 after waiting a total of $200 + 500 = 700$ million years. That's the same as saying that the half-life of X is 700 million years, since after that time half the atoms of X have split, and half haven't.

This describes U-235, which has a half-life of slightly over 700 million years and is used in nuclear power plants. It's never used in its pure state, but in concentrations of from 3% to 5%, and additional diluting elements such as carbon rods are used to help control what could otherwise be a hard-to-manage chain reaction. (From a manufacturing standpoint, U-235 is never diluted down from 100%, but is concentrated up from raw uranium ore, which itself contains only about 0.72% U-235.)

Now the solar system is "only" about 4.5 billion years old, so how could a material whose half-life is the better part of a billion years ever serve as a nuclear fuel? The answer has to do with the sheer number of atoms in a sample of everyday proportions. A little over a quarter-pound of U-235 (235 grams) contains Avogadro's number of atoms. This number, about 6×10^{23}, is easy enough to write, but its size is stupendous:

- To finish counting this number at the rate of 5 million each second would take about 4 billion years, nearly the age of our earth and the solar system.

- If each atom expanded to the size of a corn kernel, an Avogadro number of them would cover the entire U.S. to a depth of more than 9 miles.

- Avogadro's number is more than 100 trillion times greater than the number of people in the world.

One moral: extremely large numbers can dramatically decrease waiting time. For example, a driver may go for years without an auto accident, but nationwide, accidents occur on average about once every 15 seconds.

In uranium, such great numbers have an added effect: the decay of a U-235 atom releases other particles that can destabilize nearby U-235 atoms and significantly boost the overall rate of fission.

A Random-Integer Game

Suppose we choose exactly two integers i and j between 0 and 1,000. The sample space for this experiment is $S \times S$, where S is the set of integers from 0 to 1,000. What is the probability that the sum of the two already equals or exceeds 1,000? This condition is $i + j \geq 1,000$, shown as the shaded area in the picture at the top of the next page.

The square represents the $1001^2 = 1,002,001$ points of $S \times S$. Assuming the shaded area includes the entire diagonal, the number of points in this area is $1 + 2 + \cdots + 1001 = \frac{(1001+1)(1001)}{2} = 501,501$ points. The shaded part thus covers more than half of $S \times S$, so our probability is more than 50%. Therefore exactly two integers is the most common outcome, making (a) the answer to the problem. Since all probabilities sum to 100%, the probabilities of requiring exactly 3, or exactly 4, $\cdots$ are less than 50%. Although two tries is the most common, the *average* number of tries is larger. This is because ending after three or more tries occurs over 49% of the time. These larger numbers of tries boost the average number of tries. As it turns out, that's slightly more than 2.7.

The sample space idea works on analogous problems. For example, if you randomly

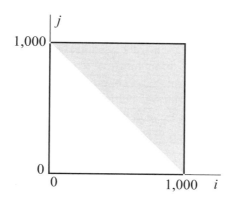

choose two integers between 0 and a million, then most of the time their sum will equal or exceed a million. It also works for real-number analogues. For example, keep picking real numbers in the interval [0, 1] until their sum equals or exceeds 1. Once again, two picks will do the trick most often. One could also ask how frequently you'll require exactly *three* real-number picks. The above picture then becomes a cube with a shaded portion consisting of points satisfying $x + y + z \geq 1$. The cube picture on p. 23 depicts this, where we take the bottommost point as the origin. Since the corner $x + y + z \leq 1$ represents a sixth of the cube, $x + y + z \geq 1$ covers five-sixths of it, so the probability of reaching or exceeding 1 in three tries is $\frac{5}{6}$. Since the probability is already $\frac{1}{2}$ for doing this in two tries, the probability for requiring exactly three tries is $\frac{5}{6} - \frac{1}{2} = \frac{1}{3}$. One can similarly show that the probability that you need exactly four tries is $\frac{1}{8}$, which we can suggestively write as $\frac{4!-1}{4!} - \frac{3!-1}{3!}$. For five tries, it's $\frac{1}{30} = \frac{5!-1}{5!} - \frac{4!-1}{4!}$. Using these factorials, it is easy to see that all these probabilities do in fact sum to 1.

Skewed Left or Right?

If you record heights in the general male population to the nearest half-inch, their histogram is approximately symmetric. The further out you go on the right side of the horizontal height-axis, the fewer people you'll generally encounter there. As a subpopulation, professional basketball players mainly come from the right side of the general population plot. Nonetheless, it's easier to come across players 6′ 5″ than players 7′ 1″. The curve's peak is toward the left side, so the curve is skewed right, making (c) the answer.

Bottles versus Boxes

The answer is (b). When you weigh groups of 24 bottles (cases of four six-packs), outliers get averaged in with the rest of the bottles. If the filler is working properly, it is rare to get many extreme overfills or underfills. That can happen, but not often. Therefore most crates will weigh in at closer to the average. The bunching around the mean tends to give the curve a bell shape. Even if the filler created a bell-shaped distribution with single bottles, the box-weight distribution curve would still be bell-shaped, only tighter, meaning the measurements are more closely grouped around the peak of the curve. The reason is the same: the occasional one-bottle outliers get averaged in with the weights of the remaining bottles in the case, and that average is what's recorded.

One measure of how tightly the measurements are clustered around the mean is the standard deviation. The Central Limit Theorem implies that if the standard deviation of single-bottle weights is s, then the standard deviation of the box weights is $\frac{s}{\sqrt{24}}$. This generalizes from $\sqrt{24}$ to $\sqrt{n}$ for n bottles. There's nothing special about bottles—the Central Limit Theorem applies to a wide variety of statistics such as annual income, auto ages, IQ, and so on.

Throwing Darts

The answer is (c). We can inductively find the volume of a unit n-ball by using calculus. One natural way is to use the slicing method. To begin the induction, you can start with an ordinary unit ball $x^2 + y^2 + z^2 \leq 1$, which can be regarded as a succession of disks $x^2 + y^2 \leq r^2$ of radius r varying from 0 to 1 and back down 0 as z increases from -1 to 0 to $+1$. More generally, an n-ball $\sum_{i=1}^{n} x_i^2 \leq 1$ can be sliced into a succession of $(n-1)$-balls $\sum_{i=1}^{n-1} x_i^2 \leq r^2$, with radius r varying from 0 to 1 to 0 as x_n increases from -1 to 0 $+1$. The volume of the $(n-1)$-balls is known, so summing up these slices of known volumes completes the induction step. Here is what you will find:

- When n is even, let $m = \frac{n}{2}$. Then

$$B(n) = \frac{\pi^m}{m!}.$$

- When n is odd, let $m = \frac{n-1}{2}$. Then

$$B(n) = 2^n \pi^m \frac{m!}{n!}.$$

In the first case, the numerator increases exponentially, while the denominator increases *factorially*. That's also true in the second case, because $2^n \pi^m = 2^n \pi^{\frac{n-1}{2}}$ can be regarded as the numerator, and $\frac{n!}{m!}$ can be looked at as the denominator, where m is about half of n. Even this "partial factorial" growth eventually beats exponential growth, because no matter how large a number A you choose, in going from A^k to A^{k+1}, you always multiply by the same number A, while with this factorial setup, the multiplier increases with k. For any A, our partial factorial eventually catches up and surpasses A^k. This is why the volume size will turn around and decrease.

Probability Answers

Monte Carlo

I ran ten trials of a million dart throws each, and obtained an average of 30.207. The maximum value of the ten runs was 30.267 and the minimum, 30.163. It therefore appears that 30.2, or (d), is a reliable answer.

The following Maple code is given just to illustrate the logic for simulating a large number of throws at a $6 \times 6 \times 2$ box enclosing the torus. In Maple, the command rand() returns a random 12 digit non-negative integer. In the last line, SUM/1000000*72 is the probability of success times the box volume of 72.

SUM:=0;

for i from 1 to 1000000 do

x:=6*evalf(rand()/1000000000000)−3:

y:=6*evalf(rand()/1000000000000)−3:

z:=2*evalf(rand()/1000000000000)−1:

if (x+y+z<2 and (sqrt(x*x+y*y) - 2) ^ 2 + z*z<1) then SUM:=SUM+1 end if:

end do:

evalf(SUM/1000000);

evalf(SUM/1000000*72);

Monte Carlo and Average Distance

In the unit square $[0, 1] \times [0, 1]$, randomly picking two points (x, y) and (u, v) is equivalent to randomly choosing four real numbers in $[0, 1]$ and assigning them to x, y, u, v. Computing $\sqrt{(u-x)^2 + (v-y)^2}$ 100,000 times and taking the average gives an answer closest to .52, making (b) the answer.

Monte Carlo and the Needle Toss

The answer is close to .79, making the answer (d). To use the Monte Carlo method, convert tossing a needle to throwing a dart. Assuming the parallel lines are vertical, the needle's resting position is given by the x-coordinate of its left endpoint and angle shown in this picture:

This picture reveals a few things. First, since the parallel lines are vertical, the y-coordinate doesn't matter. Second, the lines are spaced one unit apart, so we can assume that the x in $(x, 0)$ ranges between 0 and 1. And clearly the angle θ can be taken to lie

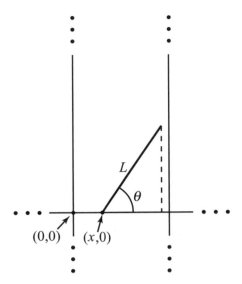

between $-\frac{\pi}{2}$ and $\frac{\pi}{2}$. Our dart board is therefore a $1 \times \pi$ rectangle. Finally, the picture reveals that the needle touches a line if and only if the bottom leg of the right triangle intersects the vertical line $x = 1$. This translates into $L \cos \theta \geq 1 - x$, our condition for success. Of course the success rate depends on what L is! To solve this "inverse problem" we guess at a value for L, run a Monte Carlo simulation and see what the probability of success is, then make a better guess. After a few successive approximations, we arrive at a value of L close to .7854. (Can you show that the answer is exactly $\frac{\pi}{4} = .785398 \cdots$?)

Short Sticks Revisited

Randomly pick two points in the unit square as in the solution to "Monte Carlo and Average Distance." Then increment SUM whenever $(u - x)^2 + (v - y)^2 < 0.25$. For 100,000 or 1,000,000 cycles, you'll get an answer close to 48.2%, so (a) is the answer.

Mechanics Answers

How Much Work?

The answer is (b). The law $W = Fd$ is commonly presented as a scalar law, but it is really a law involving vectors, and is more accurately written as W=**F**·**d**. Force **F** is a vector, and displacement **d** is, too. **F**·**d** is their dot product, so in the statement "Work equals force through distance," *through* means the component of **F** in the direction of **d**. If θ is the angle between **F** and **d**, then that component is $|\mathbf{F}| \cdot \cos\theta$. Although at each instant there's a force in the tetherball rope holding the ball in orbit, this force is always perpendicular to the instantaneous direction of the ball's travel around the circular orbit. That is, θ is 90°, so $\cos\theta$ is zero. Therefore $|\mathbf{F}| \cdot \cos\theta$ is zero and no work is done.

A similar argument shows that if frictional losses are ignored, then no work is done when heavenly bodies revolve around each other.

Toy Car

The work done in speeding up the car gets converted to kinetic energy, which is $\frac{1}{2}mv^2$. The weight w of a mass m is the force of gravity on that mass, $w = mg$. In the foot-pound-second (fps) system, we can assume g is 32 ft/sec², so the mass of the toy car is $m = \frac{w}{g} = \frac{1}{32}$. Therefore $\frac{1}{2}mv^2 = \frac{1}{2} \cdot \frac{1}{32} \cdot 10^2 = 1.5625$ ft-lb., making (c) the answer. The problem says the applied force is constant, but it doesn't say how large that force F is. That's easy to determine, however: the work done is F through the 10 ft., that is, 1.5626 ft-lb= $F \cdot 10$, so $F = 0.15626$ lb.

Grandfather Clock

The answers are (a) and (d). The frequency of the clock's pendulum is proportional to $\sqrt{\frac{g}{L}}$ (Galileo's first big scientific discovery). On the moon, g is only a sixth of what it is here, so the frequency on the moon is $\frac{1}{\sqrt{6}} \approx 41\%$ of its frequency here. With every tick, the drive gear rotates by a small amount, the chain therefore unwinds a bit and the weight descends a tiny distance, typically less than a thousandth of an inch. The slower ticking rate on the moon means that the hands turn more slowly, making (a) an answer. The weight descends more slowly, too, so (d) is also right.

Since on the moon the driving force (the weight) is a sixth of that on earth, and because we're assuming the clock does tick there, the smaller force is enough to overcome frictional effects there. Since there's no air on the moon, the pendulum needn't worry about air resistance. Compared to what it does on earth, the pendulum apportions its available energy differently, putting most of it toward overcoming friction in the gears and at the point of suspension.

Gravity's Horsepower

The answer is (a). Since we're simply dropping the brick, both the force (the weight) of the brick) and the path are directed downward, so work reduces to the scalar law $W = Fd$. The power (work W per time t) generated by gravity is then $\frac{Fd}{t}$, and we can write this as $F \cdot \frac{d}{t}$. Now $F = mg$ is a positive constant since m and g are. But because g is constant, the brick's speed $\frac{d}{t}$ increases at a constant rate. Power is the product of these two quantities F and $\frac{d}{t}$, so it continually increases, too.

Spinning Turntable

The answer is (e). The experiment begins with the penny resting on the disk, so by Newton's first law the total force on the penny is zero. The platter is frictionless, so as it begins to rotate, no additional forces are brought to bear on the penny. It therefore continues simply sitting there.

A Running Start

The answer is (a). Red has double the total energy of Green. Here's why: when Red is still at the top, it already has kinetic energy due to its speed of 5 ft/sec. But in addition, Red has *potential* energy at the top, exactly the same amount that Green had at the top and which converted to kinetic energy as it slid to the bottom, giving Green its speed of 5 ft/sec. So at the top, Red, in addition to its kinetic energy from its speed of 5 ft/sec, will have an additional, equal amount of kinetic energy once it slides to the bottom. That's why Red's total energy is double Green's. Since kinetic energy is $\frac{1}{2}mv^2$, Red therefore has twice the v^2 of Green. Green's v^2 is 25, so Red's v^2 is 50, and Red is therefore traveling with a speed of $\sqrt{50} \approx 7.07$ ft/sec.

Notice that the faster Red zips along at the top of the plane, the less speed it gains by going down it. For example, if its speed is initially $v = 100$ ft/sec, then adding Green's speed-squared of 5^2 to 100^2 gives 10,025, so Red speeds up less than $\frac{1}{8}$ ft/sec to $\sqrt{10,025} \approx 100.1249$ ft/sec.

Squeezing the Earth

The answer is (e)—a whopping 800 lbs. To see why, apply Newton's universal law of gravitation,

$$F = G \frac{m_1 m_2}{r^2}.$$

The mass of the earth and the person each remains fixed, but r is halved. Thus $\frac{1}{r^2}$ changes to

$$\frac{1}{\left(\frac{r}{2}\right)^2} = \frac{4}{r^2},$$

quadrupling the person's weight.

Paring Down the Earth

The answer is (b). In addition to the change in radius in the last question, we have a mass change, too. The earth's volume is reduced by a factor of $(\frac{1}{2})^3 = \frac{1}{8}$, and since it is uniform, the smaller earth has $\frac{1}{8}$ of its original mass. Therefore the force between the earth and person—that is to say, the person's weight—is multiplied by both 4 and $\frac{1}{8}$. The original 200 lbs decreases to $200 \text{ lb} \cdot 4 \cdot \frac{1}{8} = 100$ lb.

Galileo's Cannonball

The answer is (c). To see why, use the displayed formula on p. 112 to get a formula for the range. Since the cannonball lands at $(x, 0)$, set $y = 0$ and solve for x. The trivial solution is $x = 0$, and the nontrivial one is

$$x = \frac{\tan \theta \cdot 2 \cdot v_o^2 \cdot \cos^2 \theta}{g}.$$

This can be rewritten as

$$x = \frac{(2 \sin \theta \ \cos \theta) \cdot v_o^2}{g}.$$

The double-angle formula $2 \sin \theta \cos \theta = \sin 2\theta$ then tells us that the range is

$$\sin 2\theta \cdot \frac{v_o^2}{g},$$

a constant times $\sin 2\theta$. Obligingly, $\sin 2\theta$ has a maximum at 45°, and the graph of $\sin 2\theta$ is symmetric about the vertical line through this maximum. Therefore, firing at the angle $45° + \alpha$ results in the same range as firing at $45° - \alpha$.

Muzzle-Speed

The answer is (b). This comes from the range formula worked out above:

$$\text{Range} = \sin 2\theta \cdot \frac{v_o^2}{g}.$$

Since $(2v_o)^2 = 4v_o^2$, the range is quadrupled.

Muzzle-Speed, Again

The answer is (a). Once fired, the projectile's horizontal velocity component remains constant. If the muzzle speed is doubled, then so is the horizontal speed. From the last question, we know that doubling the muzzle speed quadruples the range. At twice the horizontal speed, it takes twice as long to cover four times the distance.

Muzzle-Speed, One More Time

The answer is (c). The maximum height Y is attained when x is at the half-range point. So set

$$x = \frac{1}{2} \cdot \left[\frac{2 \sin \theta \cdot \cos \theta \cdot v_0^2}{g} \right],$$

and substitute into the range formula on p. 287. Simplifying gives

$$Y = \frac{v_0^2 \cdot \sin^2 \theta}{2g}.$$

Since θ and g are kept constant, doubling v_0 quadruples Y.

Greatest Range

The answer is (e). The volume of a sphere is $V = \frac{4}{3}\pi r^3$, so to halve that, just halve r^3. Therefore r itself shrinks to

$$\frac{r}{\sqrt[3]{2}} \approx 0.7937r.$$

In Newton's law of gravitation, the gravitational force at the surface of the compressed earth is then obtained from

$$F = G \frac{m_1 m_2}{(0.7937r)^2}.$$

The ratio of this strength to the strength on the unsqueezed earth is

$$G\frac{m_1 m_2}{(0.7937r)^2} \div G\frac{m_1 m_2}{r^2} \approx 1.59,$$

so the original gravitational force g is multiplied by about 1.59. Since g appears in the denominator of

$$\text{Range} = \sin 2\theta \cdot \frac{v_0^2}{g},$$

the 1000-foot range is divided by 1.59, making the maximum range on the compressed earth about 630 feet.

How Much Faster?

The answer is (c). A spring is ideal if the strength of force needed to stretch or compress the spring a distance a from its natural length is ka, where k is the Hooke constant of the spring. The work (energy) required to stretch or compress the spring a distance a from its natural length is therefore $k \int_0^a x \, dx = \frac{ka^2}{2}$. Doubling the compression from a to $2a$ therefore requires four times the energy, which is then stored as potential energy in the spring. When the spring is released, this quadrupled potential energy gets converted into quadrupled kinetic energy $\frac{1}{2}mv^2$, so the speed v is doubled.

Pulling Strings

The answer is (e). As in the solution to the last problem, the energy needed to stretch a distance a is proportional to a^2. For the one-inch deflection, in both the left and right hand springs, the deflection is $a = \sqrt{18^2 + 1} - 18$. Therefore in appropriate units, the total energy stored in the two string halves is

$$E = \left[\sqrt{18^2 + 1^2} - 18\right]^2 \approx .00077.$$

For the two-inch stretch it is

$$E^* = \left[\sqrt{18^2 + 2^2} - 18\right]^2 \approx .01227.$$

The *extra* amount of energy needed to stretch from one to two inches is $E^* - E$. That works out to approximately .01150, which is about $14.9265 \cdot 0.00077 \approx 15E$.

Where Does the Arrow Land?

The answer is (e). From the last question, we know that compared to the one-inch position, pulling the arrow to the two-inch position gives it about $1 + 15 = 16$ times the stored energy. When the arrow is released, the energy is converted to kinetic energy, propelling the arrow to a greater launch speed v_0. This 16-times-greater kinetic energy is $\frac{1}{2}mv_0^2$, so v_0^2 is itself 16 times greater. Setting $\theta = 45°$ and $y = 0$ in the displayed formula on p. 112 leads to

$$\text{Maximum Range} = \frac{v_0^2}{g}.$$

The original range of 100 yards increases to about 1600 yards, which is $\frac{10}{11}$ mile.

A "Buoyancy Machine"

Since force and displacement are both vertical, work is just force times distance. The force is never very large, and the distance covered is very small. After the piston has done its work, it has at the farthest traveled to the surface of the pool of water at the bottom of the glass, and that pool is extremely shallow. If there are some 40 drops of water to a cc and the bottom of the glass has area 25 cm², then the water depth is approximately $\frac{1}{40} \cdot \frac{1}{25} = \frac{1}{1000}$ of a centimeter, less than the thickness of a human hair. So (a) is one answer. Given the minute amount of work performed, (c) is reasonable, too. If one really were intent on making a scheme like this work, one could add a little surfactant to the water to eliminate surface tension and therefore eliminate (d) as an answer.

A Pendulum's Critical Speed

The critical speed gives the mass exactly enough energy to rise 6 ft to the top. Once there, all the initial kinetic energy $\frac{1}{2}mv^2$ has become potential energy mgh. Doubling the rod's length means the mass travels upward twice as far, so there's twice as much final potential energy. Therefore the initial $\frac{1}{2}mv^2$ must double, meaning v^2 must double. Therefore v itself is multiplied by $\sqrt{2}$. That makes the answer (a), because the required initial speed is $15\sqrt{2} \approx 21.2$ ft/sec.

On the Moon

On the moon, g is a sixth of its value on earth. So at the top, the mass has a sixth of the potential energy compared to here. That means a sixth the kinetic energy is required at the bottom, and therefore a sixth of v^2. So v itself gets divided by $\sqrt{6}$. The initial speed is thus $\frac{15}{\sqrt{6}} \approx 6.124$ ft/sec, making (c) the answer.

Have You Heard of This?

You learn these basic ideas in a beginning calculus course, but in a language that artificially divides them from vector field concepts. In elementary calculus, it is the everyday function from $\mathbb{R}$ to $\mathbb{R}$ that corresponds to a vector field in multivariable calculus, and there's a simple trick to revealing their brotherhood. First, you can change any value $f(c)$ into a vector: draw a directed line segment from $(c, 0)$ to $(c, f(c))$ to obtain a vertical vector based at $(c, 0)$. (If $f(c)$ is negative, the vector goes downward.) The brotherhood? Here's the trick: rotate each vertical vector 90° clockwise about $(c, 0)$. Magically, we now have a traditional vector field in the line! The vectors point to the right where $f(x)$ is positive, and to the left where it's negative. In elementary calculus, $\int_a^b f(x)dx$ is most often presented as the signed area under the graph of $f(x)$ from a to b. After rotation, it's natural to think of $f(x)$ as a force field, and $\int_a^b f(x)dx$ as the amount of work done in going from a to b.

In multivariable calculus, the problem states that for certain force fields **F** there's a potential function ϕ for which the work done going from A to B is $\phi(B) - \phi(A)$, independent of the path from A to B. Is there a counterpart for *this* in elementary calculus? Yes, the Fundamental Theorem of Calculus. One phrasing: if $f(x)$ is continuous, then there exists an antiderivative ϕ so that $\int_a^b f(x)dx = \phi(b) - \phi(a)$. This theorem calls the multidimensional concept of potential function an "antiderivative" and guarantees such a function exists so that the work done in going from a to b is $\phi(b) - \phi(a)$. What about the multidimensional concept of "any path"? In $\int_a^b f(x)dx$, the expected path is a direct shot from point a to point b. In keeping with the multidimensional analogue, we need not go from a to b in only that one way. We could overshoot, then backtrack, and in fact move back and forth along the line many times before finally stopping at b. Inefficient to be sure, but integrating leftward between two points is the negative of integrating rightward between them, so the integral's extra contributions over all these paths cancel out, leaving us with the usual one from a to b. Therefore the two correct answers to the problem are (c) and (d).

Now What Will Happen?

The answer is (a) because in the right-angle problem, the right angle at the top never enters into any part of the solution.

Twanging Rod

The answer is (a). An important assumption in the problem is that the system is conservative—no energy is lost, not even in internal friction within the rod. Also, the steel in the rod is assumed to be uniform, so its behavior is the same no matter how much the chuck has turned. Assuming the chuck isn't rotating, begin gently pushing directly down near the free end of the rod using a horizontal knife edge. As the rod begins to deflect downward, it pushes back on the knife edge, and if at any stage you hold the edge fixed so that the deflected rod doesn't move, then the rod pushes directly upward with the same strength of force that you're applying downward. The assumption of no friction means that this remains true even if the chuck has been given a little torque so that it is slowly but freely rotating.

If that's all there were to it, we'd be done with our explanation—the rotation has no influence on the restoring force within the rod so, like a stretched string, it returns to equilibrium, goes on to deflect upward as far as it deflected downward, then goes down again, with the cycle repeating endlessly. However, the rod is almost always moving in a way akin to the spokes on a bicycle wheel—think of the wheel as spinning one way, then the other, like the to-and-fro motion of a balance wheel in a spring-driven watch or clock. A rapidly-spinning bicycle wheel acts like a gyroscope, and a gyroscope offers resistance when you try to alter its axis of spinning. With only one spoke, the effect isn't large, but it can be felt in a laboratory setting. The resistance is due to the spinning gyroscope's angular momentum. The chuck started off with a small angular momentum, and angular momentum is conserved. So some of the angular momentum of the chuck gets transferred to the gyroscope, slightly changing the gyroscope's axis of spinning. At this point, the angle of twanging alters slightly. This transfer of angular momentum is most pronounced when the rod is passing through equilibrium, for then the gyroscope is turning fastest and has the greatest angular momentum. Then the chuck's turning slows down since angular momentum is conserved. The chuck speeds up again when the rod reaches maximum deflection, where it is momentarily stopped and its angular momentum is zero.

Historically, it was this phenomenon that led to Foucault's pendulum. He was experimenting with a rod in a lathe in his basement, and noticed that when the chuck turned, the twanged rod still vibrated in a fixed plane. Foucault was not at all a highly-educated physicist, but he realized that this directional persistence meant that a pendulum would behave the same way, despite the slow turning of the earth. His first basement pendulums to show this were small and the effect questionable, but he had a knack for the experimental and soon constructed bigger, more friction-free models that clearly and dramatically demonstrated the earth's rotation.

Total Torque

The answer is (e). The L-bar is bolted to the wall, so it doesn't move. That means the total torque must be zero, otherwise it would move. This is the angular version of Newton's second law of motion.

Saturn's Rings

The answer is (a). Each chunk orbiting Saturn satisfies Kepler's Third Law. For a satellite orbiting about a central body in a circle of radius r, his law tells us that the satellite's period varies as $r^{\frac{3}{2}}$, meaning the nearer an orbiting chunk is to Saturn, the shorter its period. (This has also been confirmed by observation.) The inside of any solid ring would try to rotate faster than the outside, creating internal stresses that would tear apart the ring. A large collection of small chunks doesn't have this problem since each piece can individually attain its natural speed depending on its distance from the planet.

One might argue that the solid ring might be so strong that it *couldn't* rip apart. But that's problematic, too, because no ring would be perfectly centered about Saturn's center of mass. Part of the ring would be more strongly attracted to the planet, and as it moves closer to the planet because of that, the inequality would only grow. Things would spiral out of control and the ring would come crashing into the planet.

Electricity Answers

Staying Power

The answer is (b). Ohm's law $V = iR$ says that for a constant current i, the voltage drop between two points varies with the resistance between those two points. The fresh flashlight battery supplies $1.5V$, so applying the law to the problem's top picture shows that there is a 0.75 volt drop through each bulb. But Ohm's law, written as $i = \frac{V}{R}$, tells us that the current i varies directly with voltage V and inversely with resistance R. In the problem's bottom picture there is a 1.5 volt between each bulb's threaded side and bottom, so each of the bottom-picture bulbs drains the battery faster than either of the top-picture bulbs.

Survivors

The answer: (a), (d) and (h). A bulb dims as less current flows through it. Since the bulbs' resistance R is constant, Ohm's law tells us that the only way i can decrease is if V decreases. That happens as the battery runs down.

When the battery has voltage V, each bulb in the top picture is subjected to voltage $\frac{1}{2}V$, while in the bottom picture each bulb receives the full battery voltage V. Since bulbs 1 and 2 receive equal voltage, they go out simultaneously, making (a) an answer. Likewise, bulbs 3 and 4 always get equal voltages, so (d) is also an answer. So is (h), because the bottom arrangement draws current from the battery faster than the top arrangement. The bottom battery therefore goes dead first, so bulb 3 goes out before bulb 2.

A Current Difference?

The answer is (e). In the top picture, each bulb has .75 volts applied to it, so the current through each bulb is $\frac{.75}{R}$. The current at the arrow is the same, since the current's path never branches. In the bottom picture, each bulb has 1.5 volts applied to it, so the current through each of them is $\frac{1.5}{R}$, which is double the current through the bulbs in the top picture. In addition to this, the current at the bottom arrow *does* branch, supplying each bulb with a double-sized current. So the current at the top arrow is just a quarter of that at the bottom arrow.

Later

The answer is (a). The batteries initially have the same voltage, but the bottom one starts by delivering more current, so its voltage decreases faster, and its greater initial current diminishes more quickly. When the bottom battery is exhausted, the top one, having held on to its reserves, still has some good life in it, and the current through it will eventually be larger than the current through the bottom battery.

Voltage Change

The answer is (a). The battery creates the system circuit's overall voltage V_{total}, like a water pump lifting water to the top of a waterfall. The rest of the circuit uses this potential to do work. If R_{total} is the total resistance, then from Ohm's law, the system's overall current is $i = \frac{V_{total}}{R_{total}}$. As current flows through a resistor R such as a wire or filament, the potential decreases, once again according to Ohm's law, which can be written as $\Delta V = i \Delta R$. All the ΔR's in the loop circuit add up to R_{total}, and all the corresponding ΔV's add up to the voltage difference V_{total} between the battery's terminals.

Current Difference Again

The answer is (b). Bulb 3 is hooked up in parallel and therefore receives the full 1.5 volts of the battery, while bulb 1 shares the battery's voltage with bulb 2, each getting only .75 volts. Double the voltage difference means double the current.

Type AA

The answer: (a) and (c). Initially, the D and AA batteries have the same voltage and create the same currents, so (a) is one answer. The other answer is (c), because the AA battery contains smaller amounts of material (current-generating chemicals), causing it to exhaust more quickly.

Most Massive

The battery has more energy when it's fully charged, so by $E = mc^2$ there's a minute increase in mass. As the battery is used, energy radiates away from it as heat and light.

Electricity Answers

The amount doesn't compare in magnitude with the amount of energy radiating away when mass converts to energy in an atomic blast, but $E = mc^2$ applies to both cases. In the atomic blast, mass quickly converts to a lot of energy; in the battery, an extremely small amount of mass slowly converts to everyday amounts of energy. The answer can't be (d), since that choice says *exactly*; (a) is the right answer.

According to $E = mc^2$, there are about 9×10^{13} joules of energy in one gram. A fresh D-size alkaline battery contains about 75,000 joules of electrical energy, so it would take roughly a billion such batteries to equal the atomic energy in one gram, the weight of a dollar bill. A scale able to detect a difference of a ten-billionth of a gram, 10^{-10} g, would suffice. Today, there exist scales able to measure very tiny particles with an accuracy of a zeptogram, 10^{-21} g.

A Question of Polarity

You can determine the magnet's polarity by thinking of current going from the positive battery terminal to the negative one and applying the right-hand rule to one full turn of the wire coil. You'll consistently get the same direction of field lines within the iron core, and it's those lines that determine the polarity of the magnet. Doing this in the upper left picture gives field lines within the iron core. The only other choice having this orientation is at the bottom right, making (c) the answer.

A Coil Around a Coil

Using the right-hand rule shows that magnetic field lines point from left to right inside both the inner and the outer coils. That very nearly doubles the number of lines and therefore the strength of the magnet, so (a) is the answer.

A Fatter Wire

The batteries supply the same voltage, but the fatter wire offers less resistance so it has more current flowing through it. The picture on p. 126 says that the strength of the magnetic field lines is proportional to the current. The greater strength is picked up by the iron core, making the fat-wire magnet stronger. Therefore (a) is the answer.

Forgetting to Turn It Off

This is the electrical analogue of Newton's $F = ma$ appearing in the box on p. 130: $V = L\frac{di}{dt}$. According to Newton, if we continue to apply a constant force to a mass, it will accelerate indefinitely. In the electrical analogue, as long as the voltage V is maintained, the current in our ideal setup will continue to increase. That makes (b) the answer.

Building a Large Magnet

This crazy creation is actually a passable description of what happens inside a permanent iron magnet, so the answer is (a). Each iron atom is actually a minuscule magnet, because it has four electrons in its outer shell which effectively spin around the nucleus to create a "current" and magnetic field. In a permanent iron magnet, the iron atoms are lined up sufficiently well so that the strengths of these little magnets add up by the trillions. Our imitation approximates this.

A Larger Capacitor

The "Mechanics versus Electrical Analogies" list on p. 130 says that the frequency of an undamped mass-spring oscillator varies as $\sqrt{\frac{k}{m}}$, where k is the spring stiffness and m is the mass. The list also tells us that k corresponds to $\frac{1}{C}$, and that m corresponds is L, so the frequency of the analogous electric oscillator varies as $\sqrt{\frac{1}{LC}}$. Doubling the capacitor's plate diameter quadruples its area, so its capacity increases by a factor of 4. Quadrupling C in $\sqrt{\frac{1}{LC}}$ means the oscillator frequency is halved, making (e) the answer. Large plates give the charges more room to spread around, resulting in decreased charge density on the plates. There is therefore less repulsive force to move charges through the wire. In the mechanical analogue, a weak spring doesn't try as hard to accelerate the bob. Everything moves at a more leisurely pace, implying a slower rate of oscillation.

More Turns

Doubling the number of turns doubles L, so from the last problem, we see that the value of $\sqrt{\frac{1}{LC}}$ gets divided by $\sqrt{2}$, making (d) the answer.

Warm Resistors

Ohm's law $i = \frac{V}{R}$ tells us that doubling the voltage V doubles the current i. Referring to the box on p. 130, we argue by analogy. V is like mechanical force F, and i is like speed, or distance per time $= \frac{d}{t}$. Therefore $V \cdot i$ corresponds to $\frac{Fd}{t}$, or work per time. Since both V and i are doubled, the work per time, or power, is quadrupled. The work is expressed as heat from the resistor, so four times as much heat is given off per unit time. Heat flows because of temperature difference, an application of the larger truth Ohm

Electricity Answers

taught us. (See the box on p. 135.) Therefore to quadruple the steady-state heat flow away from the resistor, the temperature difference must also quadruple. So the difference is 4°, making (d) the answer.

Four Charges

The answer is (b). To see why, think of the equator of a sphere as being a rubber band, and on this rubber band make four equally-spaced ink dots to depict the charges. Fix three of the dots and move the fourth one north. This stretches the rubber band, so it means the distance has increased, and increasing mutual separation is what mutually-repelling charges try to do. With one of them displaced northward, we see that the charges are now vertices of a tetrahedron. Now let all four charges be free to move on the sphere. They're equal in strength, so all charges play an equal role in the final tetrahedron. It is the regular tetrahedron that perfectly fills this prescription. Its symmetry means that each charge feels the repulsive forces of the other three equally, and all four remain locked in a symmetric stalemate.

Energy Loss

The answer is (b). Example: when you turn off a flashlight, you increase the resistance of the circuit so much that essentially zero current flows. Therefore the rate of energy drain is almost zero.

The Shrinking Capacitor

Opening the irises decreases the capacitor plate size. It takes more work to squeeze a given charge onto a smaller plate since the electrons more strongly repel each other the closer together they are. In moving a stream of electrons through a wire and into a plate, they encounter back pressure from electrons already in the plate. The pressure is greater if the plate is small, since the charges are more densely packed with electrons wanting to escape back into the wire. Moving electrons onto the plate against an increased force takes more work. By conservation of energy, you can't cheat and make the plate sizes smaller without performing necessary additional work, so the answer must be (a). As an analogue, if you fill an air-compressor tank with a certain quantity of air, you encounter back-pressure sooner with a small tank than with a large one. If you've added the quantity of air to a big tank, you can't then squeeze the tank to a smaller size without doing extra work.

Heat and Wave Phenomena Answers

A Cooling Rod Problem

Choose units so the temperatures on the long and short rods are $T(x,t) = e^{-\alpha t}\sin(x)$ and $U(x,t) = e^{-\beta t}\sin(5x)$ for some α and β. Putting these into the heat equations

$$T_t = T_{xx}, \quad U_t = U_{xx}$$

gives

$$-\alpha e^{-\alpha t}\sin(x) = -e^{-\alpha t}\sin(x)$$

and

$$-\beta e^{-\beta t}\sin(5x) = -25e^{-\beta t}\sin(5x),$$

from which we conclude that $\alpha = 1$ and $\beta = 25$. Therefore

$$T(x,t) = e^{-t}\sin(x), \qquad U(x,t) = e^{-25t}\sin(5x).$$

The $-25t$ tells us that in the short rod the amplitude of the sine curve decreases 25 times faster than in the long rod. So if it takes 60 minutes for the long rod's midpoint to cool from 100°C to 50°C, then it takes $\frac{60}{25} = 2.4$ minutes for the short rod to do the same, making (c) the answer.

How Hot is Hottest?

The answer is (e). We saw from the last problem that the one-foot rod has a one-arch half-life of 2.4 minutes, so that rod experiences 25 half-lives in one hour. Therefore after an hour, the initial midpoint temperature of 100°C has dwindled to $\left(\frac{1}{2}\right)^{25}$ of 100°C. That's about 3 millionths of a degree.

Does the Peak Walk?

The answer is (c). Let's begin with a little intuition. Although heat drains through both ends of the rod, the peak is closer to the right end and the random-walking heat packets will reach there before they reach the left end. Once they go beyond the right end, they're captured by the ice block and are out of the game. That cuts off the supply of jumping beans which with probability .5 would have jumped to the left to maintain a symmetric peak. Instead, the right side is denied this heat, so the temperature profile gets eaten away on that side.

We can construct an easy-to-visualize model by reversing heat and cold: construct the tall peak out of ice, and put a heat lamp at each end of the rod. The heat on the right will make that side of the peak melt faster, and after a while the initially tall peak melts down to a more stubby one whose high point has moved away from the intense heat on the right.

We can quantify this peak movement in both the discrete and continuous levels. Here's the idea.

In the discrete case, divide the rod's interval $[0, 1]$ into a large number of equal, small Δx-intervals or cells, and regard the temperature in any cell as the number of miniature jumping beans in it. Each jumping bean represents a tiny unit of heat, so the number of them in a cell measures the amount of heat in it. We can model heat flow by assuming that from one time interval to the next, half the jumping beans in each cell jump to the right and half jump to the left. At time $n + 1$ a cell gets its entire supply of jumping beans based on what its immediate left and right neighbors possessed at time n. In the rightmost cell, however, those beans jumping to the right are irretrievably absorbed by the fixed $0°$-temperature, and similarly for the leftmost cell. The rightmost cell feels the lack of heat donation by its ice-cold right neighbor, and the heat-impoverished cell can't donate many beans to its left neighbor. So this heat-poverty radiates leftward. The temperature of any cell at time $n + 1$ is the average of the temperatures of its two immediate neighbors at time n. The simplicity of this approach makes it easy to write a little program to automate the process. This picture

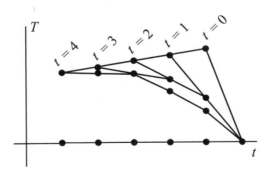

illustrates one case, where the temperature rises all the way to the rightmost graph point at time $t = 0$. The point just below the t in "$t = 1$" represents the peak at that time. It has already moved left, and has moved yet further at $t = 2$.

For the continuous version of the peak-walking phenomenon, a Fourier approach is appropriate. Assume the Fourier solution is

$$T(x,t) = \sum_{n=1}^{\infty} A_n e^{-n^2 t} \sin nx.$$

(See **Fourier solutions** in the Glossary, in particular the first displayed equation on p. 346.) In our case, $A_1 \neq 0$. The reason is that if $f(x)$ is the initial temperature $T(x, 0)$, then the peak of $y = f(x)$ has positive area under it, so $\int_L f(x) \sin x\, dx$, used in computing A_1, is nonzero. So the slowest-decaying term is the one-arch sine curve. The n^2 in $-n^2 t$ means that for $n > 1$, the $e^{-n^2 t}$ terms go to zero faster than $e^{-1^2 t}$, so after a time the original peak has spread out and looks more like a one-arch sine curve, its amplitude decreasing as heat spreads out and escapes through the ends of the rod.

Going to Zero

With the endpoints held at zero, only the fundamental building-block functions $\sin nx$ do this, making the answer (e).

How Far?

Adding an identical weight means the top spring feels an additional 1 pound. The top spring is ideal, so that 1 pound makes it and the weight attached to it descend another 4 inches. So (c) is the answer.

Hefty Block, Wimpy Spring

An ideal spring satisfies Hooke's law, so the graph of weight versus extension is a straight line. It doesn't matter where on the line you are: increasing the weight by one ounce always makes the spring stretch one more inch. So it takes only 12 ounces ($\frac{3}{4}$ pound) to get the block to the ground. A two-year-old would put that block very firmly on the ground, making (a) the right answer.

Spring Stiffness

The answer is (e). Because the 1 pound weight stretches the 12-inch spring 1 inch, each 6-inch section of it gets stretched $\frac{1}{2}$ inch. Similarly, a half pound weight stretches the 12-inch spring $\frac{1}{2}$ inch, so each 6-inch section of it stretches $\frac{1}{4}$ inch.

Spring Stiffness, Again

Applying Hooke's law $F = -k\Delta x$ tells us that a unit stretching force elongates the spring by $\frac{1}{k}$. With a unit force, the spring with $k = 2$ stretches $\frac{1}{2}$ and the one with $k = 3$ stretches $\frac{1}{3}$. When you join them end to end and apply that unit force, they both feel it and accordingly stretch a total of $\frac{1}{2} + \frac{1}{3} = \frac{5}{6}$. That is, the longer spring's $\frac{1}{k}$ results in a stretching of $\frac{5}{6}$. Therefore k is $\frac{6}{5} = 1.2$, making the answer (c).

This generalizes to ideal springs with stiffness k_1 and k_2. Applying a unit force makes the longer spring's $\frac{1}{k}$ equal to $\frac{1}{k_1} + \frac{1}{k_2} = \frac{k_1+k_2}{k_1 k_2}$, and for n springs k_i in series, $\frac{1}{k} = \frac{1}{k_1} + \cdots + \frac{1}{k_n}$. In this respect, springs in series behave like resistances in parallel. Such a statement begs for the other shoe to drop! Here's a picture depicting springs in parallel:

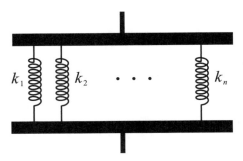

Do springs in parallel behave like resistances in series? Can you prove your answer?

Switching Off Gravity

The answer is (a). The pictures say something about the frequency, but they don't tell us a thing about the mass of the bob. The oscillation frequency is proportional to $\sqrt{\frac{k}{m}}$, where k is the stiffness of the spring and m is the bob's mass. Knowing the frequency tells us what the *ratio* of k to m is, but nothing about either k or m. The spring could be three times stiffer and the bob three times more massive, and the frequency would be the same. If m is very small, then when it was initially put on the weightless spring it would have stretched the spring just a little bit, establishing an equilibrium position where the mass could hang at rest. If m is large, the initial stretch would have been larger and the equilibrium position lower.

When gravity is switched off, no mass is lost but the weight goes from $w = mg$ to $w = m \cdot 0 = 0$. Since the mass has lost its weight, the equilibrium point moves up to where the bottom end of the spring was before any point-mass was suspended from it. Exactly where that is we don't know, but we do know that the bottommost point of the resulting motion is where the bob was when gravity disappeared, and we also know that the equilibrium point has become higher, with a resulting increase in amplitude. Only choice (a) satisfies both these conditions.

Method versus Madness

The answer to the first question is (b). In the first method, the middle spring is neither stretched nor compressed. It goes along for the ride, but since it exerts no force on either mass it is, in effect, not even there. Therefore the only force acting on the left mass comes from the left spring, and the only force acting on the right mass comes from the right spring. If each spring has stiffness constant $k = 1$ and mass $m = 1$, then each mass vibrates with angular frequency $\sqrt{\frac{1}{1}} = 1$. In the second method, if the masses are squeezed toward each other, the left spring pulls on the left mass, and the middle spring pushes on it. So two springs are trying to make it move leftward. Similarly, the middle and right springs both apply force on the right mass, trying to move it rightward. There's more force acting on each mass compared to what's going on in the first scenario. The second scenario makes each mass get to its equilibrium position faster. That's a shorter time to make a quarter-cycle, meaning a higher frequency.

The answer to the second question is (a). Whether we symmetrically squeeze the masses toward each other, or symmetrically pull them further apart, the middle point of the middle spring never moves. There could just as well be a wall at the middle, with the middle spring cut in half and each cut end glued to the wall. From the last problem we see that if each original spring has stiffness 1, then the cut-in-half spring has stiffness 2. With each mass feeling force in the same direction on both sides, this amounts to a spring of stiffness 3 acting on each mass. Since neither mass has changed, the angular frequency is now $\sqrt{\frac{3}{1}} \approx 1.732$. Therefore the faster frequency is less than double the slower one.

This solution is direct and physically meaningful, and can serve as an introductory handshake to the general theory of coupled mass-spring systems. In the general theory, Hooke's constant, a scalar, gets generalized to a stiffness matrix, and the pleasant, easy-to-see motions—the normal modes—correspond to eigenvectors, with the normal mode frequencies corresponding to eigenvalues. The components of any eigenvector $\mathbf{v} = (a_1, \cdots, a_n)$ can be plotted out as a function whose domain consists of the eigenvector's subscripts i. A typical point of the finite graph is (i, a_i). These graphs rise and fall precisely like sine functions and can be normalized to lie on the usual graphs of the sine functions of Fourier series. A good reference for these ideas is [10], Chapters 10 and 11—especially section 11.3.

A Ceiling Under Stress

The answer is (b). The spring is weightless, so when it was attached to the ceiling unstretched and uncompressed it added no stress to the ceiling. As soon as a weight was hung from it the spring stretched and the ceiling felt it. The spring determines what force the ceiling experiences, and the force is determined by whether the spring is stretched or compressed from its natural length. When it's compressed, the ceiling experiences an upward force and when it's stretched, the ceiling feels a downward force. Only when the spring is at its natural length, unstretched and uncompressed, does the ceiling feel nothing. The spring is at its natural length when the weight is 1 inch above equilibrium, since the spring originally stretched 1 inch to get to that point.

Adding Waves

The answer is (d). The waves are $A \sin x$ and $B \sin(x + \phi)$ for some A, B and ϕ. The ϕ horizontally translates the graph of $B \sin x$ The translation is called a *phase shift* and ϕ, the *phase angle*. Expanding $\sin(x + \phi)$ using the sum-of-angles formula gives

$$A \sin(x) + B \sin(x + \phi) = [A + B \cos(\phi)] \sin(x) + B \sin(\phi) \cos(x).$$

The right-hand side has the form

$$\alpha \sin(x) + \beta \cos(x)$$

and we can show that this is $\rho \sin(x + \psi)$ for appropriate ρ and ψ, which will justify the answer (d).

Expanding $\rho \sin(x + \psi)$ gives

$$\rho \left[\sin(x) \cos(\psi) + \cos(x) \sin(\psi) \right],$$

and we can now find ρ and ψ so that the last two displayed expressions are the same. How to do this? Simply equate coefficients—that is, find ρ and ψ so that

$$\alpha = \rho \cos(\psi) \quad \text{and} \quad \beta = \rho \sin(\psi).$$

But these are just the equations for converting from rectangular to polar coordinates! This picture

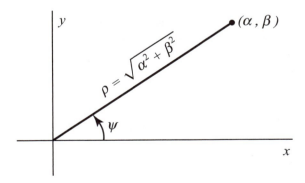

pretty much says it all: $\rho = \sqrt{\alpha^2 + \beta^2}$ and $\psi = \arctan(\beta/\alpha)$. So

$$A \sin(x) + B \sin(x + \phi)$$

is an example of
$$\alpha \sin(x) + \beta \cos(x),$$
which is the pure, translated sine curve
$$\rho \sin(x + \psi),$$
with
$$\rho = \sqrt{\alpha^2 + \beta^2}, \quad \psi = \arctan(\beta/\alpha).$$

Moral: Take two sine curves of the same frequency. You can vary their amplitudes and/or slide each horizontally as much as you please, but their sum will still be a sine curve of the same frequency. The new amplitude and phase angle can be found using the picture.

Losing Energy

The answer is (d). The system loses energy through working against air resistance, which generates heat. As this heat radiates, the system's energy decreases monotonically. This rules out (a) and (b).

The frictional effect of air is greatest when the bob is moving fastest. You experience this when you put your hand out of a moving car: the faster the car is going, the more air resistance you feel. In a full cycle of the bob, the energy loss is fastest when the bob is going the fastest as it passes through the equilibrium point. As the bob reaches its rightmost or leftmost points it is, for an instant, motionless, and no energy is leaving the system. This is described by picture (d).

Not only does your hand feel air resistance, so does the car. It and the air it's plowing though both warm up slightly from the heat generated by the friction. The chemical energy in the car's gasoline is converting into that wasted heat-energy. That's why, to improve gas mileage, auto designers try to reduce energy-consuming wind friction. (Even so, the super-sleek Bugatti Veyron 16.4 gets no more than 50 miles to *an entire tank of gas* when going at its top speed of slightly over 250 mph.) Commercial aircraft fly high to get into thin atmosphere to decrease drag. As a really extreme case of air resistance, consider asteroids: most of them enter our atmosphere at several miles per second and the great air resistance makes them vaporize well before hitting the ground.

Losing Energy, Again

Static friction means there's a minimum force needed to get the object moving. The block's motion stops for an instant every time the graph hits a maximum or minimum. In picture (a), the extrema tend towards zero, meaning the restoring force when the box is stopped tends to zero. But the damped oscillation continues forever, so there could not be static friction in (a)—if there were, the block would eventually stop oscillating. The picture in (b), however, does fit the physics.

It's a stretch, but (c) could qualify, too, provided the ratio of stickiness to force is high enough. For example, on a table top, press down a long strip of tape sticky on both sides, and quickly drag along the strip the eraser end of a pencil, but then slow down. At some point, the tape grabs the eraser. The location in picture (c) where the graph suddenly goes horizontal describes that phenomenon.

Friction and Frequency

The answer is (a). As the mass oscillates, the system loses energy through friction with the oil, and the energy drain from friction is faster when things move faster. This proportionality often implies exponential growth or decay. In this case, it's the amplitude of oscillation that decays exponentially as the system runs down. The real part of λ translates into exponential growth or decay, while the imaginary part translates into oscillation.

Substituting $e^{\lambda t}$ into $m\ddot{x} + D\dot{x} + kx = 0$ and simplifying gives

$$m\lambda^2 + D\lambda + k = 0,$$

from which we get

$$\lambda = \frac{-D \pm \sqrt{D^2 - 4mk}}{2m}.$$

For $D^2 < 4mk$, λ has imaginary part $\frac{\pm\sqrt{4mk - D^2}}{2m}$, so the oscillation has frequency $\frac{\sqrt{4mk - D^2}}{2m}$, a constant. The rate never varies, making (a) the answer.

Going, Going, Gone!

The answer is (b). A simple counterexample puts a nail through this one! Suppose the fluid is so thick it's like tar. After one day, the mass has crawled somewhat closer to the origin. During all this time, the spring has been holding on to its potential energy, releasing it ever so slowly. The total energy drain from the system? Slo-o-o-w. (Compare this problem with "Energy Loss" on p. 141.)

Change of Scene

The answer is (c). We've seen before that the frequency of the mass in an undamped mass-spring system varies as $\sqrt{\frac{k}{m}}$. Notice what's *not* in this formula—there's no gravitational strength constant g in it. If you take the apparatus to the moon, then the equilibrium point of the mass will rise because the smaller gravity there doesn't pull the mass downward as much. But the frequency? For our undamped system, it's a tango exclusively between mass and spring stiffness.

Change of Scene, Again

As in the last problem, the frequency of the mass-spring system depends on m and k, but not on the gravitational strength of the earth or moon. The equilibrium point of the mass would be higher on the moon since the mass weighs less, but a little thought shows that the two springs of stiffness k_1 and k_2 behave like a single spring of stiffness $k = k_1 + k_2$. The frequency is $\omega = \sqrt{\frac{k}{m}}$, so using a different spring constant k changes the frequency while changing g does not. The answer is therefore (c).

Pendulum versus Mass-Spring

The answer is (e). The boxed-in pendulum can make only small-angle oscillations, so Galileo's law tells us that the pendulum's frequency is approximately $\sqrt{\frac{g}{L}}$, where g is gravitational strength and L is the length of the pendulum's weightless rod. Since gravity on the moon is one-sixth of earth's, the frequency of our pendulum on the moon is $\frac{1}{\sqrt{6}}$ of its frequency here on earth. So on the moon, our pendulum's frequency decreases from twice as fast as the mass-spring to only $\frac{2}{\sqrt{6}} \approx 0.816$ as fast—that is, about 81.6% of the mass-spring's frequency.

"Fixing" Our Moonbound Pendulum

The answer is still (e). Increasing the bob's mass does nothing to $\sqrt{\frac{g}{L}}$ because mass doesn't appear in the formula.

A Shorter Period? Or Longer?

Move the bob to the top and try to balance it as well as you can, akin to balancing a pencil on its point. The closer you get to a perfect balance, the longer the bob will take to move either to the right or left. Falling down is just part of the entire period, so a large swing means an increased period, making the answer (a).

Which Runs Faster?

With each tick, the weight falls the same small distance in both clocks, so the heavier weight loses double the energy per tick. In either clock, a part of this energy goes into giving a little kick to the pendulum to overcome air resistance and maintain its amplitude. Assuming that the heavier weight delivers a bigger kick, this creates a larger amplitude. As we learned from the last problem, a pendulum's period increases with amplitude. Therefore the answer is (a)—the heavier weight makes the clock run slower.

Inclined Mass-Spring

The answer is (a). If the mass is m and the spring's stiffness is k, then the frequency is proportional to $\sqrt{\frac{k}{m}}$. By varying θ, the component of gravity pulling on the mass-spring varies too. This apparatus gives a clever way of changing gravitational strength for this experiment right here on earth. (See the boxes on pp. 119 and 120.) But g doesn't appear in the mass-spring frequency formula, so the experiment's fancy tap dance does nothing to change the system's frequency.

Wave Motion? Version 1

The answer is (c). To eliminate (b)—and therefore (a)—look at this ellipse with semi-axes a and b in a circle of radius a:

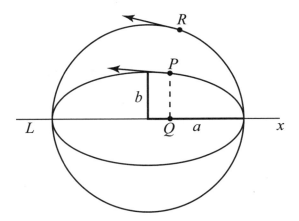

When P gets to its highest point, it and its shadow Q are both moving horizontally, so both have speed v there. If Q's motion were simple harmonic with amplitude a and top speed v, then that motion would be that of the vertically-projected shadow on L of a point R running around the circle at constant speed. That constant speed is also v, because at the top, R is going just as fast as P, whose speed is v. Therefore P and R must have the same constant speed. However, the trip around the circle is longer than that around the ellipse, so their shadows must get out of synch. That means Q's motion cannot coincide with the one possible candidate for simple harmonic motion.

Wave Motion? Version 2

The answer is (c). An argument like the above once again eliminates (b) and therefore (a). This picture

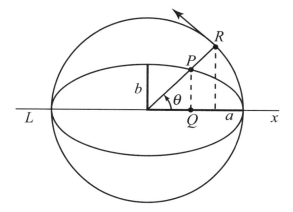

depicts R going around the circle of radius a at a constant speed. R's vertically projected shadow on L therefore moves in simple harmonic motion. The projected motion of both P and Q pass through the extreme point $(a, 0)$ simultaneously, so they have the same frequency and amplitude. If P's shadow also executed simple harmonic motion, both motions would be identical. But the picture shows at once that the projections don't agree, so cannot be identical.

Wave Motion? Version 3

The answer is (c). In the picture at the top of the next page, assume the semi major axis is a. Then simple harmonic motion along the x-axis is described by $x = a \sin \omega t$ for some frequency ω. This motion is symmetric about the origin O, and at equal distances from this center the speeds are equal. However, Kepler's second law tells us that the planet sweeps out equal areas in equal times, so when the planet is near the sun, where the pie-shape is more stubby, the planet must move faster to match the area swept out by the planet when it's further from the sun. For example, here

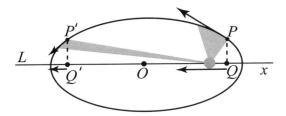

we see that the shadow Q of P is moving faster than the shadow Q' of P', where Q and Q' are at equal distances from the center of motion. So the motion can't be simple harmonic.

Is Energy Conserved?

The answer is (a). Your friend chose two points whose velocities cancel, but his "ditto for the other points" is bunk. Let's pick a vertical pair of points slightly to the left of the vertical pair that he picked:

The upper of the two points is moving down since its wave is moving to the right. The lower point is also moving down, because its wave is moving leftward. You can check that an analogous pair slightly to the right of his pair are both headed up.

Before and after the instant when the string is flat, the total energy is a mix of kinetic (from string particles' motion) and potential (from the string being stretched from a straight-line shape). At the instant the string is flat, all the energy is in kinetic form. At all times the total energy remains constant.

Which Has More Energy?

The answer is (a). Suppose the Hooke constants of the weak and strong springs are k_1 and k_2 with $k_1 < k_2$ and that the respective amount of compressing is x_1 and x_2. Since the block is being pressed equally hard by the two springs, Hooke's law $F = -kx$ tells us that $k_1 x_1 = k_2 x_2$. Our assumption $k_1 < k_2$ implies $x_1 > x_2$. The springs' stored potential energies are $\frac{1}{2} k_1 x_1^2$ and $\frac{1}{2} k_2 x_2^2$, and multiplying $x_1 > x_2$ by the equality $k_1 x_1 = k_2 x_2$ leads to $\frac{1}{2} k_1 x_1^2 > \frac{1}{2} k_2 x_2^2$. That is, the weak spring stores more energy.

A Low Orbit

The magic bullet here is Kepler's third law, which says that the period is proportional to $a^{3/2}$, a being the semi-major axis of the satellite's elliptical orbit. One consequence: The lower the orbit, the shorter the period. The answer is (b). To see why, stand on the equator and hold the ball one inch above the ground. When you let go, it of course promptly falls to the ground! Its speed when you hold the ball is, of course, one orbit per day. Because it falls to the ground, that speed isn't fast enough to keep it in that orbit. To keep it orbiting, you need to speed it up substantially, and that results in a shorter period.

The period isn't even half a day, since GPS satellites circle the earth in approximately that time, and they are some 12,625 miles above earth, so the lower one-inch orbit must have a period less than that. Kepler's third law allows us to use the GPS facts to find the golf ball's period. The orbits of both golf ball and satellite are nearly circular, so we can phrase Kepler's law as $P = kr^{3/2}$, where r is the orbit's radius. The earth's radius is about 3,960 miles, and at 12,625 miles above earth, the GPS orbit radius comes to 16,584 miles. We know the GPS period, so we can find k; we also know the imaginary golf ball's orbiting radius—it's just the earth's. Kepler's law then gives us the orbiting time: the golf ball circles the earth once every 84 minutes, which at the equator is about 17,000 mph. If the earth actually spun that fast, then near the poles objects would weigh about what they do now, while at the equator, it would be substantially easier to launch them into space.

Kepler's third law helps to put some everyday knowledge into perspective: the inner planets have shorter orbiting times than the outer ones. Mercury, the first rock from the sun, orbits in 88 earth days, or about $\frac{1}{4}$ year; the third rock (us) takes a year, while the dwarf planet Pluto takes over 248 years.

Spike

The answer is (b). Look at the initial shape as the sum of two shapes: a sine wave, and the horizontal dashed-line shape supplied with a downward vee at the midpoint between the two walls. By the superposition and independence principles, the shape at a slightly later time t is the sine shape at t, plus what has become of the downward blip then. At that time, the sine graph's amplitude is halved, while the blip has separated into two half-height blips, one moving to the right and the other to the left at equal speeds, an application of d'Alembert's wave solution stated in the chapter's Introduction on p. 148. This is depicted in (b)'s picture.

Spike — Temperature Version

The answer is (c). Just as in the wave version of this question, the initial shape is the sum of a pure sine arch and a downward spike. The shape at any later time is, by the superposition and independence principles, the sum of what each of the two shapes have become then. We know how these shapes change: the sine arch decreases in amplitude, remaining an arch. The downward temperature spike? That cold spot is sandwiched on both sides by lots of heat, creating a high temperature gradient that makes heat rush into the spike area. So the cold blip warms far sooner than the whole arch cools down to the half-life picture shown in (a), (b) and (d)—leaving (c).

Can This Happen?

The answer is (a). To simplify things, assume that the rod is represented by the unit interval [0, 1]. Denote the rod's temperature at point x and time t by $T(x, t)$.

The "yes" answer to the problem means that for some T, the temperature graph of x versus T uniformly translates upward as time progresses. This suggests that T should have the form $T(x, t) = f(x) + t$ for some function $f(x)$, the t-term making the temperature graph rise. To keep the endpoint temperatures equal to t, we need $T(0, t) = T(1, t) = t$. This implies that $f(0) = f(1) = 0$. A candidate for $f(x)$ is the quadratic $ax(1-x)$, $a \neq 0$.

Does this work? Is there some a so that $T(x, t) = ax(1-x) + t$ performs the required magic?

Begin with the heat equation. Differentiating gives $T_{xx} = -2a$, and $T_t = 1$. For these to be equal, choose $a = -\frac{1}{2}$. Our candidate is therefore

$$T(x, t) = \frac{1}{2}x^2 - \frac{1}{2}x + t.$$

This does everything we want:

- The heat equation is satisfied: $T_{xx} = T_t$ becomes $1 = 1$;
- Both endpoint temperatures rise in tandem: $T(0, t) = t$, $T(1, t) = t$;
- The t-term makes the temperature of all points increase at the same rate.

Beyond Hooke

For an ideal spring, pulling or pushing equally hard on it stretches or compresses it the same amount, and that tells us that changing the sign of x reverses the sign of F—that is, $F(-x) = -F(x)$. The only function listed having this property is the one with only odd powers, so the answer is (c).

By way of perspective, any polynomial with only odd powers has this property, as does any function with a power series representation involving only odd powers, such as $y = \sin x = x - \frac{x^3}{3!} + \frac{x^5}{5!} - \cdots$. A function having the property $f(-x) = -f(x)$ is called *odd*. Geometrically, the graph of an odd function is symmetric about the origin.

There are also *even* functions: they satisfy $g(-x) = g(x)$, and their graphs are symmetric about the y-axis. A polynomial with only even powers satisfies $g(-x) = g(x)$, as does a function with a power series representation having only even powers, such as $y = \cos x = 1 - \frac{x^2}{2!} + \frac{x^4}{4!} - \cdots$. Any polynomial is the sum of an odd and an even polynomial. Can you prove that *any* function $y = h(x)$ is the sum of an odd and an even function? Can you prove that the derivative of any differentiable odd function is even, and that the derivative of any differentiable even function is odd?

Exponential?

The answer is (b). To see why, note that the quotient of two exponential functions is also exponential:

$$\frac{Ae^{k_1 t}}{Be^{k_2 t}} = \frac{A}{B} e^{(k_1 - k_2)t}.$$

We assume that B isn't zero. If $e^{-t} + e^{2t}$ were exponential, then after dividing by e^{-t}, the result would be exponential, too. But this quotient is just $1 + e^{-t}$, which *isn't* exponential—a decaying exponential goes to zero, while this goes to 1.

Does This Chord Stay Musical?

Assume the strings act like ideal springs obeying Hooke's law. Then the tension in any string is proportional to how much it was stretched from its natural unstretched, uncompressed length. A string with a high spring constant will require just a little lengthening to bring it up to pitch, while a string with low constant requires much more. From the geometry, sliding the four strings means that they all lengthen by the same amount. That could drastically raise the frequency of a high-k string, while not affecting it much for a low-k string. Two strings of differing k could be tuned to the same pitch, but stretching them both the same amount would destroy this unison and could create a sour sound. The answer to the problem is therefore (b).

The Leaking Tank Answers

How Long?

Suppose the tank is one foot tall. Then in 10 sec the filled tank's level goes down 3 in, at which point the pressure at the hole is less. Therefore it takes longer than 10 sec to fall another 3 in, making (b) the answer.

In general, let r ($0 < r < 1$) be the proportion of water remaining in the tank after some unit of time has elapsed, and call that time the tank's r-life. If the initial level A decreases to Ar between $t = 0$ and $t = 1$, then at any time t, any level α decreases to αr between t and $t + 1$. So between $t = 1$ and $t = 2$, the level drops from Ar to $(Ar)r = Ar^2$. From $t = 0$ to $t = n$, the level goes from A ($= Ar^0$) to Ar^n.

In this problem, the proportion remaining after 10 minutes is .75, so the amount left after another 10 minutes is $.75^2 = .5625$. Therefore after 20 minutes, the tank's level hasn't quite decreased to 50%, so again, the half-life is longer than 20 minutes.

Again: How Long?

The answer is (a). If after 30 sec the amount remaining is 50%, then after another 30 sec, the amount left is half of 50%, or 25%. That is, 75% of the water has been lost.

How Fast?

The tank's flow rate is proportional to the water height, so if 1 cc/sec flows out when the tank is half full, then 2 cc/sec flow out when it's full, making the answer (c).

Two Philosophies Compared

The box appearing at the bottom of p. 181 gives a hint. Assume units are chosen so that the initial height is 1, and so that the spongy-plug tank's flow rate is $\dot{y} = y$. Therefore when $y = 1$, then also $\dot{y} = 1$. Substituting these into the Torricelli tank equation $\dot{y} = C\sqrt{y}$ shows that the proportionality constant C is 1. At all later times, the height in both tanks is less than 1. At any height $a < 1$ in the spongy-plug tank, the tank is discharging at the rate $\dot{y} = a$; at the same level $y = a$, the Torricelli tank is discharging at the rate

$\dot{y} = \sqrt{a}$. Because $a < 1$, $\sqrt{a} > a$. So at all levels, the Torricelli tank is emptying faster, making the answer (b).

Two Movies

The answer is (a). Look at this labeled version of the two-tank arrangement:

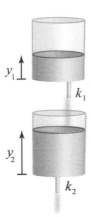

$\dot{y}_1$ and $\dot{y}_2$ are the rates of change of the water heights. In the top tank, the rate of change of y_1 is $\dot{y}_1 = -k_1 y_1$, where k_1 is a positive constant. This says y_1's rate of increase is negative (the level is falling) and depends jointly on y_1 (the greater the head, the faster the level falls) and the constant k_1. In the bottom tank, liquid flows both in and out. The rate of inflow is $+k_1 y_1$, since liquid from the top tank flows into the bottom one. The leak rate in the bottom tank is $-k_2 y_2$, so the overall rate of level increase in that tank is $\dot{y}_2 = +k_1 y_1 - k_2 y_2$.

The constants k_1 and k_2 influence how fast the levels change. Suppose we were to double k_1 and k_2 to $2k_1$ and $2k_2$. That doubles the rate at which y_1 falls, since then $\dot{y}_1 = -2k_1 y_1$. Same for y_2, because $\dot{y}_2$ is then $2k_1 y_1 - 2k_2 y_2$. If we were watching a movie, the doubled constants would double its speed. Similarly, if the k_i were halved, the movie would be the original one run at half speed. Therefore the basic question is, *what determines the k_i?* In a physical system, they would be obtained experimentally. Here are a few things that influence their values:

- Hole size. Increasing the cross-section of the hole increases that tank's k.
- Tank base size. The larger the base, the smaller the corresponding k. For example, doubling the base's area halves the constant.
- Gravitational constant. The larger g is, the more pressure there is at the outlet and the faster the flow.
- Viscosity. The more viscous the liquid is, the more slowly it flows.

The liquid's density doesn't affect the constants because gravity accelerates each particle of matter the same amount.

The Leaking Tank Answers

We now see why the answer is (a). Each tank contains only water or only oil. This means that in the right tank, if replacing water by oil multiplies that tank's constant k by some constant $r < 1$, then compared to the water-movie, the oil movie progresses at only r^{th} the rate. Therefore speeding up the right tank by the factor $\frac{1}{r}$ will synchronize the two movies.

Two Movies, Again

The answer is (a). The reasoning is the same as in "Two Movies," except that now it is not viscosity but rather the decreased gravity on the moon that uniformly decreases both k_i. One can therefore similarly speed up the moon-picture to match the earth-picture.

Who Wins?

The answer is (g). In the left tank, the weight of the 10-inch water column directly above the hole is the force pushing water through the pure resistor. In the middle tank, there's the additional weight of the 10-inch underwater column directly over the hole, but by Archimedes' principle there's an equal upward force of buoyancy canceling the extra weight. The net force at the resistor is therefore the same as for the left tank. And the right tank? For it, the underwater column of the middle tank is just air whose weight we ignore, so it does not cancel the buoyancy in the middle tank. Also, the column's weight seen in the left tank is missing. So at the hole in the right tank, there's a force of the same strength as in the left or middle tank, but directed upward. Therefore all three water levels initially approach L at identical rates. There is nothing special about 10 inches—all three columns not only start, but *continue* to approach L at identical rates. Therefore the distance-from-L histories of all three tanks are identical.

How Many Half-Lives?

The answer is (c). The question is, how long does it take to cool *to* 75°C? After one-half of a half-life, it has cooled to

$$(\frac{1}{2})^{\frac{1}{2}} = \frac{1}{\sqrt{2}} \approx .707$$

of its original temperature, so it's cooled to about 70.7°C. It therefore takes less than that time to cool to 75°C.

To find the exact number x of half-lives write

$$\left(\frac{1}{2}\right)^x = .75$$

and take logarithms to get

$$x = \frac{\log .75}{\log .5} \approx .415.$$

A Musical Analogue

The answer is (a). On both modern pianos, and pre-Bach harpsichords and organs, counting down 12 successively-lower keys brings you to the next lower octave. Even for the ancient Greeks, two plucked strings, identical except for length, sounded an octave apart when the string lengths had a ratio of 2:1. Today, we realize that such a setup creates frequencies in a 1:2 ratio. So although today we fill in the space in each octave differently from the way the ancients did, and differently from less ancient pre-Bach musicians, one octave down amounted then, and still amounts today, to half the frequency—one half-life.

Down One Key

The answer is (b). Any note has 100% of its own frequency, so the next semitone down has $100\% \times 2^{-\frac{1}{12}} \approx 94.4\%$ of that frequency, a decrease of about 5.6%.

Raising Frequency

The answer is (b). Counting up n semitones from any note multiplies its tempered frequency by $2^{\frac{n}{12}}$. Raising a frequency by $33\frac{1}{3}\%$ means multiplying the frequency by $\frac{4}{3}$, so we're looking for n that satisfies $2^{\frac{n}{12}} = \frac{4}{3}$. Taking the logarithm of each side gives $\frac{n}{12} \log 2 = \log \frac{4}{3}$, or

$$n = 12 \times \frac{\log \frac{4}{3}}{\log 2} \approx 4.98.$$

This is closest to five semitones, called a fourth.

A Full Piano Keyboard

The answer is (b), because 88 keys represents only 87 semitone *intervals*. Now 84 successive semitone intervals represent $\frac{84}{12} = 7$ half-lives. The extra 3 semitone intervals contribute another $\frac{3}{12} = \frac{1}{4}$ of a half-life, giving a total of $7\frac{1}{4}$ half lives.

A Model for Relative Humidity

The answer is (b). If in going from noon to evening the relative humidity increases from 20% to 85%, then the rightmost picture could represent the situation during the chilly night when the air has cooled so much it can no longer hold all the moisture it did earlier. The spilled water in the picture represents dew that has condensed out of the air.

Two Interconnected Tanks

The answer is (d). The large hole has double the diameter of the small one, therefore four times the cross-sectional area, which means a quarter the resistance. Water heights are steady when both flow rates are equal, so water height times resistance must be the same in both tanks. Therefore the top tank's water column must be four times that of the bottom tank. The tanks have the same base area, so the top tank contains four times as much water as the bottom tank. With 10 gallons between them, that means there are 2 gallons in the bottom tank and $4 \cdot 2 = 8$ gallons in the top one.

Three Interconnected Tanks

The answer is (d). Water heights are steady only when all three flow rates are equal, so water-height times resistance must be the same in all three tanks. Water heights of 6, 3 and 2 work, because $6 \cdot 1 = 3 \cdot 2 = 2 \cdot 3$.

Tanks Can Model Current Flow

The answer is (a). Look at the water analogue: maintaining the same charge on each capacitor plate is like keeping the same amount of water in each tank. But keeping the amounts the same and shrinking the bases makes the columns grow taller, stretching the two vertical height-vectors in the picture. This causes the water pressure at the resistor to increase and speed up the flow rate. For a compressed-air analogy, compress 10 cubic feet of air in a cannister down to 9 cubic feet. When you open a small valve, some air will at once begin to exit, but at a rather leisurely rate. But compress that 10 cubic feet all the way down to one cubic *inch*! Open that same small-sized valve, and air will rush out with a deafening scream.

Pulling on the Plates

Opposite charges on the plates attract. If the charges are kept fixed, then pulling the plates apart takes work. That means the capacitor with further-apart plates requires more energy to charge up—that is, its capacitance is smaller. Physically, opposite charges on the two plates attract each other, partly neutralizing the tendency of repelling charges on a plate to send those charges back up the attached wire. When the plates are further away, the attractive force is less, making it harder to keep the charges on the plate. One can also argue by analogy with our tank model: capacitance corresponds to tank base area, and a tank having a small base requires more work to fill with a gallon of water, since we have to lift much of that gallon to a greater height. The answer is (c).

Tanks Can Model Heat and Temperature

The answer is (b), because we are doing more than changing the size of the tank's base—we're adding water to it, too. Heat is added during the compression stroke, because work is done in compressing the air. At the molecular level, as the piston decreases the volume, air molecules in its path receive a kick and this increases their average speed. Average speed is heat, so the air gets hotter as it's compressed. Of course this work is far more than compensated for when the diesel fuel's chemical energy converts to mechanical energy in the power stroke.

Tank With Constant Water Supply

Let R be the rate that water discharges from the hole when the tank is full. If water enters the tank at a rate equal or greater than R, then start the tank completely full—any excess water will flow over the tank sides and the tank remains full. If water enters at a rate r less than R, then r is sandwiched between the maximum discharge rate R and its minimum, which is 0. Of course the tank can discharge at any rate between 0 and R. Choose that level with discharge rate r. The problem's answer is therefore (a).

Increasing the Water Supply

The answer to the first question is (c): $\dot{y} = t - y$ says that the slope $\dot{y}$ at any point (t, y) is $t - y$, and at the point $(1, 1)$, this is $1 - 1 = 0$.

The answer to the second question is (a). Differentiating $\dot{y} = t - y$ with respect to t gives $\ddot{y} = 1 - \dot{y}$. We just found that $\dot{y}$ is zero at $(1, 1)$, so that makes $\ddot{y}$ equal to $1 - 0$ there. The positive second derivative says that the solution curve is concave up at $(1, 1)$.

Linear Algebra Answers

A Matter of Distance

A sketch of all four lines encloses a diamond-shape symmetric about the origin, making the answer (e). But there's another way to solve this problem that leads to a powerful tool. In the problem's equations, factor the left-hand side into a dot product. For example, $3x+4y$ is the product $(3,4) \cdot (x,y)$. Geometrically, $(3,4) \cdot (x,y) = 0$ says that the vectors (x,y) and $(3,4)$ are perpendicular, and therefore all (x,y) satisfying $(3,4) \cdot (x,y) = 0$ form the line L through the origin that is perpendicular to $(3,4)$. This pictures gives the idea:

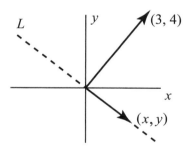

What about when the constant on the right isn't 0? From the top box on p. 203 we know that $\mathbf{v} \cdot \mathbf{w} = \|\mathbf{v}\| \cdot \|\mathbf{w}\| \cos\theta$, where θ is the angle between $\mathbf{v}$ and $\mathbf{w}$. Therefore $3x+4y=5$ becomes $\|(3,4)\| \cdot \|(x,y)\| \cdot \cos\theta = 5$ which, since $\|(3,4)\| = 5$, is

$$\|(x,y)\| \cdot \cos\theta = 1.$$

This is illustrated by the figure at the top of the next page. This last equation describes the line L of vectors (x,y) orthogonally projecting onto the point P in that figure—that is, the line through P perpendicular to the vector $(3,4)$.

Everything we've just said holds more generally. The equation $ax+by=c$ of any line in the plane can be written as $(a,b) \cdot (x,y) = c$, or as $\|(a,b)\| \cdot \|(x,y)\| \cdot \cos\theta = c$. Therefore the distance D of the line from the origin, which is $\|(x,y)\| \cdot \cos\theta$, becomes

$$D = \frac{c}{\|(a,b)\|}.$$

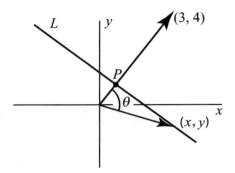

It is both handy and powerful to regard (a, b) as the "coefficient vector" of $ax + by = c$. One can always divide the equation through by $\|(a, b)\|$, normalizing the coefficient vector to unit length, and then the constant on the right side is the distance of the line to the origin.

For this problem's four lines, all the coefficient vectors have the same length because $\|(a, b)\| = \sqrt{a^2 + b^2}$, making the signs of a and b irrelevant.

There is a designer aspect to this method. Given a unit vector **u** in the (x, y)-plane, one can directly write the equation of the line perpendicular to **u** and at any desired distance D from the origin: $\mathbf{u} \cdot (x, y) = D$. The distance is signed, because a negative D puts the line into the opposite half-plane. This picture encapsulates the idea for $\mathbf{u} = (a, b)$:

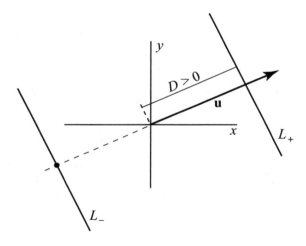

The equation of L_+ is $\mathbf{u} \cdot (x, y) = ax + by = D$, and the equation of L_- is $ax + by = -D$.

A Matter of Distance (3D Version)

In the last problem, there's nothing special about dimension two. For example, it generalizes to 3-space, making it very easy to get the answer. The distance between the origin

Linear Algebra Answers

and the plane $ax + by + cz = d$ is

$$D = \frac{d}{\|(a, b, c)\|},$$

making the answer (d).

In keeping with what might be called the principle of minimum astonishment, this method unastonishingly generalizes to hyperplanes in $\mathbb{R}^n$: $a_1 x_1 + \cdots + a_n x_n = d$ is perpendicular to the coefficient vector $(a_1, \cdots, a_n)$, and this hyperplane is at the signed distance $D = \frac{d}{\|(a_1, \cdots, a_n)\|}$ from the origin.

What's its Area?

The answer is (a). Thanks to some good news, there's a simple way to find the area of any triangle in a plane from the coordinates of its vertices. We can always assume that one of its vertices is the origin—if not, translate the triangle so that one vertex *is* there. Let the other translated vertices be (a, b) and (c, d). Up to sign, the area is half the magnitude of a determinant:

$$\text{Triangle area} = \frac{1}{2} \begin{vmatrix} a & b \\ c & d \end{vmatrix}.$$

In our case, that's $\frac{1}{2} \times (100 \cdot 1 - 2 \cdot 2) = 48$, exactly an integer.

To see where this formula comes from, consider any two vectors (a, b) and (c, d) that are not scalar multiples of each other, and look at the parallelogram determined by them:

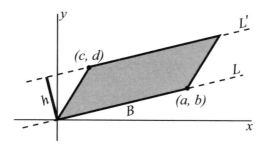

The parallelogram's area, base × altitude, is Bh in the picture. What are B and h in terms of (a, b) and (c, d)?

- B is obviously $\sqrt{a^2 + b^2}$.

- We obtain h from the method used in the last two questions. In the parallelogram, this is the distance from the origin to L'. Since L' has the same slope $\frac{b}{a}$ as L and passes through (c, d), its equation is $y - d = \frac{b}{a}(x - c)$, which can be written as $-bx + ay = ad - bc$. Up to sign, the distance of L' from the origin is therefore $\frac{ad - bc}{\sqrt{a^2 + b^2}}$.

This brings us to a *coup de grâce*: up to sign,

$$Bh = \sqrt{a^2 + b^2} \cdot \frac{ad - bc}{\sqrt{a^2 + b^2}} = ad - bc = \begin{vmatrix} a & b \\ c & d \end{vmatrix}.$$

Therefore the area of the triangle with vertices $(0, 0)$, (a, b), (c, d) is the absolute value of $\frac{1}{2} \begin{vmatrix} a & b \\ c & d \end{vmatrix}$.

By the way, in translating the triangle to place a vertex at the origin, the choice of vertex is arbitrary—the other two work just as well. In general, all three 2×2 matrices will be different, but up to sign they will have the same determinant. This provides a nice numerical check. (Can you prove that up to sign, all three determinants are in fact equal?)

Interestingly, this problem can be solved by a quite different route, using Pick's Theorem (Georg Alexander Pick 1859 - 1942). This theorem says:

> Let P be a polygon whose vertices are lattice points and whose boundary is a sequence of connected, non-intersecting line segments. Suppose that the number of lattice points in the interior of P is I, and that the number of lattice points on P's boundary is B. Then the area of P is
>
> $$I + \frac{B}{2} - 1.$$

In our triangle, there are 47 interior lattice points: $(3, 1), (4, 1), \cdots, (49, 1)$. There are four lattice points on the boundary: $\{(0, 0), (2, 1), (100, 2), (50, 1)\}$. Therefore Pick's Theorem tells us that the triangle's area is $47 + \frac{4}{2} - 1 = 48$. (Can you use Pick's Theorem to show that no lattice triangle can ever be equilateral?)

What's the Volume?

Use the information in the box at the bottom of p. 203. Proceeding as in the previous problem, translate all four vertices so that, say, $(1, 1, 1)$ ends up at the origin. The other vertices are then $(0, -2, -2)$, $(-2, 0, -2)$ and $(-2, -2, 0)$. The absolute value of the determinant of these three vectors is 16 and is the volume of the parallelepiped defined by them. As when cutting off a corner of a cube, the pyramidal corner's volume is a sixth of the parallelepiped's volume. The pyramid therefore has volume $\frac{16}{6} = \frac{8}{3}$, making the answer (d). As in the previous problem, there's nothing special about sending $(1, 1, 1)$ to the origin, and any of the four vertices can be chosen. Although the four 3×3 matrices look different, they'll have the same absolute value, providing a three-way check on the answer.

Stretched From Both Ends

The answer is (b). The rubber band is evenly stretched, both before and after pulling. This can be modeled by a straight-line function sending 1 to 0 and 10 to 12. The "1 to 0" determines what happens to the left end, the "10 to 12" says what the right end does, and "straight line" describes the even stretching of the rubber band. This is just a fancy way of asking for the equation of the line through $(1, 0)$ and $(10, 12)$. In this picture, the x-axis is the before axis and the y-axis is the after axis:

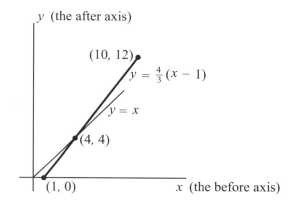

The slope of the line through $(1, 0)$ and $(10, 12)$ is $\frac{12-0}{10-1} = \frac{4}{3}$, so its equation is

$$y = \frac{4}{3}(x - 1).$$

A point ending where it started also satisfies $y = x$. Eliminate y to get $x = \frac{4}{3}(x-1)$, which has solution $x = 4$. Therefore (b) is the answer.

Which Way Do They Point?

The answer is (c). Look at points P and Q as vectors based at the origin. At time t, $P = (2, 0)$ has moved to $(2, -t)$, and $Q = (0, 1)$ has moved to $(-t, 1)$. P and Q, as basis vectors of the plane, describe by this movement a time-dependent linear transformation $X \to XA(t)$. Because $P = (2, 0)$ moves to $(2, -t)$, $P/2 = (1, 0)$ moves to $(1, -\frac{t}{2})$. The matrix of this transformation is therefore

$$A(t) = \begin{pmatrix} 1 & -\frac{t}{2} \\ -t & 1 \end{pmatrix}.$$

One eigenvalue is $\lambda = 1 - \frac{t}{\sqrt{2}}$, and its corresponding eigenvector is $\left(1, \frac{1}{\sqrt{2}}\right)$. The other eigenvalue is $\lambda = 1 + \frac{t}{\sqrt{2}}$, with corresponding eigenvector $\left(1, -\frac{1}{\sqrt{2}}\right)$. The line through an eigenvector and the origin has important significance: under the linear transformation, points on this line stay on it, sliding either toward or away from the origin. (See **eigenline**, eigenvalue, eigenvector in the Glossary.) We can call this an *eigenline*, and in this problem, computation shows that you get the same two eigenlines for every nonzero time t. The eigenline L through the origin and $\left(1, \frac{1}{\sqrt{2}}\right)$ intersects the rubber band in the point $\left(\frac{2}{1+\sqrt{2}}, \frac{\sqrt{2}}{1+\sqrt{2}}\right) \approx (0.8284, 0.5858)$. If we mark this point on the rubber band, then as t varies, we'll see that this mark always stays on L. Although there are two eigenlines, the problem says "always," so the answer is in fact (c). A little later on, the rubber band intersects the other eigenline.

Pyramid Peak

In a cube 200 ft on a side, let one face be the square base of the pyramid, and let the pyramid peak be at the center of the cube. This gives the pyramid an altitude 100 ft. Six copies of it surround the cube's center, one for each of the cube's six faces. Since the total steradian measure around any point is 4π, the solid angle at the peak is a sixth of this, making the answer (b)

Cross Product

The answer is (a). The cross product of two vectors in 3-space generalizes to a cross product of $n-1$ vectors in n-space. Here's the idea: in $\mathbb{R}^n$, write the standard basis as

$$\epsilon_1 = (1, 0, \cdots, 0)$$
$$\vdots \qquad \vdots$$
$$\epsilon_n = (0, \cdots, 0, 1)$$

and let $\mathbf{v}_i = (a_{i1}, \cdots, a_{in})$ $(i = 1, \cdots, n-1)$ be $n-1$ vectors in $\mathbb{R}^n$.

Then the cross product of the $\mathbf{v}_i$ is

$$\begin{vmatrix} \epsilon_1 & \cdots & \epsilon_n \\ a_{11} & \cdots & a_{1n} \\ \vdots & \vdots & \vdots \\ a_{(n-1)1} & \cdots & a_{(n-1)n} \end{vmatrix}$$

This determinant is in fact a vector $\mathbf{w}$ in n-space, since each term in its expansion is a scalar times some vector $_i$. The vector $\mathbf{w}$ is orthogonal to each $\mathbf{v}_i$ — that is, $\mathbf{v}_i \cdot \mathbf{w} = 0$ for $i = 1, \cdots, n-1$. To see this, write $\mathbf{v}_i$ as $(a_{i1} + \cdots + a_{in}\epsilon_n)$ and then observe that

$$\mathbf{v}_i \cdot \mathbf{w} = \begin{vmatrix} a_{i1} & \cdots & a_{in} \\ a_{11} & \cdots & a_{1n} \\ \vdots & \vdots & \vdots \\ a_{(n-1)1} & \cdots & a_{(n-1)n} \end{vmatrix}$$

In the determinant, the first and i^{th} rows are the same, so the determinant is zero.

A Special Cone

The answer is (c). The description "right circular cone tangent to each coordinate plane" implies symmetry among the three coordinates x, y and z. Tangency means that each coordinate plane intersects the cone in a line, and symmetry implies this must be a 45°-line in its plane. For example, the cone intersects the (x, y)-plane in the line $y = x$. This tells us that the point $(1, 1, 0)$ is in the cone, and this point satisfies exactly one of the equations—the one in (c).

This single point can do even more: it lets us actually derive the equation in (c). The line of symmetry running down the cone's center passes through the origin and $(1, 1, 1)$. To determine the angle θ between that line of symmetry and any line in the cone, take the dot product of $(1, 1, 1)$ with $(1, 1, 0)$. The dot-product observation in the first box on p. 203 then leads to the cone's equation.

Discriminant versus Area

Since the answer is one of the four choices, we can arrive at the right one by comparing the choices with the area πr^2 of a circle, which is a special ellipse. Write $x^2 + y^2 = r^2$ as $\frac{x^2}{r^2} + \frac{y^2}{r^2} = 1$, which means $A = C = \frac{1}{r^2}$ and $B = 0$. Then $4AC - B^2 = \frac{4}{r^4}$, and this varies as the reciprocal-squared of the circle's area. Therefore (d) is the answer.

How Much of the Heavens?

The three triangle vertices define a spherical triangle on the sphere $x^2 + y^2 + z^2 = 6$ since the vertices are all at a distance $\sqrt{6}$ from the origin. Then divide the triangle's area by the sphere's area of $4\pi r^2 = 24\pi$.

Since the triangle is relatively small, we can get close to the exact answer by replacing it with an ordinary plane triangle. It's easy to see that each triangle side is $\sqrt{2}$. Since the area of an equilateral triangle of side s is $\frac{\sqrt{3}}{4}s^2$, its area is $\frac{\sqrt{3}}{2}$. Dividing this by the sphere's area gives $\frac{\sqrt{3}}{2 \cdot 24\pi} \approx .0115$, making (a) the answer.

The spherical triangle bulges a bit, so the exact answer is a little more than 1.115%, and we can use the method outlined in the box on p. 206 to find it. Since the triangle is equilateral, it's enough to find just one angle. Set

$$\mathbf{v}_1 = (2, 1, 1), \quad \mathbf{v}_2 = (1, 2, 1), \quad \mathbf{v}_3 = (1, 1, 2).$$

Then

$$\mathbf{w}_1 = \mathbf{v}_2 \times \mathbf{v}_3 = (3, -1, -1), \quad \mathbf{w}_2 = \mathbf{v}_1 \times \mathbf{v}_3 = (1, -3, 1), \quad \mathbf{w}_3 = \mathbf{v}_1 \times \mathbf{v}_2 = (-1, -1, 3),$$

so

$$\cos \theta = \frac{\mathbf{w}_1 \cdot \mathbf{w}_2}{\|\mathbf{w}_1\| \cdot \|\mathbf{w}_2\|} = \frac{5}{11}.$$

The exact answer is then

$$\left[3 \arccos\left(\frac{5}{11}\right) - \pi \right] \div 4\pi,$$

which is approximately $.01235 = 1.235\%$. As predicted, this is slightly more than 1.115%.

Degrees in a Spherical Triangle

Every spherical triangle T has a complementary triangle T^* whose sides are the same as those of T:

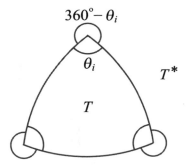

If the angles of T sum to $\theta_1 + \theta_2 + \theta_3 = 180° + \phi$, then the angles of T^* sum to

$$(360° - \theta_1) + (360° - \theta_2) + (360° - \theta_3) = 3 \cdot 360° - (\theta_1 + \theta_2 + \theta_3) =$$

$$3 \cdot 360° - (180° + \phi) = 900° - \phi.$$

As T's area approaches zero, so does ϕ. Therefore the sum $900° - \phi$ of the angles of T^* approaches $900°$, making (e) the answer.

Shearing an Ellipse

The principal axes of an ellipse in standard form lie on the coordinate axes. The box just below the problem tells us that adding the term Bxy to the conic equation $Ax^2 + Cy^2 = 1$

uniformly rotates the direction of each principal axis. Now a nonsingular linear transformation changes any conic equation in standard form to one of just this form, so B rotates the coordinate-axes a common amount to the new principal axes, meaning that they remain at right angles. Any shear keeps the ellipse's equation quadratic, so the ellipse stays an ellipse. Therefore (d) is the answer.

Geometry in a Mixed Term

We can read off the answer using the relation $\tan 2\theta = \frac{B}{A-C}$ given at the end of the first box on p. 208. For this ellipse, $A - C = \frac{1}{4} - 1 = -\frac{3}{4}$, so $\tan 2\theta = -\frac{4}{3}B$. Increasing B makes $\tan 2\theta$ go increasingly negative, so θ does, too. Therefore the rotation is clockwise, making (a) one answer.

The equation $\tan 2\theta = \frac{B}{A-C}$ can be written as $B = -\frac{3}{4}\tan 2\theta$. As B approaches $+\infty$, 2θ decreases toward the finite value $-\frac{\pi}{2}$, but not beyond that. So as B increases, the rate of rotation must slow down, making (e) the other answer.

There's something else significant happening as B increases: as the principal axes begin their clockwise rotation, the ellipse stretches out more and more, its curvature at the ends of the minor axis decreasing all the while. When B reaches 1, the ellipse has stretched out "infinitely far," the curvature has decreased to zero, and the ellipse has degenerated to the two parallel lines $y = -\frac{1}{2} \pm 1$. As B increases beyond 1, the curvature continues to decrease, and the two lines begin curving *away* from each other, becoming the two branches of a hyperbola. See [4] an instructive applet, as well as Chapter 9 in the book for more information and perspective on the geometric role played by Bxy.

How Long are the Axes?

The answer is (c), the only choice that works for an ellipse in standard form, $\frac{x^2}{a^2} + \frac{y^2}{b^2} = 1$. In that case, $A = \frac{1}{a^2}$, $B = 0$ and $C = \frac{1}{b^2}$. The answer, $4(A + C)$, is then $4(\frac{1}{a^2} + \frac{1}{b^2})$ and substituting this into the problem's quartic and simplifying gives

$$s^4 - (a^2 + b^2)s^2 + a^2b^2 = 0.$$

Applying the quadratic formula to this gives

$$s^2 = \frac{a^2 + b^2 \pm \sqrt{(a^2 - b^2)^2}}{2}.$$

The four solutions are then $s = \pm a$ and $s = \pm b$. Again, [4], Chapter 9 provides more information and perspective on this formula.

Comparing Times

The answer is (a). Looking at the diagram shows that at any time, there is less water in the P case than in the Q. In the steady-state diagram, think of the line segment $P_1 P_2$ as the hypotenuse of the small right triangle below it having vertical and horizontal legs, and do likewise for the line segment $Q_1 Q_2$. The two triangles are congruent, so the increases in water levels (the horizontal legs) in the top tank are the same in both cases. The same is true of the level decreases in the bottom tank (the vertical legs). Therefore in both cases, each tank undergoes the same level change, but with less water in the P case, water pressures are less, water flows are slower and tank levels change more slowly than in the Q-scenario. It therefore takes longer to go from P_1 to P_2.

Smaller Drain Holes

The answers are (a) and (f). Cutting each k_i in half decreases the rate at which the process progresses, which would be like decreasing gravity (diminishing the driving force), or replacing water with maple syrup (increasing viscosity). Now consider the connection mentioned toward the end of this chapter's Introduction: the size of an eigenvalue λ is a measure of how fast the experiment progresses.

In all these cases, the eigenvalues decrease. If slowing down time is the only effect, then the steady-state heights are the same and the relative rates at which they get there are also the same, suggesting on physical grounds that there's no effect on the eigenvectors. One can see this algebraically by replacing each k_i by $\frac{1}{2}k_i$ and looking at the general two-tanks solution on p. 199. The original eigenvalues 0 and $k_1 + k_2$ get replaced by 0 and $\frac{1}{2}(k_1 + k_2)$, so one eigenvalue has decreased in magnitude. The original eigenlines have slopes -1 and $\frac{k_1}{k_2}$, and these are unchanged.

A Leak to the Environment

One answer is (a): a leak in either tank means that both tank levels go to zero with time, and this can happen only if both eigenvalues are negative. This also implies (d).

Behavior of Water Heights, Version 1

From the two-tank solution displayed on p. 199, we see that in each tank the level approaches the steady state level (the first term) as a decreasing real exponential (the second term). Since any real exponential function is monotone, choices (b) and (c) are eliminated, making (a) the answer.

Behavior of Water Heights, Version 2

The answer is (b). Suppose the little hole leaking water to the environment is in the bottom tank and has constant k_3. Mimicking the solution method in this chapter's Introduction, we arrive at the quadratic equation $\lambda^2 + (k_1 + k_2 + k_3)\lambda + k_1 k_3 = 0$. Its discriminant is $(k_1 + k_2 + k_3)^2 - 4k_1 k_3$, which after expanding and collecting terms can be written as $(k_1 - k_3)^2 + k_2^2 + 2k_1 k_2 + 2k_2 k_3$. Since each k_i is positive, so is the discriminant, which tells us that the solutions to the quadratic are real. The constant term $k_1 k_3$ is the product of the two roots, and since it is positive the roots have the same sign. From the physics of the system, they can't be positive, so both are negative. (This fits in with things algebraically: in $\lambda^2 + b\lambda + c$, the sum of the roots is $-b$, which in our case is $-(k_1 + k_2 + k_3)$. This is indeed negative since each k_i is positive.) The difference of two exponential functions need not be monotone, as this plot of $y = e^{-0.5t} - e^{-t}$ from $t = 0$ to $t = 10$ shows:

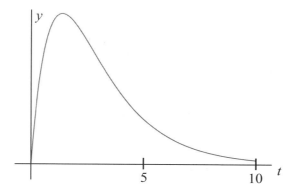

As an example of a physical situation which behaves like this, put water in the top tank, none in the bottom, and make both the holes in the bottom tank small. As the experiment begins, water from the top tank flows into the bottom tank, whose level then rises. This rise can continue for only a limited time, since that hole to the environment eventually drains all the water from the system. The result is a plot similar to the above.

Behavior of Water Heights, Version 3

The answers are (b) and (c). We can use the same approach for getting water heights as we did in this chapter's Introduction, but now there are three heights y_1, y_2, y_3. The determinant equation in λ simplifies to

$$\lambda \left(\lambda^2 + (k_1 + k_2 + k_3)\lambda + (k_2 k_3 + k_1 k_3 + k_1 k_2) \right) = 0.$$

On physical grounds, we could have predicted that one root is 0 since the system is conservative. What about the discriminant of the quadratic factor? After simplification, it becomes

$$k_1^2 + k_2^2 + k_3^2 - 2(k_2 k_3 + k_1 k_3 + k_1 k_2).$$

If we set this to zero, it defines a cone in (k_1, k_2, k_3)-space—in fact, the same cone we encountered on p. 205. The k_i are positive, so the relevant part of the cone is in the first octant of (k_1, k_2, k_3)-space. At points inside the cone, the quadratic expression

$$k_1^2 + k_2^2 + k_3^2 - 2(k_2 k_3 + k_1 k_3 + k_1 k_2)$$

is negative, while at points in the first octant outside it, it's positive. Therefore, choosing hole sizes corresponding to points outside the cone leads to real zeros of the quadratic $\lambda^2 + (k_1 + k_2 + k_3)\lambda + (k_2 k_3 + k_1 k_3 + k_1 k_2)$. On physical grounds, we see that the two roots must be negative. Initially dividing the water between the two tanks implies behavior (b) for this conservative system, eliminating (a). However, hole sizes corresponding to points *inside* the cone lead to a negative discriminant and therefore complex λ. That leads to terms of the form $e^{i\theta} = \cos\theta + i\sin\theta$ which, when starting heights are put into the general solution, produce real solutions involving sines and cosines. So in this case the water levels oscillate! The oscillation terms are multiplied by a decreasing exponential, so the amplitudes of oscillation approach zero.

Here's a geometric way to put this behavior into perspective. The intersection of the cone with the plane $x + y + z = 1$ is a circle in that plane, and intersecting the first octant with the plane defines a region enclosed by an equilateral triangle, the circle tangent to each side of this triangle. Since all the k_i are positive, the ray from the origin $(0, 0, 0)$ through (k_1, k_2, k_3) must intersect within the triangular region. Oscillation occurs when that intersection is in the disk, but not outside. In fact, oscillation is fastest at the disk's center and tapers off to zero at its circular edge. Here's the suggestive picture. The darker the gray, the faster the oscillation:

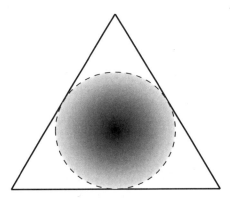

Since the circle corresponds to the quadratic factor's discriminant

$$k_1^2 + k_2^2 + k_3^2 - 2(k_2 k_3 + k_1 k_3 + k_1 k_2),$$

the discriminant serves as an oscillation threshold, similar to the way the discriminant of $M\lambda^2 + D\lambda + k$ does for the mass-spring system having equation $M\ddot{y} + D\dot{y} + k = 0$ with mass M, damping D and spring constant k. When D is small, the discriminant is negative and leads to $e^{i\alpha t} = \cos\alpha t + i\sin\alpha t$ and oscillation. A zero discriminant implies critical damping, and a positive discriminant means overdamping. See [5] for more on oscillations in three tanks.

Ellipses and Hyperbolas

We can use the double-angle formula for $\tan 2\theta$ to convert $\frac{B}{A-C}$ to an expression in $m = \tan\theta$. The double-angle formula is

$$\tan 2\theta = \frac{2\tan\theta}{1 - \tan^2\theta} = \frac{2m}{1 - m^2},$$

and setting this equal to $\frac{B}{A-C}$ gives $B(1 - m^2) = 2m(A - C)$, which becomes

$$Bm^2 + 2(A - C)m - B = 0.$$

Therefore (e) is the answer.

Interpolation

Compare their values at the point π. Now $\sin\pi = 0$, while $p(\pi) = \pi - \frac{\pi^3}{6} + \frac{\pi^5}{120} > 0$, making the answer (b). The approximation strategies are different in that the power series asks for higher and higher order of goodness of fit at some one point, while the interpolation polynomial simply passes through the specified points, with no attention given to such things as ensuring that slopes agree. So its total goodness of fit is spread over several points, with less required at each point. Yet more spread out are Fourier approximations, where the fit is over an entire interval $[a, b]$. It can happen that such an approximation $g(x)$ agrees *nowhere* with the original function $f(x)$, and at the same time the approximation can have a very small total deviation from $f(x)$ over the interval, measured by $\int_a^b [g(x) - f(x)]^2 dx$.

Adding Equal Weights

The answer is (b). One surprise here is that the top and bottom halves of the string of weights act like mirror images of each other. To see why, assume that the weights are cubes without boxes, and redraw the picture as on the next page, separating top and bottom halves by cutting the middle cube.

With the middle cube cut in half, we can create the downward stretching by starting with no gravity and gradually increasing it. Weight 1 then travels down a certain distance. If the spring just below weight 1 were completely rigid, weight 2 would descend the same distance. But that spring *isn't* rigid, so weight 2 descends a little farther than that. Therefore weight 2 travels down farther than weight 1.

What about the bottom half? Since all springs are identical, Hooke's law tells us that the spring attached to the floor is pushed down by weights 3 and 4 just as hard as the ceiling spring is pulled down by weights 1 and 2. Since weight 3 is riding on a spring, it descends more than weight 4, and by the same additional distance that weight 2 did.

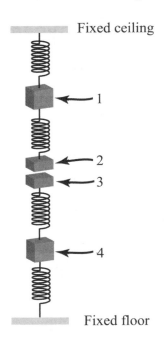

Whether the half-cubes are separated by an inch, a thousandth of an inch, or 0 inches, they each do their own thing, with weights 2 and 3 dropping the same distance, which is greater than for weights 1 or 4. Conclusion: the middle weight, thought of as weights 2 and 3 with zero separation, moves down the most, making (b) the answer.

Notice that assuming only that weights 2 and 3 are equal and positive still implies that they move down the farthest. Also, a little thought shows that all these arguments extend to a string of any number of equal weights. In fact, for an even number of weights, there's nothing to cut in half. To summarize, if the number of weights is odd, the middle one descends the most, and if the number is even, the middle *two* descend by the same amount, which is more than any of the others.

This solution is only qualitative. Although it tells us that the middle weight moves down the most, it doesn't tell us how far that or any other weight moves. And if not all the weights are equal, then our method tells us almost nothing! But the problem itself is important, almost crying out for some method able to solve the general problem. There *is* such a method, and it's based on the least energy principle saying that in equilibrium, the masses are located so that the total energy is a minimum. In our system, energy arises from two sources: the potential energy of the masses coming from their height (think mgh); and the potential energy stored in the springs by their being stretched or compressed. Certainly the mgh energy goes down when the masses drop. But as they drop, the springs get stretched or compressed, making *their* energy rise. In this mixture of energies, there is some configuration in which the total energy is the smallest, and we use this general approach to obtain a quantitative answer. Though we carry it out for our problem of three equal masses and equal springs, the setup suggests how to proceed in general, with n unequal masses and unequal springs.

Linear Algebra Answers

It's easy to write down the total energy. Look at this picture:

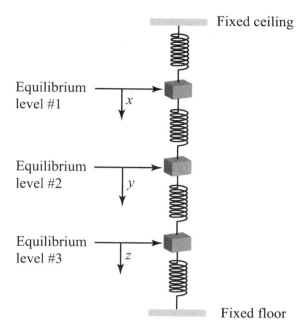

Assume all weights are 1, all spring constants are 1, and that the gravitational strength g is 1. Then stretching or compressing a spring a distance D requires $\int_{s=0}^{D} 1 \cdot s \, ds = \frac{1}{2} D^2$ amount of work. The spring then has that energy stored in it. In the picture, x, y and z are the distances the springs move relative to a fixed background. Therefore from top to bottom, the changes in the four spring lengths are x, $(y - x)$, $(z - y)$ and z. The total energy stored in all four springs is thus

$$\frac{1}{2}x^2 + \frac{1}{2}(y - x)^2 + \frac{1}{2}(z - y)^2 + \frac{1}{2}z^2.$$

Since each weight mg is 1, the picture shows that the total decrease in potential energy due to the masses dropping is $x + y + z$, giving

$$E = \frac{1}{2}x^2 + \frac{1}{2}(y - x)^2 + \frac{1}{2}(z - y)^2 + \frac{1}{2}z^2 - x - y - z$$

as the total energy $E(x, y, z)$ of the system. To find where this attains a minimum, set

$$E_x = E_y = E_z = 0.$$

Performing these differentiations results leads to the system

$$2x - y + 0 = 1$$
$$-x + 2y - z = 1$$
$$0 - y + 2z = 1.$$

The solution is
$$x = \frac{3}{2}, \quad y = 2, \quad z = \frac{3}{2}.$$
This corroborates what we saw earlier: the middle weight moves down the most, and the top and bottom weights move down equally far. But now we know just how far each one moves.

Glossary

angular momentum: (See **momentum**.)

Archimedes' principle: A body partly or completely immersed in a fluid is buoyed up by a force equal to the weight of the displaced fluid.

autumnal equinox: (See **equinox**.)

base (counting): In positional notation, any real number is expressible as a linear combination, possibly infinite, of integral powers of an integer base $B > 1$ with coefficients chosen from $\{0, \cdots, B-1\}$. Among finite linear combinations, the representation is unique. For infinite linear combinations, the representation is unique provided we omit redundant representations ending in an infinite string of $(B-1)$s.

bell curve: A hypothetical data spread that is approximated by the graph of outcomes of a **binomial experiment** with very large n. The curve roughly describes many common statistical distributions such as heights or weights of U.S. adult males or females, SAT scores, the weights of a large number of pennies, and so on. The x-value of a bell curve's peak is the data set's average value μ. Here is the curve's equation:

$$y = \frac{1}{\sigma\sqrt{2\pi}} e^{-\frac{(x-\mu)^2}{2\sigma^2}}.$$

The distance from μ to the x-value of either inflection point is the data set's **standard deviation** σ. Here's the curve, showing a possible location of μ. σ is the distance from there to an inflection point:

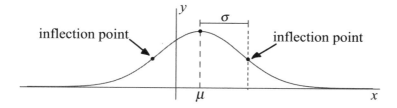

If σ is small, as for yardsticks off an assembly line, the curve will be compressed horizontally, resulting in a sharp peak. But if σ is large, as for lengths of hardwood flooring, the graph is stretched out, making for a gentle hill. See the box on p. 100.

binary: Base two.

binomial experiment: An experiment with a fixed number n of independent trials, exactly two possible outcomes for each trial, and the same probability of outcome in each trial. "Independent" means that the outcome of any trial does not affect the outcome of any other trial. Some examples of binomial experiments are:

- Flipping a fair coin 50 times. The possible outcomes for each trial are heads and tails.

- Flipping an unfair coin 100 times. The possible outcomes for each trial are heads and tails.

- Randomly choosing, one at a time, 15 ping-pong balls from a basket containing 25 red and 35 blue ping pong balls, thoroughly stirred. After each draw, the drawn ping-pong ball is returned to the basket and the contents restirred. The possible outcomes for each trial are red and blue.

- Rolling a die 25 times, each time recording the face value as even or odd.

Here are two experiments that are *not* binomial:

- Randomly choosing, one at a time, 15 ping-pong balls from a basket containing 25 red and 35 blue ping-pong balls (thoroughly stirred). Once out of the basket, a ball stays out of the basket. This is not a binomial experiment because the probability of drawing a red ball changes from one draw to the next.

- Flipping a penny 101 times and, beginning with the second flip, recording whether or not the outcome is the same as that of the previous flip. The outcomes for each trial are "same" and "different." This isn't a binomial experiment because the outcomes are not independent.

Brownian motion: The jittery motion of microscopic particles due to molecules of surrounding fluid bombarding them. Examples are seen in smoke particles suspended in air or tiny dust or pollen particles floating on water. It was in 1827, about when Ohm published Ohm's law, that the botanist John Brown was doing what botanists typically do: inspecting parts of plants. Under a microscope he saw pollen grains floating on a drop of water, shivering and wandering around. Were they alive, wiggling like spermatozoa? To check, he repeated the experiment with microscopic bits of dust. They, too, jittered. The perplexed botanist never could figure out what was going on.

Astonishingly, in those days—just over a century ago—scientists were *still* debating a 2000-year-old-question: are atoms and molecules real, or are they just a useful crutch for the imagination inherited from ancient Greek philosophers? In 1905, Einstein turned his mind to the problem. It seemed to him that Brownian motion offered concrete evidence that atoms and molecules were actual. If a molecule was moving fast enough when it hits a grain of pollen, the grain could lurch. (Think of a fast-moving cannonball crashing into an automobile.) Since the grain is surrounded by millions of such cannonballs, the hapless pollen grain would continually lurch this way and that, just what one sees under a microscope.

What set Einstein apart from so many others is that he included *testable predictions* with his theories. Accompanying his explanation of Brownian motion, he showed that one could make a number of fundamental calculations: the number of molecules hitting a single pollen grain per unit time, how fast the water molecules were moving, Avogadro's number, and even the size of various molecules. Later, the physicist Jean Perrin confirmed Einstein's predictions, and in so doing contributed to earning his 1926 Nobel prize in physics. In one stroke, Einstein offered testable evidence that the kinetic theory of fluids was correct and that Brown's phenomenon is the result of what molecules do when they're heated up to a liquid or gaseous state: they gain speed and therefore momentum. As a bit of perspective, the average speed of a water molecule in air is over 1400 mph, nearly double the speed of sound. It is this relatively great momentum that succeeds in making the small particles jitter.

capacitor: A storage device for electrical energy generally, it consists of a pair of electrically conducting plates that can be oppositely charged. To keep opposite charges apart and thus maintain the stored energy, the plates are separated by an electric insulator such as a vacuum, air, glass, plastic or a ceramic.

central limit theorem: This theorem says three things. First, it tells us that even if the set of individual measurements from a large population doesn't form a bell curve, the set of n-at-a-time measurements does, if n is fairly large. Second, it says that if the original set of measurements has mean μ, then so does the set of n-at-a-time measurements. Third, if the original curve has standard deviation σ, then the n-at-a-time curve has standard deviation $\frac{\sigma}{n}$.

Here's an example. Suppose we measure the heights of several thousand adults. This will probably form a curve with two peaks rather than one, since on average men are taller than women. Instead of using individual heights of adults, randomly select many subgroups of, say, 100 of those adults and find the average height of each group. If we use these averages as our set of measurements, then since there will usually be roughly equal numbers of men and women in a group of 100, that will tend to average out the height difference between men and women, resulting in a curve with just one peak. In addition, the averaged points will tend to be bunched closer together than individual heights because the heights of very tall and very short people—people increasing the standard deviation—are averaged, and doing that blunts the extremes. In this example, the central limit theorem says that the standard deviation of the new curve is just a tenth of the original. This is true if we average 100-at-a-time heights of adult females, or the weights of boys in the third grade, or weights of new pennies, or lengths of old yardsticks or U.S. incomes—this is very general.

coefficient vector: The linear function $a_1 x_1 + \cdots + a_n x_n$ can be factored as the dot product $(a_1, \cdots, a_n) \cdot (x_1, \cdots, x_n)$, and $(a_1, \cdots, a_n)$ is called the *coefficient vector*. This is handy because for the hyperplane $a_1 x_1 + \cdots + a_n x_n = c$, the coefficient vector

$a = (a_1, \cdots, a_n)$ and c together say what the hyperplane looks like: it's perpendicular to a and at a signed distance $c\|a\|$ from the origin. "Signed" means that when c is positive, the plane cuts the ray extending a, and if c is negative, it lies on the other side of a.

cone: A cone is a union of lines through a common point V, called the cone's *vertex*. Often the lines are in $\mathbb{R}^3$, but since line is a vector space concept, the definition works in any vector space. The notion of *ray* makes sense in any $\mathbb{R}^n$ ($n > 0$), and sometimes *cone* is meant to be a union of rays all starting from V. (Think of an ice cream cone.) A cone defined by rays rather than lines is a *nappe* of the corresponding full cone. The cone defined using the negatives of those rays is the other nappe.

conservation of angular momentum: See **conservation of momentum**.

conservation of energy: The law of conservation of energy says that the total amount of energy in an isolated system never varies. To illustrate, assume a fixed coordinate system and imagine a box surrounding the entire solar system. No mass, light, heat or anything else is permitted to enter or leave the box, making what's inside this box an isolated system. Within it, a multitude of energy-exchanges constantly take place: the sun, through $E = mc^2$, is continually converting mass to radiant energy that eventually bounces off the walls of our box. Plants on earth convert some of the radiant energy to chemical energy, a form of potential energy. Animals use plants as fuel for their bodies, sometimes converting to kinetic energy, sometimes to mechanical potential energy if they climb a hill, and into radiant energy by creating body warmth. Asteroid collisions, the receding of the moon from the earth, a car skidding to a stop, wind, rain, an eagle's screech, the hot midday sun heating up a brick building—in everything happening within this box-universe, energy is moved about or changed from one form into another. But do the bookkeeping, and the totals within the box remain precisely constant.

conservation of momentum: In this book we use **linear momentum** and **angular momentum** in the classical sense, and both are conserved. The total (ordinary, straight-line or linear) momentum, as well as angular momentum remains constant in any force-isolated system—that is, one in which no external forces act on the system. Since measuring the velocity of a mass depends on which coordinate system (or moving frame) we're in, we assume throughout that we are in a non-accelerating coordinate system. (See **momentum**.) Both momentum and angular momentum are vector quantities, and if we add the momentum vectors for every particle in the force-isolated system we get one grand, total momentum vector. Conservation of momentum says that the single total momentum vector remains unchanged through time. All the collisions, reflections, smashes, twists, turns and bumps can change countless trillions of vector summands within any grand total, but the grand totals themselves never vary.

convex: A set S in $\mathbb{R}^n$ is called *convex* if the line segment connecting any two points in S lies entirely in S. $\mathbb{R}^n$ is itself convex, so any subset T in $\mathbb{R}^n$ is contained in at least one convex set. The definition implies that the intersection of convex sets is convex, so the intersection $H(T)$ of all convex sets containing T is convex. $H(T)$ is called the *convex hull* of T.

cross product: The *cross product* **v**×**w** of two vectors **v** and **w** in euclidian 3-space is a

Glossary

certain vector perpendicular to each of **v** and **w**. It There are two equivalent definitions of **v**×**w**, one geometric and one algebraic:

- **geometric definition.** Assume **v** and **w** emanate from the origin, let P be the plane containing **v** and **w**, and let L be the line perpendicular to P passing through the origin. Take a standard slotted screw (one that goes into the wood when you turn the screwdriver clockwise), and align it along L so that its slot is parallel to **v**. When you turn the screw through an angle ($\leq 180°$) so that the slot becomes parallel to **w**, the screw advances. That advance is the direction of **v**×**w**. Its length is equal to the unsigned area of the parallelogram in P having **v** and **w** as two of its sides. In aligning the screw with L, we have not said which way the screw points. For example, if L is vertical, does the head go on top, or at the bottom? A bit of contemplation shows that it doesn't matter—the screw will advance the same way whichever choice you make.

- **algebraic definition.** In the usual right-hand rectangular coordinate system, let

$$\mathbf{i} = (1,0,0), \quad \mathbf{j} = (0,1,0), \quad \mathbf{k} = (0,0,1).$$

("Right-hand" means that using the above screw, turning **i** into **j** makes the screw advance in the direction of **k**). If $\mathbf{v} = (v_1, v_2, v_3)$ and $\mathbf{w} = (w_1, w_2, w_3)$ then **v**×**w** is the determinant

$$\begin{vmatrix} \mathbf{i} & \mathbf{j} & \mathbf{k} \\ v_1 & v_2 & v_3 \\ w_1 & w_2 & w_3 \end{vmatrix}.$$

The order of **v** and **w** matters. The geometric and algebraic definitions each imply that **v**×**w** = −**w**×**v**.

discriminant: In this book, *discriminant* refers to discriminant of a quadratic polynomial in two variables. An ellipse, parabola or hyperbola can be defined by a real polynomial equation of the form $Ax^2 + Bxy + Cy^2 + Dx + Ey + F = 0$, and the *discriminant* of this polynomial is $B^2 - 4AC$. This discriminant is well named, for it does indeed discriminate among these three kinds of nondegenerate conics: if $B^2 - 4AC < 0$, then equation determines an ellipse; if $B^2 - 4AC = 0$, the equation describes a parabola; and if $B^2 - 4AC > 0$, the equation defines a hyperbola. Note that D, E and F don't enter into the discriminant; D, E and F can translate and/or magnify a conic, but they never unsymmetrically stretch it or change its type. See [4, pp. 149-150] for more on this.

eigenline, eigenvalue, eigenvector: In linear algebra, the eigenvalue problem is this: given a square matrix A, find nonzero vectors **v** such that **v**A is a scalar multiple of **v**. The factor λ by which **v** is multiplied is called an *eigenvalue* of the matrix A, **v** is called an *eigenvector* of A corresponding to this eigenvalue, and we call the set of all scalar multiples of any eigenvector an *eigenline* of A. If **v** is an eigenvector, then **v**$A = \lambda$**v**, and, importantly, λ**v** can be written as λI, where I is the identity matrix. This means we can write **v**$(A - \lambda I) = \mathbf{0}$, where $\mathbf{0}$ is the zero vector. This has a nontrivial solution only when $\det(A - \lambda I)$ is zero, giving the familiar polynomial equation in λ for finding eigenvalues.

equinox: Either one of the two times of the year when the length of day equals the length of night, in the sense that the sun spends as much time above the horizon (day) as it does below (night). They occur around March 20 and September 23. In the northern hemisphere, the equinox close to March 20 is the vernal (spring) equinox. That same day in the southern hemisphere is the autumnal equinox. Likewise, in the northern hemisphere, September 23 is the autumnal equinox and in the southern, the vernal. See the box on p. 63.

Euclid's postulates: In modern form, they are:

- I A straight line segment can be drawn joining any two distinct points.

- II Any straight line segment can be extended indefinitely in a straight line.

- III Given any straight line segment, a circle can be drawn having either endpoint as the circle's center, and the segment as the circle's radius.

- IV All right angles are congruent.

- V If two lines are drawn that intersect a third so that the sum of the inner angles on one side is less than two right angles, then the two lines intersect each other on that side when extended sufficiently far.

force field: A curve, surface, or all or part of three-space together with a force vector at each point. Force fields arise from masses creating gravity, charges creating electrostatic forces, moving charges creating magnetic forces, and so on.

Foucault's pendulum: An **ideal pendulum** swinging in a plane and suspended from a frictionless gimbal. A circle of blocks or bowling pins is often added to the system, though this is not part of a Foucault pendulum. One can think of the suspension as a frictionless ball at the top end of the pendulum's cable, the cable in turn passing through a hole in a plate on which the ball rests. The frictionless arrangement decouples the pendulum from the earth's rotation, and the **symmetry principle** informs us about the two basic points of view: an observer in the fixed plane of swing sees the earth rotate, and therefore sees a rotating circle of bowling pins, each pin getting knocked down as it moves into the plane of the swinging pendulum. Symmetrically, an observer on earth sees a fixed earth and a slowly rotating plane of swinging. As this plane slowly rotates, the pendulum in it knocks down the pins standing in a fixed circle.

The pendulum cannot be completely isolated from the earth since there's always some friction at the point of suspension, as well as with the surrounding air. There are several ways to minimize or counteract friction, including using a heavy, dense ball, a long connecting cable and a ring of timed magnetic boosters positioned either a little below the suspension point or hidden under the floor to maintain the swing's amplitude.

Fourier solutions: The Fourier solution of the cooling rod problem provides a nice illustration of the general method. Assume that the rod is a uniform conductor, say a round copper rod of diameter $\frac{1}{4}$ in covered with insulation except at its end points, which are kept at 0°C. The rod starts off with some initial temperature $T_0(x)$ at each of its points x. To keep everything simple, we assume T varies continuously throughout the length of the rod.

It turns out that a cluster of remarkable and beautiful properties fit together to solve this problem. The function $T_0(x)$ is a possibly infinite sum of pure sine functions that are zero at the endpoints. Let's assume that the rod stretches along the x-axis from 0 to π. Then each function has the form $A_n \sin nx$, where n is an integer, so this says that $T_0(x) = \sum_{n=1}^{\infty} A_n \sin nx$. One can regard the $\sin nx$ as the basic building blocks which are then combined by an infinite linear combination. The pieces $A_n \sin nx$ have two properties, each remarkable in its own right.

First property. As time goes on, each temperature distribution $A_n \sin nx$ keeps looking like that sine shape, but its amplitude A_n approaches zero. Pictorially, think of a large rubbery piece of the (x, y)-plane, and then on this stretched sheet plot one of the sine curves. As time elapses, let the plane gradually shrink toward the x-axis. The plot stays sine-shaped, but approaches the x-axis.

Second property. Each point of the plot approaches the x-axis as a decreasing exponential $T = T_o \, e^{-Kt}$, the very sort of function solving the zero-dimensional problem. Furthermore, the same K works for every point in the rod. It's as if you put a row of one thousand lilliputian cups of coffee out in the freezing cold at 0° C. Their initial temperatures, as you scan from left to right, follow the profile of one of our sine curves, with the endmost cups of coffee nearly freezing. Though each cup independently cools, at any instant the thermometers always plot out the initial sine temperature curve, but with its amplitude diminishing toward zero as time goes on.

So for any such sinusoidal temperature curve, each point of the rod acts as if it is cooling down completely unaware of its neighbors, though of course because it's within the rod, the point is constantly exchanging heat with its neighbors. Remarkably, each term $A_n \sin nx$ of $T_o(x)$ cools down pointwise this way: $e^{-K_n t} (A_n \sin nx)$.

What is K_n, the constant for an n-arch sine curve? Look at these two building-block temperature curves, a one-arch and a two-arch sine curve:

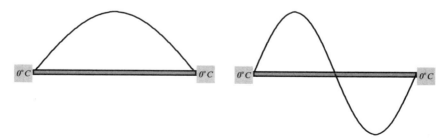

We take K_1 to be the rod's K, and suppose that this K_1—the constant for the curve on the left—is 1. Is K_2 also 1 for the curve on the right? Let's argue this way: the two-arch curve crosses the x-axis at the midway point, and the only way the curve changes in time is that its amplitude approaches zero. So it always crosses the x-axis at the midpoint, meaning the temperature is always zero there. Effectively, there's a chunk of ice at the midpoint, so any point in the right picture is more closely surrounded by ice than the corresponding point in the left picture. Result? Any point in the right picture feels that 0 °C more: warmer temperatures drop faster, meaning that heat flows away from it faster and cold temperatures rise more quickly, with heat flowing to these points from the points staying at 0°C. Since the K-constants measure how fast heat flows, it follows that $K_2 > K_1$. It turns out that $K_1 = 1^2$ and $K_2 = 2^2$. In general, $K_n = n^2$, so we now know that the nth building block $A_n \sin nx$ approaches 0°C pointwise according to $e^{-n^2 t}(A_n \sin nx)$.

And now there comes yet another remarkable fact: we are guaranteed not only that there exist sine functions $A_n \sin nx$ summing up to the initial temperature profile. Additionally, *each one of these basic sine-temperature functions does its own thing, entirely independent of what any other sine-temperature function $A_m \sin mx$ may be doing.*

These observations supply enough information to solve the problem, for we can track what each $A_n \sin nx$ does as time progresses. To get the temperature distribution $T(x,t)$ throughout the rod at any time t, just add up the components to get the solution,

$$T(x,t) = \sum_{n=1}^{\infty} A_n e^{-n^2 t} \sin nx.$$

Up to this point we've assumed that the physical properties of the rod yield a $K = K_1$ of 1. What about a real-world rod, one made, say, of copper, iron or glass? There are three physical constants of the material that affect how fast the cooling takes place. They are thermal conductivity κ, density δ and specific heat σ. Most metals transmit and absorb heat easily, and a high κ is one reason why if you step on a metal grating at room temperature with your bare feet, it feels cold—the metal, at about 70°, quickly absorbs heat from your feet, at about 98°. And there's a lot of material per unit volume (a high δ). However, your discomfort doesn't last long, since metals usually have a lower capacity to hold heat (a smaller σ), and a little of your body heat will warm them up quickly. Silicates like sand and glass are different, having lower conductivity and higher specific heats. (Brick buildings can radiate heat stored during a hot day well into the evening.) The three constants combine to give an overall cooling rate: $C = \frac{\kappa}{\delta \sigma}$. This constant appears in the exponential part of the solution and controls the rate at which the experiment proceeds. For a rod made of material of constant C, we then have $K_n = Cn^2$, and general solution

$$T(x,t) = \sum_{n=1}^{\infty} A_n e^{-n^2 Ct} \sin nx.$$

The constant C is larger for metals and smaller for glass. In a "cooling-race" between a copper rod and a glass rod having identical length, diameter, initial temperature of 100°C, and with ice kept at the ends, the copper rod easily wins, cooling much faster. Movies of their temperature profiles through time would be the same, up to speed, with the copper-rod movie running faster. At the end of the day, most of the melted water is lying at the ends of the glass rod, not the metal one! At 100°C, the glass rod contained considerably more heat than the metal rod.

We have said nothing about actually computing the coefficients A_n. For our rod of length π and initial temperature distribution $f(x)$, the formula is

$$A_n = \frac{2}{\pi} \int_{x=0}^{\pi} f(x) \sin nx \, dx.$$

As a bit of intuition, for large n and continuous f, any single arch of $\sin nx$ above the x-axis is followed by one below the x-axis, and because n is large, $f(x)$ doesn't vary much, meaning our integral of $f(x) \sin nx$ can't build up in size. For any arbitrarily small $\epsilon > 0$, $0 < |A_n| < \epsilon$ for all n sufficiently large.

Heat and wave phenomena share a kinship through the equations governing them. As with heat, we can get the gist of things by looking at a special one-dimensional case, transverse oscillations of a string stretching on the x-axis from 0 to π. On a horizontal string, vertical displacements play the role of temperature. The string is fixed at each end, so there is no displacement there. That's analogous to keeping the ends of the heat-conducting rod at 0°C.

The building block sine functions are the same, too. Instead of the sum $\sum_{n=1}^{\infty} A_n \sin nx$ representing the initial temperature distribution on the rod, it represents initial transverse displacement of the string. Instead of lining up a thousand cups of coffee in the cold, suspend from a ceiling a line of a thousand separate tiny mass-spring oscillators. Just as we assumed a uniform constant K for the cooling cups, we assume a uniform Hooke constant k for the springs and equal, small masses. And just as the temperatures of the thousand cups formed a sine curve zero at the ends, we push up or pull down on each tiny mass so that the resulting transversely-displaced masses trace out a sine curve. Taking the cups of coffee out into the cold corresponds to releasing the thousand masses from rest. They all begin heading toward zero displacement, just as all the cups of coffee head for zero temperature.

Each of the cups of coffee in the cold cooled, unaware and unconcerned about its neighbors. Yet the overall sine curve simply decreased in amplitude, modeling an actual rod in which points constantly exchange heat with its neighbors. Similarly, each tiny mass-spring vibrates separately, unaware of its neighbors, yet the overall sine curve begins by just decreasing its amplitude, modeling an actual string in which points constantly exchange forces with its neighbors. There's a difference, however: a string vibrating with no loss of energy will vibrate forever with undiminished amplitude and does not head to a steady-state of zero. Superficially, this may seem quite different from exponential decrease, but the two behaviors are mathematically close. To see the connection, compare

e^{-1kt} with e^{-ikt}. With the change from 1 to i, we go from decreasing exponential (the real part of e^{-1kt}) to oscillation (the real part of $e^{-ikt} = \cos kt - i \sin kt$).

In the cooling rod, we saw that each term $A_n \sin nx$ of $T_0(x)$ varies through time as $e^{-K_n t}(A_n \sin nx)$, for some K_n. The string analogue faithfully replaces exponential decrease by oscillation: each term $A_n \sin nx$ of the initial displacement function $Y_0(x)$ varies through time as $\cos k_n t \, (A_n \sin nx)$ for some k_n. For the cooling rod, we argued physically that when n is larger, cooling progresses faster since K_n grows quadratically as $K_n = n^2 K_1$. For the vibrating string, when n is larger, vibration likewise progresses faster, but here k_n increases linearly as $k_n = nK_1$. We can argue physically that in the picture of two sine curves, the one on the right can be produced on a violin by lightly touching the string at its midpoint as you bow the string. That creates a sound one octave higher, or double the frequency. Touching the string at a point one-third of the way along its length creates a sound having triple the frequency of the whole length. For an ideal string having no physical thickness, touching it at its $\frac{1}{n}$-point produces a frequency n times the whole-string frequency. (The whole-string frequency is itself called the *fundamental*.)

As for heat, we are guaranteed not only that there exist sine functions $A_n \sin nx$ summing up to the initial wave displacement function, but also that *each one of these basic sine-displacement functions does its own thing, independent of what the others do.*

As in the case of heat, we can solve the wave problem by tracking what each $A_n \sin nx$ does as time progresses. To get the displacement distribution $Y(x, t)$, add the components to get

$$Y(x,t) = \sum_{n=1}^{\infty} A_n \cos nt \cdot \sin nx.$$

We have so far assumed that the physical properties of our string all combine so that $C = 1$, meaning $\cos Cnt$ is $\cos nt$, as above. But what about a real-world string like a steel piano string? For one thing, the string's length certainly affects how fast its fundamental mode vibrates. Continue to assume the string stretches on the x-axis from 0 to π. There are now just two other physical constants of the string that affect the string's fundamental frequency. They are the string's tension τ and its linear density δ (mass per unit length). Everyone knows that by tightening up a string on a piano, violin or guitar, the pitch goes up. And for a given length and tension, a lighter string vibrates faster than a heavy one, as can be seen by comparing the thin, treble strings on a piano with its fat, heavy bass strings. The two constants combine to give an overall rate at which vibration proceeds: $C = \sqrt{\frac{\tau}{\delta}}$. Altogether, we have

$$Y(x,t) = \sum_{n=1}^{\infty} A_n \cos Cnt \cdot \sin nx.$$

Assuming that the string starts its motion from rest, the formula for the coefficients A_n

applies here, too.

For more on Fourier series, see [1], [8] or [9].

Galileo's law of projectiles: In a uniform gravitational field, a projectile launched non-vertically and traveling without friction follows a parabolic path.

Gauss curve: See **bell curve**.

Gauss elimination: A stepwise procedure for solving a system of linear equations

$$a_{i1}x_1 + \cdots + a_{in}x_n = c_i \quad (i = 1, \cdots, r \leq n).$$

There are two halves to the process: forward and backward. The forward half customarily goes from top to bottom, and then the backward half goes from bottom to top.

- **Forward.** If $a_{11} = 0$, then re-index the variables so that $a_{11} \neq 0$. For $i = 2, \cdots, r$ multiply equation i by $-\frac{a_{11}}{a_{i1}}$ whenever $a_{i1} \neq 0$, making the coefficient of x_1 equal to $-a_{11}$ in each of these $r - 1$ equations; then add the first equation to each of them. After doing that, only the first equation has an x_{11} in it. The matrix of coefficients and right-hand constants has been changed so that it now has the shape

$$\begin{pmatrix} a_{11} & a_{12} & a_{13} & \cdots & a_{1n} & c_1 \\ 0 & a_{22} & a_{23} & \cdots & a_{2n} & c_2 \\ 0 & a_{32} & a_{33} & \cdots & a_{3n} & c_3 \\ \vdots & \vdots & \vdots & \vdots & \vdots & \vdots \\ 0 & a_{r2} & a_{r3} & \cdots & a_{rn} & c_r \end{pmatrix},$$

with zeros as indicated in the first column. For simplicity, we've continued to write the other matrix entries as simply a_{ij}.

Repeat with a_{22}. That is, strip off the top row and the left column, and do the same on the smaller matrix. Continue until we can't repeat any more. When we have gone as far as possible, the matrix will have this staircase form:

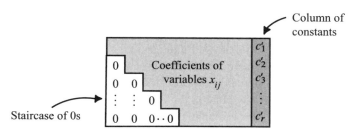

The corresponding r equations have this same staircase form, too.

- **Backward.** Starting with the last equation of the staircase, solve for any one of the $n - r + 1$ variables $x_r, \cdots, x_n$ in terms of the other $n - r$ variables—say, solve for x_r. Substitute this (which is some linear combination of $x_{r+1}, \cdots, x_n$) into the equation above it, then solve for x_{r-1}. The staircase form makes this easy. Repeat until you get to the top equation. At the end, each of $x_1, \cdots, x_r$ is expressed in terms of $x_{r+1}, \cdots, x_n$.

This illustrates a fundamental process in mathematics: going from the *implicit* to the *explicit*. For example when $r = 2$ and $n = 3$, the above equations can represent two planes in $\mathbb{R}^3$ intersecting in a line. The line is defined implicitly by the two equations, and it is Gauss elimination that leads us from that form to an explicit representation of the line. By setting the left-over variable x_3 equal to a parameter, say t, each point (x_1, x_2, x_3) of the line gets an explicit representation

$$x_1 = A_1 t + B_1,$$

$$x_2 = A_2 t + B_2,$$

$$x_3 = t.$$

Another example is $r = 1$ and $n = 3$. But even here the spirit is the same. Assume the equation defines a plane in $\mathbb{R}^3$. It has dimension two, and correspondingly, there are two left-over variables that can be made into parameters, say s and t. One then has

$$x_1 = A_1 s + B_1 t + C_1,$$

$$x_2 = s,$$

$$x_3 = t.$$

A less trivial example is $r = 7$ and $n = 10$. The seven equations can define a three-space in $\mathbb{R}^{10}$, and each component x_j of a point $(x_1, \cdots, x_{10})$ gets represented explicitly as $x_j = A_j s + B_j t + C_j u + D_j$, $(j = 1, \cdots, 10)$.

What about when $n = r$? Assuming the determinant of the matrix is nonzero, Gauss elimination solves n equations for n variables, so it makes explicit the solution that was originally implicit. (In $\mathbb{R}^3$, the equations represent three planes intersecting in a point.) The number of left-over variables is zero, the number of parameters is zero, the dimension of the solution space (the point) is zero, and in analogy to the above $x_j = A_j s + B_j t + C_j u + D_j$, we have simply $x_j = A_j$, a constant. But this case belongs to the same club and has precisely the same spirit as its more general cousins.

This spirit turns out to be a veritable gold mine of riches and can be generalized almost beyond recognition. For one, there are nonlinear analogues. As an illustration, a function $F(x_1, x_2, x_3) = 0$ can define a surface in $\mathbb{R}^3$, and under appropriate conditions, throughout a neighborhood of any point in this surface, the surface locally has an explicit representation $x_j = f_j(s, t)$. For example, the sphere $x^2 + y^2 + z^2 - 2z = 0$ is tangent to the (x, y)-plane at the origin, and the lower half of the sphere near the origin can be explicitly represented using the parameterization

$$x = s,$$

$$y = t,$$

$$z = \sqrt{1 - s^2 + t^2} + 1.$$

genus: The only topological surfaces appearing in this book are those of connected objects made from modeling clay, and these turn out to be connected compact orientable surfaces without boundary. Topological transformations of them correspond to bicontinuous transformations (see **topology**), and it happens that any of these surfaces is topologically equivalent to the surface of a ball with a finite number of attached handles, each looking like a teacup handle. The number of handles is the *genus* of the surface. Not only is this a topological invariant, it's a *classifier* within this world of connected compact orientable surfaces. That is, any two such surfaces of the same genus are topologically equivalent to each other. The world of topological objects is vastly larger than modeling-clay surfaces, however, and decades of effort have gone into extending the notion of genus to other topological spaces and of finding numbers and other objects that do for such spaces what the genus does for modeling-clay surfaces: completely classify them.

For our modeling-clay surfaces, there are several equivalent definitions of genus. For example, the genus is also the greatest number of disjoint circles that can be drawn on the surface such that after the loops are removed, what's left remains in one piece, not falling into two or more pieces. Another equivalent definition: draw edges on the clay surface to make it a topological polyhedron with F faces, E edges and V vertices. It turns out that $V - E - F$ is always even, and the genus may be defined as $1 - \frac{V-E-F}{2}$.

gravitational locking: A general phenomenon between pairs of celestial bodies, in which differences in gravitational attraction due to the bodies' finite sizes alters their spins until one or both of the bodies forever hides a face from the other. The most familiar example is our moon, which is gravitationally locked to earth. Our moon always faces us, the other half always hidden from earth's view. The reason is that the moon spins about its own axis at the same angular rate that it orbits us.

This is not peculiar to our moon. It also happens for both of Mars' two moons, Phobos and Deimos. Although they orbit the red planet at different rates, their far sides both stay hidden from it. The same thing occurs with all the moons of Jupiter that Galileo discovered: Io, Europa, Ganymeade and Callisto. And of Saturn's numerous moons, 16 of them behave this way. This is true even for Pluto's moon, and in this case yet more is true: Pluto *itself* acts like this! That is, a camera on its moon Charon would always see the same side of Pluto. The planet is locked to its moon and the moon is locked to its planet.

Clearly, something basic is going on, here—these are not mere accidents. What's the story?

The story is that looking at planets as ideal point-masses may be fine for calculating their orbits around the sun, but planets and their moons are not at all simply points. Each has a finite size whose shape can be changed by outside gravitational forces, and each spins about its own north-south axis. These properties work together with gravity to make *all* moons attempt to hide a face. They may not always succeed because other competing gravitational forces may intrude and ruin the plan, but if a sufficiently isolated planet-moon pair is left alone long enough, the moon will finally hide a side. In fact, if you could supply an unlimited amount of time, perhaps many billions of years, then the heavier planet will hide a face too, as Pluto already does.

What is it that makes this mechanism tick? Take a look at this picture depicting a planet on the left and a smaller moon on the right:

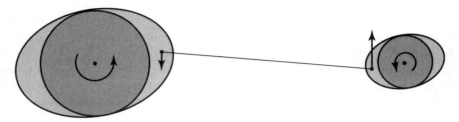

This is a "before" picture: time has not yet locked the moon into showing just one face.

Gravitational attraction has stretched both bodies, this effect being tremendously exaggerated in the drawing. Why stretched? The attracting force is greater between nearer points and less strong between more distant ones. If the planet on the left is earth, then the lightly shaded area could symbolize water that has reshaped itself in response to the force difference. We see a bulge of ocean on the right side of the earth. The solid earth itself has moved rightward, leaving another oceanic bulge on the left, and these two oceanic bulges mean we experience two tidal cycles each day. As the earth spins, land masses sticking above the ocean surface plow through the water, the bulge staying where it is. As an analogue, walking from the shallow end of a pool toward the deeper end corresponds to a rising tide. The water might rise from your knees to your waist but *you* are one doing the moving! Because the earth's swimming pool is so huge and the land mass's walk is just the rotation of the earth, few people think of it this way.

And the moon? It gets elongated by gravitational force differences. Although the earth pulls much harder on the moon than the moon pulls on us, the reshaping is less pronounced on the moon because the moon is a huge rock mass and doesn't flow like water. The tremendous pull by the earth nonetheless forces the rock to demonstrate some of its elasticity.

In the picture, each ellipse shape is rotated counterclockwise a few degrees. That is induced by the rotation of each body. Gravity creates the elongation, but the elongated mass has inertia simply because it *is* mass, so the bulge can't immediately reconfigure itself as the earth rotates. It travels a bit in the direction of the body's rotation. The constant pull by the other body keeps the bulge in place, tilted a few degrees.

The line segment drawn between the bodies symbolizes the net attractive forces between the bodies. Its endpoints are on the semi-major axes of the ellipses, and are placed on the near sides because the gravitational attraction is stronger there, predominating over the weaker forces between further-away points. This line tilts because the bulges tilt, with the result that there's a vertical force component at each endpoint. Both bodies are spinning in the same direction in our picture, so in each case the vertical component *opposes* the turning direction so that moon experience a braking effect.

This vertical component is larger on the moon because the more massive earth pulls harder on the moon than the moon pulls on earth. The large braking force is applied to the lighter body. Result? The moon's rotation slows down faster than the earth's, and with a slower rotation, the moon's bulge has more time to snap back. That means the bulge isn't tilted as much, so there is a smaller vertical force component on the moon and the braking decreases. It continues to decrease until the bulge has no reason to be ahead (or behind), and that occurs exactly when the moon rotates just fast enough so that one side always faces us! In an "after" picture, the moon's ellipse would be untilted.

Here are a few additional observations:

- The earth is not gravitationally locked to the moon, so the moon is now braking our rotation. The mechanism is not unlike a disk brake in a car—think of the solid earth as the spinning disk in a disk brake, and the ocean as stationary brake pads rubbing against the disk surface (ocean bottom). This action converts kinetic energy to heat and slows down the spin rate. The moon is able to slow us down only about 23 seconds every million years. But a billion is a thousand million, and over the last billion years, the earth has lost about a quarter of its angular rotational speed. Back then, a day was only 18 hours long. The moon will probably never gravitationally lock to the earth, however. The earth is just too massive and the braking too weak. The moon is racing against the clock, for in just a few billion years, our sun will swell into a red giant and melt or engulf both earth and moon.

- As the earth's spin slows down, its angular momentum decreases. By conservation of angular momentum, this loss must be made up somewhere within the pair. It mostly goes into the moon's orbital motion around us, hiking the moon up into an ever-larger orbit. Every time the moon circles the earth (once a month), it spirals out about an eighth of an inch farther from us.

- In the solar system, we see gravitational locking only between planets and moons. Although in theory a planet could eventually become locked to the sun, none are. The closest to this is a 3:2 rotation-orbit resonance between Mercury and the sun, Mercury spinning three times around its axis while making two orbits around the sun. However Upsilon Andromeda b, an extra-solar planet, *is* locked to its sun. One side always faces its sun and the other never sees it, creating a tremendous temperature difference. The hot side of the planet is forever lava-hot, while the other is forever cold, perhaps below freezing.

- In our solar system, there's just one example of a planet locking onto moon— Pluto always faces Charon. Even more remarkable would be a sun locking onto a planet. Once again, exploring extrasolar planets reveals the nearly-unthinkable—

the star tau Bootis, some 50 light years from us, has a planet orbiting it once every 3.3 days, the same time it takes for the star to spin about its own axis. This locking means that the far side of this star never sees the planet.

great circle: A great circle on a sphere is the intersection of the sphere with a plane through the center of the sphere. All great circles on a given sphere are largest-possible circles on that sphere, each having as diameter the diameter of the sphere.

half-life: For a process in which a quantity Q exponentially decays to zero, the *half-life* is the time required for Q to reduce to $\frac{Q}{2}$. The half-life does not depend upon the size of Q or when you begin measuring. It takes the same time for Q to decay to $\frac{Q}{2}$, no matter how large or small Q is. If Q decreases as QA^{-t} for some $A > 1$, then the half-life works out to $\frac{\ln 2}{\ln A}$.

Heron's formula: Triangles of side lengths a, b and c are all congruent, so there ought to be a formula for the area of any triangle in terms of a, b, c. There is:

$$A = \frac{1}{4}\sqrt{(a+b+c)(-a+b+c)(a-b+c)(a+b-c)}.$$

This is credited to Heron of Alexandria since it and its proof appear in his book *Metrica*, written around 60 CE. It is suspected that Archimedes before him already knew this formula, since *Metrica* is essentially a resource book of facts generally known in those times. Heron's formula is more commonly stated in an equivalent form using the triangle's semi-perimeter $s = \frac{a+b+c}{2}$:

$$A = \sqrt{s(s-a)(s-b)(s-c)}.$$

hexadecimal: Base sixteen.

Hooke's constant: This is the positive constant connecting the force exerted by a spring with the spring's stretch or compression. It's the k in **Hooke's law**, $F = -kx$, and is a measure of the spring's stiffness or strength. A watch hairspring has a small Hooke's constant, while the spring on a garage door has a large one. By assuming x small relative to the spring's length, $F = -kx$ models a real-world spring very well. It is often convenient to idealize the spring by assuming that $F = -kx$ applies exactly, and then we have an *ideal spring*. Unless stated otherwise, we assume springs in this book are ideal.

Hooke's law: The relation $F = -kx$. x parameterizes a one-dimensional range of deformations of an elastic object away from its undeformed state. The picture on p. 145 shows some examples. The number k is **Hooke's constant**, and F is the reaction force with which the object pushes against the applied force that created the deformation. Hooke's law usually models reality faithfully only when x is small in relation to the elastic object's size.

horsepower: A unit of power equal to 550 ft-lbs per second. It is close to 745.7 watts. The connection with wattage is no accident since it was James Watt who conceived the unit *horsepower* and coined the term. In the early days of the industrial revolution, Watt's

Glossary

steam engines began to replace horses, the usual source of industrial power then. A typical horse served as the power source of a mill grinding grain or sawing wood. In a typical setup, the horse walked in a circle about 12 feet from the hub, thus traveling some 75 feet per revolution, and on average that took a bit less than half a minute. Watt figured that a healthy horse would pull with a force of about 180 pounds, and found that the horse pulled with a power of approximately 32,580 ft-lbs per minute. Rounding up to 33,000 ft-lbs per minute makes this a multiple of 60, which is why one horsepower is defined as $\frac{33,000}{60} = 550$ ft-lbs per second.

ideal mass: A theoretical mass concentrated to a point. With the mass all at a point, it cannot be deformed and therefore can't store potential energy through deformation the way a spring does. In this respect, an ideal mass is opposite from an **ideal spring**, which is massless and can store energy only through deformation. These two ideal extremes lead to the teacups analogy explained in the boxes on pp. 153 and 154.

ideal pendulum: A pendulum with a point-mass bob and rigid, massless rod. When swinging, the ideal pendulum has no driving or dissipative forces, so its total energy stays constant. No pendulum is ideal, but one can be approximated by using a small, dense bob, thin rigid rod, a knife-edge suspension system and by encasing the pendulum in a vacuum chamber to reduce air resistance.

ideal spring: A theoretical, massless spring for which Hooke's law holds exactly.

inductor: A coil of insulated conducting wire.

Kepler's laws:

I Each planet moves in an elliptical orbit with the sun at one focus.

II The line segment from the sun to a planet sweeps out equal areas in equal times.

III The period of a planet is proportional to $a^{3/2}$, where a is the length of the semi-major axis of the planet's elliptical orbit.

Leyden jar: The Leyden (*lie*-dun) jar was the first man-made electric capacitor or storage device. It was invented around 1746 by the Dutch physicist Pieter van Musschenbroek (a professor of physics at the University of Leiden) and independently by Ewald Georg von Kleist of Pomerania about a year earlier. Although electrostatic generators had been around for a century and static electricity had long fascinated serious scientists and laymen alike, static electricity was evanescent. The Leyden jar, however, could store high-voltage quantities of electric charges for hours and could be discharged on demand.

Though the Leyden jar represented a breakthrough in the history of electricity, it is a very simple device. A plain-vanilla version consists of a glass jar wrapped with two separate pieces of conducting foil such as tinfoil, one piece on the outside, the other on the inside. The glass serves as an insulator, keeping charges separate. The inside foil is connected to the outside by an insulating stopper with a metal rod running through it and connected to a metal chain touching the foil inside. Keeping the outside foil grounded, the inside foil can be charged to a high voltage using an electrostatic generator. The stored charge suddenly discharges when the outside foil is electrically connected to the rod sticking up from the stopper. Benjamin Franklin was fascinated by Leyden jars and connected several of them in a row to combine charges, thus getting one huge discharge. In this connection he coined the term "battery." The first atom smasher, built in 1930 at Cambridge University, used this idea. Its banks of connected Leyden jars could store as much as a million volts.

mass-spring: A paradigm model illustrating basic interactions between mass, an energy-storage device (the spring), and possibly friction. Both mass and spring are assumed ideal, and the mass is usually assumed to move in a straight line.

- If the system is conservative, meaning no energy loss, then when the mass is set into motion it will forever oscillate sinusoidally with no decrease in amplitude.

- If the system is nonconservative but the friction is less than a certain threshold value, then when the mass is set into motion it will forever oscillate sinusoidally but with amplitude exponentially decreasing to zero. The mass approaches in this way its equilibrium point where it remains once it's at rest.

- If the system is nonconservative and if the friction is greater than a threshold value, then when the mass is set into motion it will monotonically approach its equilibrium point.

The above qualitative description can be made quantitative. We assume that the mass m and the spring with constant k are ideal. We also assume that the sum of all resistance forces is *ideal*, in that it varies linearly with the velocity of the mass. If the signed displacement of m from equilibrium is $x(t)$, then the ideal resistance or slowing-down force can be written as $F = D\dot{x}(t)$, $D \geq 0$ being the *damping factor* and the dot signifying differentiation with respect to t. This term is added to the undamped equation $m\ddot{x} + kx = 0$, essentially that on p. 145 in the Introduction to *Heat and Wave Phenomena*. With no other outside forces, the sum of the three forces gives this equation for the ideal damped mass-spring:

$$m\ddot{x} + D\dot{x} + kx = 0.$$

Assuming a solution of the form $x(t) = e^{\lambda t}$ and substituting it into the equation leads to the quadratic $m\lambda^2 + D\lambda + k = 0$ whose roots λ_1 and λ_2 are equal to $\frac{-D \pm \sqrt{D^2 - 4mk}}{2m}$. These roots determine the nature of the motion. When the discriminant $D^2 - 4mk$ of the quadratic is negative, the roots are complex and lead to the general solution

$$x(t) = e^{-\frac{D}{2m}t} \cdot (\alpha \cos \omega t + \beta \sin \omega t),$$

where $\omega = \frac{\sqrt{4mk - D^2}}{2m}$. The constants α and β are determined by the initial position and velocity of the mass.

- When the system is conservative, the damping D is zero, the roots of the quadratic are pure imaginary and the mass oscillates with constant amplitude. With $D = 0$, the oscillation frequency reduces to $\omega = \sqrt{\frac{k}{m}}$.

- If the system isn't conservative but $D^2 - 4mk < 0$, then the mass oscillates with frequency

$$\omega = \frac{\sqrt{4mk - D^2}}{2m} < \sqrt{\frac{k}{m}}$$

and the graph of $x(t)$ versus t is pointwise multiplied by the modulating factor $e^{-\frac{D}{2m}t}$. This makes the oscillation amplitude approach zero.

- When $D^2 - 4mk = 0$ the oscillation is called *critically damped*. This is a threshold value for the discriminant, a bridge between oscillation and non-oscillation. Here, $\lambda_1 = \lambda_2$, and the general solution is $x(t) = (\alpha + t\beta)e^{-\frac{D}{2m}t}$, where α, β are determined by the initial position and velocity of the mass.

- When $D^2 - 4mk > 0$ the system is *overdamped*. The solution is a sum of two decaying exponentials:

$$x(t) = \alpha e^{\frac{-D + \sqrt{D^2 - 4mk}}{2m}t} + \beta e^{\frac{-D - \sqrt{D^2 - 4mk}}{2m}t}.$$

Although the solution's graph goes to zero monotonically, it does not decay like a single exponential. That is, the sum isn't equivalent to a single decaying exponential function.

momentum: In this book we use *momentum* and *angular momentum* as in classical mechanics. The *momentum* of a mass traveling with velocity **v** is the vector quantity $m\mathbf{v}$. Since measuring the velocity of a mass depends on which coordinate system we're in, we assume throughout that all measurements take place in some one, non-accelerating coordinate system. *Angular momentum* additionally assumes some point P fixed in that coordinate system, and uses the position vector **r** from P to m. With respect to P, the *angular momentum* of m is then defined to be the **cross product** $\mathbf{r} \times (m\mathbf{v})$.

How are the notions of momentum and angular momentum related? If a mass m moves in a straight line at constant speed, its momentum is constant, so its angular momentum with respect to any fixed P is also constant because the magnitude of the cross product is the area of the parallelogram defined by **r** and $m\mathbf{v}$, and that area is base times height. The base is always the same length in the same direction. And the height is just the distance from P to the line L.

But this is not the *only* way m's angular momentum can remain constant. If m circles around P in one direction and at a fixed speed, then its angular momentum remains constant, too. In fact, this and the above description fit together to describe the set of *all* motions of unit mass m having a common angular momentum. To see this, begin with a circle centered at P whose radius is the length of **r**. As m travels in the same direction with constant speed around the circle, its angular momentum remains constant. It still remains constant even if the string holding m in orbit were to break, with m flying off tangentially to the circle. Picture an oriented circle together with all its directed tangent lines—each direction agreeing with the circle's orientation; next, at each point on each of these tangent lines, base a unit vector aimed in the positive direction of that line. All unit masses having such a vector as velocity have the same angular momentum with respect to P. Any unit mass with any other velocity vector has a different angular momentum with respect to P. This picture illustrates this:

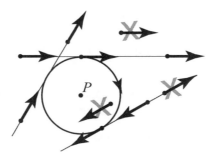

In it, all seven of the angular momenta without an **X** are equal. The ones with an **X** through them are mutually unequal and different from them.

One can push the relation between momentum and angular momentum further, by thinking of a line as a circle with center at infinity. To see the line-circle relation, start with a circle tangent to the x-axis at the origin. Keeping it tangent there, move its center vertically, making the radius grow. When the radius is astronomically large, the part of the circle near the origin looks like the x-axis, and in the limit, the circle converges to the x-axis. For any point x_0 of the x-axis and any arbitrarily small $\epsilon > 0$, there is a sufficiently large R_0 so that for any of the above circles of radius $R > R_0$, all points of the circle between the lines $x = -x_0$ and $x = x_0$ are within ϵ of the x-axis.

We mimic this to relate momentum and angular momentum. In the picture, let the horizontal tangent be the x-axis and the point of tangency with the circle, the origin. Keeping the circle tangent at the origin, move its center further and further directly downward. As the radius increases, the angular momenta increase, but they can be normalized by dividing them by the radius size. So, any unit momentum vector in the x-axis is the limit of vectors on the ever-growing circle. Furthermore, no other vector based at any other point in the plane is the limit of vectors on the growing circle. In this sense, linear momentum is a limiting case of angular momentum.

Dividing by the radius effectively throws away one length dimension L. This agrees with the dimensions of angular momentum and momentum. The dimension of angular momentum $\mathbf{r} \times (m\mathbf{v})$ is $LM\frac{L}{T} = \frac{L^2 M}{T}$, while the dimension of momentum $m\mathbf{v}$ is $\frac{LM}{T}$.

Monte Carlo method: The Monte Carlo method uses random numbers to generate a large amount of numerical data representing possible outcomes of an experiment (the experiment may range from very simple to highly complex) and then appropriately averages the outcomes to arrive at a plausible simulation. Historically, the first important Monte Carlo simulation took place in connection with the Manhattan Project to solve the problem of neutron diffusion in a fissionable material. The method created a possible "genealogical history" of a single neutron as it interacts with surrounding material. Using different neutron positions and initial velocities, this history was repeated many times to generate a statistically valid simulation. When computers grew up, so did the Monte Carlo method, and today there are applications to areas as diverse as economic risk calculation, the spread of cancer cells, designing heat shields, testing for prime numbers, special cinematic effects and even video games.

The Monte Carlo method can be used to evaluate definite integrals, since they can be looked at as finding areas or volumes. Picking two random real numbers within certain specified intervals corresponds to throwing an imaginary dart at a rectangle. To find the area of a region X defined by inequalities with a rectangle, one multiplies the rectangle area by the percentage of darts landing within X. Today's fast home computers can make this a very powerful and easy tool to use. This generalizes to multiple integrals.

It was the mathematician Stanislaw Ulam (1909–1984) of the Manhattan Project who, after many games of solitaire, decided that his own method of statistical sampling in card playing might be put to serious use. The story goes that during a long car ride with John von Neumann (1903–1957), he outlined his ideas to an initially skeptical Johnny. By the end of the trip, he had won over von Neumann, and within weeks serious work began with several of their Manhattan Project colleagues. As it happened, Ulam's uncle often borrowed money from relatives because he simply *had* to go to Monte Carlo! One of the team members, Nicholas Metropolis (1915–1999), happened to refer lightheartedly to the new technique as the *Monte Carlo*. The name stuck.

Newton's law of cooling: This can be stated as: "The rate of change of an object's temperature is proportional to the difference between its temperature and the surrounding temperature." For it to apply, the object's temperature is assumed to be uniform at all times. If the object is liquid (think of a cup of coffee in a thin metal cup, say), then keeping the liquid stirred helps to approximate this assumption. If the object is solid, then assume it's small or that the surrounding temperature isn't far away from the object's temperature.

The law is easy to translate into a differential equation: let t denote time, and let T be how much hotter the object is than the surrounding temperature. Then $\frac{dT}{dt}$ is how fast the body is warming up. Since constants of proportionality K relating physical quantities are usually taken to be positive, and because the body is cooling, Newton's law becomes

$$\frac{dT}{dt} = -KT, \quad K > 0.$$

Newton's law of gravitation: Any two point masses m_1, m_2 separated by a distance r exert on each other a gravitational force of attraction of strength

$$F = G \frac{m_1 \cdot m_2}{r^2}.$$

G is a constant of proportionality and is often called a *universal constant* because Newton's law assumes its value is the same throughout the universe. No matter where in the universe you measure the forces (they could be near opposite ends of the universe, or in the middle of a planet or a star), two masses m_1, m_2 if they're r apart, attract each other with the same strength of force. The forces themselves are vectors, pointing along the line through the two point-masses.

Newton did not discover gravity—the ancient Greeks certainly knew there was gravity. In fact, **Galileo's law of projectiles** famously assumes a uniform downward gravitational force. Newton's great insight is that this law is actually universal—the same gravity that makes a stone drop or causes a projectile finally to land is also the very same gravity that would make the moon drop to the earth like an apple from a tree, if it weren't balanced by the outward force due to its orbiting speed. In fact, Newton cleverly used Galileo's projectile to connect a dropping stone with the orbiting moon: launch a projectile with zero speed, and it drops like a stone. Launch it with ever-greater speeds, and the projectile travels further and further, finally making it into an elliptical orbit around the earth. For still greater launch speeds, the projectile can escape earth's gravity forever, traveling not in an ellipse but along a parabola or branch of a hyperbola.

Newton's law was a colossus: it unified the centuries-old and separate physics of the terrestrial and the celestial. It told us that gravity isn't just something that happens here on earth. It happens everywhere and accounts for a multitude of terrestrial and astronomical phenomena, including the paths of cannonballs, planets and comets.

It's a profitable exercise to contemplate how various scientific greats unified the world around them: Newton, Einstein, Mendeleyev, Darwin, . . .

Newton's laws of motion:

  I An object at rest will remain at rest unless acted upon by an external and unbalanced force. An object in motion continues to move at a constant velocity unless acted upon by an external and unbalanced force.

  II A force acting on a body gives it an acceleration that is in the direction of the force. The magnitude of this acceleration is inversely proportional to the mass of the body.

  III For every action force there is an equal but opposite reaction force.

nines complement: In base ten, the nines-complement of a digit n is $9 - n$. The nines complement of any integer is obtained by taking the nines-complement of each of its digits. This notion extends to any base: in base B, the $(B - 1)$-complement of a digit m chosen from $0, 1, \cdots, B - 1$ is $B - 1 - m$. As in base ten, this extends digitwise to any integer written in base B. In base two, the ones complement of any integer amounts to a toggle in which 0's and 1's are interchanged.

Ohm's law: The basic law $i = \frac{V}{R}$ connecting current i through an ideal resistor R with the applied voltage V. This law became well known after it appeared in Georg Ohm's 1827 book *The Galvanic Circuit Investigated Mathematically* (written in German). An ideal resistor is a resistor for which Ohm's law holds, so there's an immediate practical question: for real resistors like a length of copper wire, a light bulb filament or heating element, how well does Ohm's law actually hold? For small currents, it holds extremely well, i staying within billionths of one percent of its theoretical value. As you "challenge" the system by increasing voltage, the current doesn't rise linearly with V as predicted by the law, but as $\sqrt{V}$. So in using Ohm's law, care must be taken not to use voltage ranges that are too large.

There are also many non-ohmic materials such as semi-conductors for which Ohm's law fails completely, but this behavior has turned out to be a world-changing boon.

By adding precise amounts of impurities to pure silicon, for example, doped sand has been transformed into one of the most versatile materials known to man. Semi-conductor material can be made into current switches, LEDs, light sensors, high-power lasers, as well as single photon detectors that have changed the face of astronomy.

pendulum, ideal: (See **ideal pendulum**.)

photosphere: The disk-shaped part of the sun normally visible to our eyes. Actually, the sun extends millions of miles beyond the photosphere, but its gas, though glowing, is so attenuated that the light from the photosphere drowns it out. During a total eclipse, this drowning-out effect is removed and we can see the beautiful corona.

Platonic solid: A convex polyhedron all of whose faces are congruent convex regular polygons. There are exactly five Platonic solids, with three of them made up of congruent equilateral triangles: the tetrahedron, octahedron and icosahedron with 4, 8 and 20 faces, respectively. The other two are the cube with 6 square faces, and the dodecahedron, with 12 faces that are regular pentagons. Convexity is an essential part of the definition because there are nonconvex polyhedra having congruent regular polygons as faces. (See Kepler-Poinsot solids in [2]). The name "Platonic" solid makes it easy to believe that they originated with the ancient Greeks. Not so. Late Stone Age inhabitants of Scotland knew of all five, and stone models of them are today housed in the Ashmolean Museum in Oxford, England.

Why the name "Platonic"? Plato had the instinct to break up a complex system such as the physical world into elemental building blocks. His four elements were earth, water, air and fire. Although today they are not useful for science, finding irreducibles has had a long and productive history. Plato associated earth with the cube, water with the icosahedron, air with the octahedron, and fire with the tetrahedron. And the dodecahedron? It was used by "... the gods for arranging the constellations of the whole heaven." In his *Elements*, Euclid proved that there are no other regular polyhedra. The ancient Greeks, however, implicitly assumed convexity. There are actually nine polyhedra made up of congruent regular-polygonal faces. (See the box on p. 22.)

Principia: The short title for Newton's masterpiece, *Philosophiae Naturalis Principia Mathematica* (Mathematical Principles of Natural Philosophy). This three-volume work, in Latin, was published in 1687. It is divided into three books.

• Book I gives basic definitions of mass, motion (momentum), inertial, impressed and centripetal forces as well as absolute time, space, and motion, leading to the three axioms known as **Newton's laws of motion**. Using them, Newton proceeded as in Euclid's *Elements*: step by step with propositions, theorems, and problems.

• Book II considers the motion of bodies through resisting mediums as well as the motion of fluids themselves.

• Book III extends Newton's laws of motion, making them universal in scope. **Newton's law of gravitation**,

$$F = G \frac{m_1 \cdot m_2}{r^2}$$

is introduced, and Newton uses it to explain the motion of the planets and their moons, the paths of comets, the precession of equinoxes, as well as tidal action.

It can be said that *Principia* started with a chance discussion in 1684 at the Royal Society in London. Edmund Halley and the architect Sir Christopher Wren suspected that the known motion of planets had to be governed by an inverse square law, but neither could prove it. They asked their friend Robert Hooke, and Hooke boasted that he could not only prove the inverse square law, but all three of Kepler's laws, too. Wren was motivated, and offered him an expensive book as a reward for a proof within a two-month period. But Hooke was not the strongest of mathematicians, and failed to meet the challenge. Halley, too, was motivated and happened to be one of Newton's few friends. He traveled from London to Cambridge to ask Newton what the orbits of planets would be if the inverse square law of attraction were true. Newton immediately replied, "In ellipses, of course!" A surprised Halley asked how he knew that, and Newton replied that he'd worked it out some twenty years earlier. But a search among Newton's papers came up with nothing, so Newton promised to send Halley the proof. Already disappointed by Hooke, Halley returned to London a frustrated man. However some three months later Halley received a nine page paper that included a derivation of Kepler's laws from the inverse square law of gravitation and his own three laws of motion. A wildly impressed Halley immediately suggested publication. Newton, who had been stung by Hooke's criticism of his theory of light years earlier, would have no part of it. Halley, however, recognized the cosmic importance of what his friend had done. Through persistence, tact, good humor, and ultimately financial backing, he slowly won over the recalcitrant giant. Newton finally put pen to paper and, with customary thoroughness, labored for a year and a half, endlessly revising, rewriting, re-revising the short paper. It gradually grew to three volumes. The Royal Society's publication funds were at that time all but depleted, but Halley knew the world must see Newton's opus. To Halley's everlasting credit, he used his wedding dowry to cover virtually all costs of publication.

The *Principia* appeared in two additional editions during Newton's lifetime, in 1713 and 1726. It became the intellectual foundation of the modern world view, and brought him both fame and financial security. By the age of 45, Newton was recognized as one of the greatest scientists in history. Though Halley's name lives on in Halley's comet, shepherding the *Principia* from a remote possibility to concrete reality was far more important than any dirty snowball in space!

retrograde motion: The perceived reversal of direction due to the observer's moving platform. A simple example: you're riding west on a very smooth train. You look out the window only to see a westward-bound train moving on the tracks right next to you, but apparently going backwards. In fact, you're going 65 mph, while the other train is going only 55 mph, so the 55 mph train seems to be moving backwards at 10 mph. This example is something of a stretch, because although your coach may be smooth, there's a continual stream of various signals that keep telling you that your train *is* in fact moving. Historically, retrograde motion arose not from looking from one train at another, but at another planet from our planet, and the ride on our planet is incredibly smoother and signal-free than on any train! Take a look this picture of us and Mars orbiting the sun:

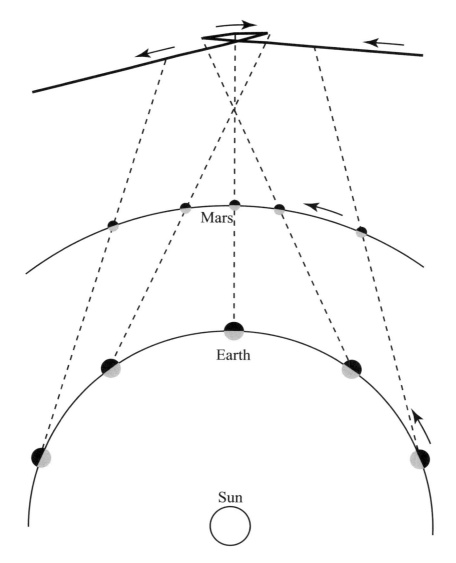

Mars is farther from the sun than we are, so from Kepler's Third Law (p. 106) it takes longer to orbit the sun than we do. The picture shows how our line of sight with Mars (the dashed lines) changes during a half year. In the third of the five snapshots, the horizontal component of earth's velocity becomes larger than that of Mars, and we see this as a reversal in Mars' direction.

saddle point: For a nonconstant, smooth, real function of two variables $F(x, y) \colon \mathbb{R}^2 \to \mathbb{R}$, a saddle point is a point P of F's graph around which the graph looks like a saddle. It is a two-dimensional analogue of an inflection point in the one-variable case, in that P is neither a relative maximum nor a relative minimum. Within a tiny ball centered at a saddle point P, the intersection of the tangent plane at P with the graph of F looks like two short line segments crossing at right angles. The algebraic condition for P to be a saddle point is $F_{xx}(P) \cdot F_{yy}(P) - F_{xy}^2(P) < 0$. Notice for the saddle surface $z = x^2 - y^2$ that $F_{xy}(0,0) = 0$, so the condition means $F_{xx} \cdot F_{yy}$ is negative at $(0,0)$—that is, $F_{xx}(0,0)$ and $F_{yy}(0,0)$ have opposite signs. This fits in with the "saddle-shapedness" of the graph and stands in analogy to the origin in the graph of $y = x^3$, which is a point of inflection, the graph having positive second derivative (is convex up) on one side of the tangent line and negative second derivative (is convex down) on the other. Geometrically, $F_{xx}(P) \cdot F_{yy}(P) - F_{xy}^2(P) < 0$ insures the right convexity behavior in the two-dimensional setting.

similarity: Two objects in euclidean space are *similar* provided they are "congruent up to size." That is, either object, after an appropriate uniform magnification or reduction, becomes congruent to the other object. And congruence itself? Two objects are congruent if one can be made to exactly coincide with the other by a finite succession of translations, rotations and possibly a reflection about a line.

solid angle: A solid angle in $\mathbb{R}^3$ is an angle formed by three or more planes intersecting at a common point, called the angle's *vertex*.

spherical excess: The sum of the interior angles of a **spherical triangle** is always more than 180°, and the amount greater than 180° is called the *spherical excess* of the spherical triangle. Gauss showed that the area of the spherical triangle is proportional to its excess. On a unit sphere, the excess of a triangle, measured in **steradians,** *is* its area.

spherical triangle: A three-sided figure on a sphere each of whose sides is an arc of a **great circle** on that sphere.

standard deviation: In this book, the standard deviation of a finite set of data $\{X_1, \cdots, X_N\}$ is computed this way: let μ be the average of $\{X_i\}$—that is,

$$\mu = \sum_{i=1}^{N} X_i.$$

Then the standard deviation of $\{X_1 \cdots X_N\}$ is

$$\sigma = \sqrt{\frac{\sum_{i=1}^{N}(X_i - \mu)^2}{N}}.$$

Standard deviation is a measure of data spread or dispersion. If the data points are close to the average, then σ is small—think of the actual lengths of a thousand yardsticks. If the data points are generally far from the computed average, then σ is large—think of the lengths of a thousand strips of hardwood flooring.

For many large populations, the graph of some measure (such as height, weight or cost) against incidence approximates a **bell curve.** The curve has a maximum of μ, and the distance from there to either point of inflection is σ. See the box on p. 100 and the picture shown under **bell curve** on p. 339. We can compute μ and σ for $\{X_1, \cdots, X_N\}$ whether or not the data set's graph happens to look bell-shaped.

Glossary

steradian: The natural unit of solid angle measurement, the steradian is the three-dimensional analogue of the radian. Just as the total radian measurement of a circle is 2π (the circumference of a unit circle), the total steradian measurement of a sphere is 4π (the surface area of a unit sphere). In the same way that the radian measure of a plane angle can be defined as the arclength the angle cuts out on unit circle, the steradian measure of a solid angle can be defined as the area it cuts out on a unit sphere.

topology: The study of properties unchanged or invariant under bicontinuous maps—that is, under continuous maps F such that the inverse F^{-1} exists and is also continuous. Two objects related by a bicontinuous map are said to be *topologically equivalent*. The topological questions in this book are restricted to closed compact orientable surfaces without boundary, surfaces such as the sphere, torus or the figures on pp. 30–31—essentially the surface of any object made from modeling clay. The **genus** of the surface is one topological invariant. Dimension and connectedness are two others. There are many good introductory references to this general geometry. See [3] or [6], for example.

Torricelli's law: This says that in an ideal, friction-free setup, water in an open tank flows from a hole at its bottom with the same speed v as it would in a vertical waterfall, where the water simply drops a distance y from the tank's water level to the hole. In appropriate units, this idealized law is often written as $v = \sqrt{2gy}$, where g is the gravitational constant. In practice, however, the water's viscosity, roughness around the hole edges, the swirling motion of the water and other energy-loss factors can easily make this velocity some 40% less. An approximation often used by engineers is $v \approx 0.84\sqrt{gy}$.

vernal equinox: See **equinox**.

Well-Tempered Clavier: Two collections of preludes and fugues by Johann Sebastian Bach (1685–1750). Each collection illustrates the advantages of his invention, tempered tuning. (See the box on p. 186.) In this tuning, the frequency ratio of any keyboard note to the one below it, is $2^{\frac{1}{12}}$—two, because the frequency ratio of any keyboard note to the octave just below it is two; and twelve, because there are twelve semitone intervals in an octave. Correspondingly, each collection consists of twelve Prelude-and-Fugues—a major- and a minor-key version of each. These are arranged going upward one semitone at a time. The first pair is a Prelude and Fugue in C major followed by a Prelude and Fugue in C minor, the next pair is one in C-sharp major followed by one in C-sharp minor, ending with the twelfth pair, a B-major Prelude and Fugue followed by a B-minor Prelude and Fugue. The two collections (Book I and II) were written approximately 22 years apart. It is hard to overestimate their influence on western music. Haydn, Mozart and Beethoven all carefully studied and learned from these great works.

work: In its simplest form, *work* is $F \cdot d$, where d is the length of a directed line segment and F is a constant applied force parallel to the line segment. This can be generalized in stages: if F isn't parallel to the line segment, take the component of F that *is*. If the path curves, then approximate the path by a series of short line segments connecting close-together points on the path and sum the work done on each segment to get the approximate total work. The exact total work is then the limit as the greatest length of the line segments approaches zero. The picture on p. 118 gives the idea: for the purposes of this definition, the actors are the curve Γ and the force vectors whose base points are on Γ. The total work done in traversing Γ is

$$W = \int_{P \in \Gamma} \mathbf{F} \cdot \mathbf{t}\, ds,$$

where $\mathbf{t}$ is the unit tangent vector to Γ at P.

The dimension of work is $\frac{ML^2}{T^2}$, where $M =$ mass, $L =$ length and $T =$ time, the same as that of kinetic energy $\frac{1}{2}mv^2$, potential energy of height mgh, and of potential energy stored in a deformed spring, $\frac{1}{2}kx^2$. From Hooke's law $F = -kx$, the dimension of spring stiffness k can be computed from $\frac{F}{x} = \frac{ma}{x}$, and is therefore $\frac{M}{T^2}$, so the dimension of energy stored in a spring, $\frac{1}{2}kx^2$, is $\frac{ML^2}{T^2}$. Dimensionally, work and all these other forms of energy are tied together. Even heat is the randomized energy of molecular motion, and each molecule has kinetic energy $\frac{1}{2}mv^2$ which has dimension $\frac{ML^2}{T^2}$. Any of these forms of energy can be converted to any other.

References

[1] Rajendra Bhatia, *Fourier Series*, Mathematical Association of America, Washington, DC, 2004.

[2] H. S. M. Coxeter, *Regular Polytopes*, Dover Publications, Inc., New York, 1973.

[3] Martin Crossley, *Essential Topology*, Springer Undergraduate Mathematics Series, London, 2005.

[4] Keith Kendig, *Conics*, Mathematical Association of America, Washington, DC, 2005.

[5] Keith Kendig, When and Why Do Water Levels Oscillate in Three Tanks?, *Mathematics Magazine*, Vol 72, No. 1 (Feb., 1999), pp. 22–31.

[6] Robert Messer and Philip Straffin, *Topology Now!*, Mathematical Association of America, Washington, DC, 2006.

[7] Clark Mollenhoff, *Atanasoff, Forgotten Father of the Computer*, Iowa State University Press, Ames, Iowa, 1988.

[8] Georgi Tolstov, *Fourier Series*, Dover Publications, Inc., New York, 1976.

[9] Transnational College of LEX, *Who is Fourier? A Mathematical Adventure*, Language Research Foundation, Belmont, Massachusetts, 1995.

[10] C. Ray Wylie, *Advanced Engineering Mathematics*, Fourth Edition, McGraw-Hill Book Company, New York, 1975.

Problem Index

Adding Equal Weights, 215, 335
Adding Waves, 159, 306
Adding a Room, 31, 242
Again How Long, 181
Air Pocket, 77, 269
Ancient "Aha", 63, 264
Another Way, 71, 267
Archimedean Spirals, 12, 227
Archimedes' Principle, 82, 271
Arctic Blast, 80, 270
Attaching the Wire, 55, 258
Average Distance-Squared, 96, 276
Average Distance, 96, 276

Beachball, 72, 268
Behavior of Water Heights (Version 1), 212, 332
Behavior of Water Heights (Version 2), 212, 332
Behavior of Water Heights (Version 3), 212, 333
Beyond Hooke, 176, 314
Body-Builder, 38, 247
Bottles versus Boxes, 100, 281
Building a Large Magnet, 138, 298
Buoyancy Machine, 116, 290
By the Way, 6, 221

Can This Happen, 176, 314
Can You Trust This Headline, 98, 278
Carbon Monoxide Detector, 79, 270
Ceiling Under Stress, 158, 305
Change of Scene, Again, 163, 309
Change of Scene, 163, 308
Chip Off the Old Block, 23, 233
Church Window Again, 15, 228
Church Window, 14, 227
Coil Around a Coil, 137, 297
Common Birthdays, 98, 279
Comparing Times, 210, 331
Cooling Rod Problem, 150, 301
Cross Product, 205, 328
Cube-Shaped Fishing Weight, 1, 217

Curious Property, 35, 245
Current Difference Again, 133, 296
Current Difference, 133, 295

Decimal versus Binary, 38, 247
Degrees in a Spherical Triangle, 207, 330
Diagonals of a Parallelogram, 20, 231
Discriminant versus Area, 205, 329
Disk in the Sky, 27, 238
Do Their Behaviors Correspond, 40, 249
Does This Chord Stay Musical, 177, 315
Does This Curve Exist, 16, 229
Does the Peak Walk, 151, 301
Down One Key, 185, 320
Dropped Penny, 94, 273
Dynamite Asteroids, 59, 262

Earth's Phases, 58, 261
Ellipses and Hyperbolas, 214, 334
Elliptical Driveway, 17, 229
Energy Loss, 141, 299
Exit Polls, 95, 274
Exponential?, 176, 315

Fatter Wire, 137, 297
Fixing Our Moonbound Pendulum, 165, 309
Flagpole Sundial, 51, 254
Float Capacity, 79, 269
Forgetting to Turn It Off, 138, 297
Foucault's Pendulum, 53, 256
Four Charges, 140, 299
Friction and Frequency, 161, 308
Friday the 13th Again, 44, 252
Friday the 13th One More Time, 44, 252
Friday the 13th, 3, 218
Full Moon Again, 57, 260
Full Moon, 57, 259
Full Piano Keyboard, 186, 320

Gain or Lose, 42, 250
Galileo's Cannonball, 112, 287
Gelosia Again, 39, 248
Genus of a 4-Cube, 32, 242

Geometry in a Mixed Term, 209, 331
Glass Skeleton, 30, 241
Going to Zero, 152, 303
Going, Going, Gone, 162, 308
Grandfather Clock, 110, 285
Gravity's Horsepower, 110, 286
Greatest Number of Stars, 51, 254
Greatest Range, 114, 288

Half and Half Again, 21, 232
Half and Half, 21, 232
Have You Heard of This, 117, 290
Hefty Block, Wimpy Spring, 153, 303
Hexadecimal versus Binary, 39, 248
High-Noon Shadows, 97, 277
Hot Buckshot Melts Through, 76, 269
Hot Buckshot, 75, 269
How Close, 4, 219
How Far, 152, 303
How Fast, 181, 317
How Grim the Reaper, 98, 278
How High is High Noon, 51, 253
How Hot is Hottest, 151, 301
How Long are the Axes, 209, 331
How Long, 181, 317
How Many Half-Lives, 184, 319
How Much Faster, 114, 289
How Much Work, 109, 285
How Much of the Heavens, 207, 329
How Often Does Lightning Strike, 36, 245
How Turns the Earth, 3, 218

Ideal Tetrahedron, 95, 274
Impact Angle, 60, 262
Inclined Mass-Spring, 166, 310
Increasing the Water Supply, 196, 322
Ingenious Student, 41, 249
Interpolation, 214, 335
Is Energy Conserved, 171, 312
Is This Possible, 51, 255

Larger Capacitor, 139, 298
Largest Cone in a Sphere, 25, 235
Largest Cylinder in a Sphere, 25, 235
Later, 133, 296
Leak to the Environment, 210, 332
Light Lumps, Massive Spheres, 59, 262
Limit or Chaos, 29, 239
Liquid Nitrogen, 81, 270
Losing Energy Again, 161, 307
Losing Energy, 160, 307
Low Orbit, 173, 312
Lunar Foucault Pendulum, 56, 259

Manned Deep Impact Mission, 61, 263
Martian Equinox, 63, 264
Mastic Spreader, 16, 229
Matter of Distance (3D Version), 202, 324
Matter of Distance, 202, 323
Melting Ice Cube, 78, 269
Method versus Madness, 156, 304
Model for Relative Humidity, 187, 320
Monte Carlo and Average Distance, 103, 283
Monte Carlo and the Needle Toss, 103, 283
Monte Carlo, 101, 282
Moonday, 56, 259
Moondial, 58, 260
Moonrise, 58, 260
Moonstruck, 58, 261
More Turns, 139, 298
Most Massive, 135, 296
Multiplication from Another Era, 35, 245
Musical Analogue, 185, 319
Muzzle-Speed Again, 113, 288
Muzzle-Speed One More Time, 113, 288
Muzzle-Speed, 113, 287

Never-Ending Decimal, 37, 245
Not a Plumb Plumb, 52, 256
Now What Will Happen, 119, 291

On the Moon, 117, 290
Once Again: How Long, 317
Opening or Closing Parenthesis, 58, 262
Our Spinning Moon, 50, 253

Paring Down the Earth, 112, 287
Pendulum versus Mass-Spring, 164, 309
Pendulum's Critical Speed, 117, 290
Perpendicular at Both Ends, 17, 230
Platonic Solid, 22, 232
Pouring Buckshot, 74, 268
Pulling Strings, 115, 289
Pulling on the Plates, 193, 321
Pyramid Peak, 205, 328
Pyramid Volumes, 23, 233

Question of Polarity, 136, 297

Rain Gauge, 71, 267
Raising Frequency, 186, 320
Random-Integer Game, 99, 280
Reality Test, 43, 251
Red and White, 94, 273
Red, White and Blue, 94, 273
Retrograde Motion, 56, 259
Right Turn, 28, 238
Ring-Shaped Fishing Weight, 24, 234

Problem Index

Round Trip, 42, 251
Running Start, 111, 286

Saturn's Rings, 121, 292
Seeing Two Sides, 97, 277
Serious Waiting Game, 98, 279
Sex and Money, 96, 275
Shearing an Ellipse, 207, 330
Short Sticks Revisited, 103, 284
Short Sticks, 96, 276
Shorter Period, Or Longer, 165, 309
Shrinking Capacitor, 142, 299
Sinking 2-by-4, 6, 220
Skewed Left or Right, 99, 281
Smaller Drain Holes, 210, 332
Smiley Face?, 64, 264
Solid Angles of a Tetrahedron, 27, 237
Special Cone, 205, 329
Spike (Temperature Version), 175, 313
Spike, 174, 313
Spin It Again, 26, 237
Spinning Turntable, 110, 286
Spiral Length, 13, 227
Spring Stiffness Again, 154, 303
Spring Stiffness, 154, 303
Squeezing the Earth, 111, 286
Star Area, 18, 230
Staying Power, 131, 295
Stereo Camera, 19, 231
Stretched From Both Ends, 204, 326
Subtracting by Adding, 2, 217
Sunspots, 52, 256
Survivors, 132, 295
Switching Off Gravity, 155, 304

Tale of Two Boats, 83, 271
Tank With Constant Water Supply, 196, 322
Tanks Can Model Current Flow, 192, 321
Tanks Can Model Heat and Temperature, 194, 321
Teachers versus Students, 95, 274
Three Children, 96, 275
Three Interconnected Tanks, 191, 321

Throwing Darts, 100, 282
Top versus Bottom, 73, 268
Total Torque, 121, 292
Toy Car, 109, 285
Travel Time, 4, 220
Twanging Rod, 120, 291
Two Interconnected Tanks, 190, 320
Two Movies, Again, 183, 319
Two Movies, 182, 318
Two Philosophies Compared, 181, 317
Type AA, 134, 296

Up or Down, 7, 221

Vaporizing Asteroid, 62, 263
Vernal Equinox, 62, 263
Visual Demonstration, 70, 267
Voltage Change, 133, 296

Warm Resistors, 139, 298
Wave Motion (Version 1), 168, 310
Wave Motion (Version 2), 169, 311
Wave Motion (Version 3), 170, 311
What About Areas, 29, 241
What Comes Next, 44, 252
What Sort of Graph, 42, 250
What Will Happen, 4, 219
What are the Chances, 94, 274
What is This, 37, 246
What's its Area, 202, 325
What's the Volume, 203, 326
Where Does the Arrow Land, 115, 289
Which Has More Energy?, 172, 312
Which Runs Faster, 166, 309
Which Way Do They Point, 204, 327
Which Way, 50, 253
Who Wins, 184, 319
Wild Sequence, 37, 246
Wooden Spheres, 74, 268
Would This Tempt You, 3, 218

Yellow Points, Blue Points, 7, 226

Subject Index

Ampère, André, 126, 131
angular momentum, 57
Archimedean spiral, 12, 13, 227
Archimedes, 67
Archimedes' principle, 6, 67, 76, 77, 79, 82, 319
Aristotle, 45, 105
autumnal equinox, 63

Bach, Johann Sebastian, 186
base, counting, 33, 34, 40
bell curve, 100
binary, 38, 39, 248
binomial experiment, 91, 92, 273
Boyle, Robert, 116
Brahe, Tycho, 105
Brownian motion, 76

Cardano, Gerolamo, 85
Central Limit Theorem, 282
coefficient vector, 324
cone, 25, 26
conservation of angular momentum, 46, 57
conservation of energy, 45
convex, 22
Copernicus, Nicolaus, 105
Coulomb, Charles, 135
cross point, 24
cross product, 205

Davy, Humphrey, 125, 126
Deep Impact mission, 61, 263
Descartes, René, 134
discriminant, 206
dodecahedron, 22

Edison, Thomas, 127, 133
eigenline, 328
eigenvalue, 198, 199
eigenvector, 198, 200, 211, 327
Einstein, Albert, 56, 134
electromagnetic radiation, 48
equinox, 63, 264

Eratosthenes, 51, 63, 264
Euclid, 9
Euclid's postulates, 107
event, 87
experiment, 86

Faraday, Michael, 126
Fermat, Pierre de, 86
Fibonacci, Leonardo, 35
force field, 117
Foucault's pendulum, 53, 257
Foucault, Leon, 53, 55, 56, 144
Franklin, Benjamin, 124

Galileo Galilei, 105, 112, 113, 309
Galileo's law of projectiles, 112, 113
Galvani, Luigi, 124
Gauss bell curve, 100
Gauss, Carl, 10
Gaussian elimination, 197
gelosia, 35, 39, 249
genus, 30, 31
great circle, 10
gyroscope, 56, 144

Hales, Thomas, 14
half-life, 320
Henry, Joseph, 126
Heron of Alexandria, 9
Heron's formula, 9
hexadecimal, 39, 40, 248
Hilbert, David, 9
Hipparchus, 10
Hooke constant, 154, 347
Hooke's law, 130, 144, 176, 198, 303, 304, 314, 315, 335
Hooke, Robert, 144
Hopkinson, Francis, 18
horsepower, 110, 286
humidity, 187
Huygens, Christiaan, 144, 165

ideal mass, 147, 154

ideal spring, 147, 154, 289

Kepler's laws, 106, 311, 313
Kepler, Johannes, 14, 105

lattice, 8
Leibniz, Gottfried, 38
Lenz, Emil, 127
Leyden jar, 124

mass-spring, 162, 164, 167, 309, 310
Maxwell, James C., 134
Méré, Chevalier de, 85, 86, 92
mho, 134
Michelson, Albert, 56
Monte Carlo method, 101
Morley, Edward, 56
Muschenbroek, Pieter van, 124

Newton's law of cooling, 143, 144, 197, 198
Newton's law of gravitation, 287, 288
Newton's law of motion, 128
Newton's laws of motion, 107, 108, 129, 286
Newton, Isaac, 106, 107, 143
nines' complement, 2, 217

Oersted, Christian, 125, 131, 134
Ohm's law, 130, 197, 220, 223
Ohm, Georg Simon, 126, 134, 135, 299
outcome, 86

Pascal, Blaise, 86
pendulum, 43, 53, 117, 144, 164–166, 285, 290, 309
Peregrinus, Petrus, 123
phase change, 75
photosphere, 77
pictogram, 33
Plato, 105

Platonic solid, 22, 233
Principia, 107
probability function, 87
pyramid, 9, 23, 27, 205, 233
Pythagorean theorem, 9

rain gauge, 71, 267
retrograde motion, 56
Ross, Betsy, 18

sample point, 86
sample space, 86, 87, 276, 280
similarity, 11
skewed, 99, 281
solid angle, 10
spherical excess, 10
spherical geometry, 10
spherical triangle, 207
standard deviation, 282
Stardust mission, 61
steradian, 10
Stonehenge, 9
sunspots, 52
symmetry principle, 48

Thales, 123
topology, 11
Torricelli, Evangelista, 181
trial, 86
Tycho Brahe, 105

vernal equinox, 63, 264
Villarceau, Yvon, 235
Volta, Alessandro, 124

Watt, James, 128, 133, 140
Well-Tempered Clavier, 186
work, 109

About the Author

Keith Kendig was born in Los Angeles and raised in Santa Monica, California. He attended UCLA, receiving a B.S. in Mathematics in 1960 summa cum laude, and was elected to Phi Beta Kappa. He received his M.S. in mathematics from UCLA a year later and his Ph.D. from there in 1965, working under Basil Gordon. He subsequently spent two years at the Institute for Advanced Study in Princeton, working with Hassler Whitney.

Kendig wrote the book *Elementary Algebraic Geometry*, published by Springer-Verlag (Vol. 44 it its Graduate Texts in Mathematics Series). In 2000 he received the MAA Lester Ford Award for the *Mathematical Monthly* article "Is a 2000-Year-Old Formula Still Keeping Some Secrets?" In 2005, his book *Conics* appeared in MAA's Dolciani Series. He is currently an associate editor of *Mathematics Magazine* and is also on the editorial board of the Spectrum Series of MAA books.

Keith and his wife Joan are vegetarians, and in 1980 coauthored a vegetarian cookbook. Keith's contribution was using linear programming to find ratios of various seeds, nuts and grains so that the amino acid profiles of the combinations most closely match those of egg. Joan then translated these optimal combinations into a wide variety of tasty meat substitutes.

Keith is a semi-professional cellist and plays in various chamber music groups in the Cleveland area.